Selected Titles in This Series

Volume

2.2 William H. Kazez, Editor
Geometric Topology
1997

2.1 William H. Kazez, Editor
Geometric Topology
1997

1 B. Greene and S.-T. Yau, Editors
Mirror Symmetry II
1997

Geometric Topology

AMS/IP

Studies in Advanced Mathematics

Volume 2, Part 2

Geometric Topology

1993 Georgia International
Topology Conference
August 2–13, 1993
University of Georgia
Athens, Georgia

William H. Kazez, Editor

American Mathematical Society • International Press

Shing-Tung Yau, Managing Editor

1991 *Mathematics Subject Classification.* Primary 57–02, 57–06.

Library of Congress Cataloging-in-Publication Data

Georgia International Topology Conference (1993 : University of Georgia)
Geometric topology : 1993 Georgia International Topology Conference, August 2–13, 1993, University of Georgia, Athens, Georgia / William H. Kazez, editor.
p. cm.—(AMS/IP studies in advanced mathematics, ISSN 1089-3288; v. 2)
Includes bibliographical references.
ISBN 0-8218-0652-1 (set : acid-free paper). —ISBN 0-8218-0654-8 (pt. 1 : acid-free paper). —ISBN 0-8218-0653-X (pt. 2 : acid-free paper)
1. Topology—Congresses. I. Kazez, William H. (William Hilal), 1954– . II. Title. III. Series.
QA611.A1G456 1993
514—dc20 96-38266
CIP

∞ The paper used in this book is acid-free and falls within the guidelines
established to ensure permanence and durability.

10 9 8 7 6 5 4 3 2 1 02 01 00 99 98 97

TABLE OF CONTENTS

PART 1

PART 2

Preface

The 1993 Georgia International Topology Conference was held at the University of Georgia in Athens, Georgia from August 2–13, 1993. 274 people from 20 countries attended the conference. The proceedings include research articles, expository articles, and problem sets. A wide variety of topics in geometric topology are covered including many which were discussed at the conference.

I would like to thank the authors for the work they have put into their papers and the referees for the fine job they have done. Many of the authors have commented that they found the referees' comments particularly useful.

My colleagues John Hollingsworth, Gordana Matić, and Clint McCrory contributed to every aspect of the conference. Simon Donaldson, Bob Edwards, Mike Freedman, Dave Gabai, and Bill Thurston suggested many of the people who spoke at the conference. Dave Gabai and Rob Kirby put a lot of energy both into organizing problem sessions at the conference and into writing up the very thorough problem sets that appear in this volume. Gail Suggs and Sharon Southwick helped with the day to day running of the conference and in the preparation of many of the manuscripts. I appreciate all of their help.

The conference would not have been possible without the very generous financial support we received from the University of Georgia (in particular, the Office of the Vice President for Academic Affairs, the Office of the Vice President for Research, the Franklin College of Arts and Sciences, and the Department of Mathematics), the National Science Foundation (grant number DMS-9300677), and the International Science Foundation. We gratefully acknowledge their support.

I appreciate the guidance, encouragement and inspiration provided by previous organizers of the conference: M.K. Fort 1961, Jim Cantrell and C.H. Edwards, Jr. 1969, Jim Cantrell 1977, and Clint McCrory and Ted Shifrin 1985.

W.H.K.

Athens, Georgia
February, 1995

Dedication

These 1993 Georgia Topology Conference proceedings are dedicated to Professor James C. Cantrell on the occasion of his retirement from the University of Georgia after 31 years of service. His direction and inspiration are responsible for the continuation of the initial 1961 Georgia Topology Conference as a major international conference held every eight years, with smaller summer topology conferences held in Athens annually for over three decades. It is therefore our pleasure to offer this most appropriate dedication in honor and recognition of Jim Cantrell's career of leadership and friendship with topologists everywhere.

AMS/IP Studies in Advanced Mathematics
Volume **2**, 1997 (Part 2)

PROBLEMS IN FOLIATIONS AND LAMINATIONS

DAVID GABAI

0. INTRODUCTION

Since Alexander's foliations proof of the PL Schoenflies theorem in 1922 [1] and Reeb's [97] discovery of a foliation on S^3 in 1948, the theory of foliations on manifolds has undergone an explosive development, its fundamental ideas and structure creeping into many aspects of mathematics. (See references in [61].) In particular the geometric topological theory of foliations and laminations in 3-manifolds has experienced a huge development, thanks in part to the works of Haefliger [65], Novikov [89] and Thurston [103]. (See the survey [50].)

I attempt to expose the frontiers of the subject through the study of outstanding open problems, associated recent advances, occasional mostly forgotten theorems, a couple of new results, and unavoidable throw away remarks. I hope that the reader has the patience and resources to look up the definitions and classical results in the references cited above and below.

Although a great deal is known about Haken manifolds, e.g. topological rigidity [105] and geometrization [104], and some secrets of 3-manifolds with taut foliations have already been revealed, the deepest mysteries of manifolds with essential laminations have yet to be exposed. On the other hand, the manifolds with essential laminations encompass a far broader class of spaces than that of Haken manifolds and possibly that of manifolds with taut foliations. How much broader, what is known and some of what needs to be known comprise much of this paper. Here is the first hint ([59] generalizing [89], [95]). If M has an essential lamination, then M is covered by $\mathbb{R}^3$, so M is irreducible with infinite fundamental group. Also the leaves of λ and the components of $M - \lambda$, π_1-inject into M.

Good expositions of the theory of foliations can be found in [21], [72] or [61]. For the theory of measured laminations in 3-manifolds see [94] or [87]. For the theory of 3-manifolds see [75] or [67]. See [59] for

---

Partially supported by NSF Grants DMS-8902343, DMS-9200584, DMS-9505253

the theory of essential laminations and essential branched surfaces in 3-manifolds.

Here are some of the basic definitions. A 2-dimensional *foliation* $\mathcal{F}$ (resp. *lamination* λ) on a 3-manifold is a decomposition of M (resp. a closed subset of M) into surfaces called *leaves* so that M is covered by charts of the form $\mathbb{R}^2 \times \mathbb{R}$ where the leaves pass through a given chart in slices of the form $\mathbb{R}^2 \times$pt. We assume that each leaf of λ or $\mathcal{F}$ is smoothly immersed in M. A lamination λ on a closed oriented 3-manifold is *essential* [59] if it contains no spherical leaf or torus leaf bounding a solid torus; furthermore if C is a *closed complementary region* (i.e., the closure of a component of $M - \lambda$ with respect to the path metric), then C is irreducible and ∂C is both incompressible and end-incompressible in C. An end compressing disc is a properly embedded (D^2 -(closed arc in ∂D^2)) in C which is not properly isotopic rel ∂ in C to an embedding into a leaf. There is an analogous definition of an essential lamination in a manifold with boundary. The transversely orientable foliation $\mathcal{F}$ of a closed orientable manifold is *taut* if there exists a transverse closed curve in M which nontrivially intersects each leaf. More generally $\mathcal{F}$ is *Reebless* if $\mathcal{F}$ contains no Reeb components. Again there are related definitions for bounded manifolds. We will abuse notation by declaring laminations to be supported in a nowhere dense subset of M. Thus a foliation is viewed as a "lamination" by Denjoying (or blowing air into) a dense set of leaves, an operation that destroys a preexisting smooth transverse structure on the foliation. Thus the notion of manifold with essential lamination is the simultaneous generalization of the Haken manifold and manifold with Reebless foliation. Finally a *genuine lamination* is an essential lamination that is genuinely not a foliation, i.e. a lamination one of whose closed complementary regions is not an I-bundle.

A foliation can be made transversely orientable by passing to the appropriate 2-fold covering space. On the other hand, for local reasons, the *generic* genuine lamination is not transversely orientable, even after passing to a covering space.

The paper is organized as follows.

7 Algebraic Topology
8 Knot Theory

Disclaimer. I generally do not assign ownership to particular problems.

Notation. If E is a topological space, then $|E|$ denotes the number of connected components. $\overset{\circ}{E}$ denotes the interior of $E, <,>$ denotes algebraic intersection number and $N(E)$ denotes regular neighborhood of E in its ambiant space.

Acknowledgments. I am thankful to the many people who contributed in various ways to this paper including Will Kazez who suggested that I write it, the participants of the foliations problem session held August 6, 1993 at the Georgia International Topology Conference and Alberto Candel, Larry Conlon, Sergio Fenley, Joel Hass and Lee Mosher. I deeply appreciate the unnamed others who corrected misstatements and/or supplied improvements of the text. Finally, to William P. Thurston who got me started.

1. Existence Problem

Problem 1.1. Let M be a closed orientable irreducible atoroidal 3-manifold with $\mid \pi_1(M) \mid$ infinite. Does M contain an essential lamination? (Atoroidal means $\pi_1(M)$ contains no non peripheral $\mathbb{Z}+\mathbb{Z}$.)

Remark 1.1. After a 30+ year effort the analogous question, when M is a Seifert fibred space, has just been completely answered [91]. Here is some history. The classical results of Milnor [82] and Wood [109] show that there exists a foliation transverse to the fibres of an S^1-bundle of Euler class e over a surface of Euler characteristic χ if and only if $|e| \leq |\chi|$. This work was generalized to Haken Seifert fibred spaces by [29], and Jankens-Neumann [77]-[78] reduced the general existence problem of foliations transverse to Seifert fibres to a conjecture that for certain triples (a_1, a_2, a_3) of numbers, id$\in$ Homeo(S^1) is not equal to the product of three homeomorphisms, each respectively conjugate to an a_i-rotation. Meanwhile Thurston's 1971 thesis showed that a circle bundle over a surface has a C^2 taut foliation if and only if it has a C^2 foliation transverse to the fibres. This was appropriately generalized (in the C^0 category) in the 1990 theses of Brittenham [7] and Claus [20] which in particular show that an essential lamination in a non Haken Seifert fibred space is isotopic to one transverse to the Seifert fibres. This reduced the general problem to the former one. Finally Ramin Naimi completed the proof by establishing in a direct and completely elementary way the Jankens - Neumann conjecture.

For completeness we mention that a compact, orientable, irreducible, toroidal (i.e. nonatoroidal) 3-manifold is either Seifert fibred or Haken by [19] or [51].

Examples 1.1. i) In the 1970's Thurston, Fried and Ghys independently observed that if M^3 fibres over S^1 with pseudo-Anosov monodromy, then the stable lamination suspends to a codimension-1 lamination λ on M^3. After either Dehn filling a boundary component or Dehn surgering (possibly trivially) the core of a closed complementary region X, the resulting region closed complementary region Y in the surgered manifold has interior $\overset{\circ}{D^2} \times S^1$ and either has compressible boundary or fibres over S^1 with fibre $D^2 - (k \text{ points of } \partial D^2)$. They observed that if this finite number k is even and the gluing map is by a $2m\pi/k$ rotation, then Y can be filled in with a bundle of monkey saddles. Furthermore if this is true for each non I-bundle closed complementary region, then λ extends to a Reebless foliation on the modified 3-manifold N. More generally, by [59], if for each non I-bundle complementary region $k > 1$, (and not necessarily even) then λ is essential in N. Here is one important special case of this construction. Suppose $\partial M = P \neq \emptyset$, e.g. M is a fibred knot complement with pseudo Anosov monodromy. The cusps of the complementary region of λ containing P are parallel to $n > 0$ copies of an essential simple closed curve $\alpha \subset P$. In this case $n(\alpha)$ is called the *degeneracy slope* $d(\lambda)$ of λ. If one does a β filling to P to obtain N, then λ remains essential in N if and only if $|<\beta, n\alpha>|> 1$.

ii) Thurston [103] contains interesting constructions of taut foliations.

iii) Each thurston norm minimizing embedded surface in a closed orientable 3-manifold is the compact leaf of a taut foliation [39]. Specializations and variations of the [39] construction can be found in [40]-[45]. Its' sequels contain other constructions (e.g. [40, 1.7], [41, 3.1]).

iv) A manifold obtained by "nontrivial" Dehn surgery along a curve in a leaf of a taut foliation yields a manifold with a taut foliation [49]. If ∂M contains a torus component P and M has a taut foliation which restricts to a foliation on P that contains a Reeb component, then except possibly for the Reeb direction, any manifold obtained by Dehn filling P has a taut foliation. Similar results hold for manifolds with essential laminations. [49] also reviews elementary methods of perturbing laminations.

v) If M is an atoroidal 3-manifold possibly with boundary, and $\mathcal{F}$ is a finite depth foliation on M, then by [58] there exists an essential lamination "almost" transverse to $\mathcal{F}$, whose complementary regions disjoint from ∂M are like those in Examples i) above. Furthermore if $\mathcal{F} \mid \partial M$ is a foliation by circles, then all the complementary region are like those in i) and the degeneracy slope $d(\mathcal{F})$ will be transverse to the circle fibration of ∂M. Since each knot in S^3 has such a finite depth foliation [41], it follows by i) that for k an atoroidal knot in S^3, off of at most two lines in Dehn surgery space, all nontrivial surgeries have essential laminations.

vi) Associated to either a punctured torus bundle over S^1 or 2-bridge knot complement M, Hatcher [71] constructs 1-parameter families of essential laminations on M, which are Reebless foliations in the case where M is a torus bundle. (Here M should be viewed as a compact manifold with torus boundary.) The restriction to ∂M is a lamination of curves of fixed slope. The slopes of a family lie in an open interval of $\mathbb{R}$. When the slopes are rational, the boundary leaves are circles and the laminations on M extend to essential laminations or Reebless foliations on the corresponding Dehn filled manifolds.

vii) Oertel [unpublished] has examples of essential laminations with transverse affine structures on some manifolds obtained by surgery on knots in S^3.

viii) Naimi and Delman independently showed that manifolds obtained by nontrivial surgery on non torus two bridge knots contain essential laminations. [90] generalizes Hatcher's construction in a 2-fold way to show that to each nontrivial, non meridinal simple closed curve $\gamma \subset \partial N(k)$ there exists an essential lamination in $S^3 - \mathring{N}(k)$ which on $\partial N(k)$ restricts to a lamination by circles parallel to γ. Naimi both constructs new laminations carried by new branched surfaces and shows that Hatcher's branched surfaces carry more than first thought.

ix) Delman [23] on the other hand finds a lamination contained in $S^3 - \mathring{N}(k)$ which remains essential after all nontrivial surgeries. Delman has named such laminations *persistent* and has generalized his construction to a class of Montesinos knots which contains all those knots whose associated rational numbers have all odd denominators [24]. (The interesting such case is when the Montesinos knot contains exactly 3 tangles, for by [93] nontrivial surgeries on ones with >3 tangles are Haken.) In particular he finds persistent laminations in the complements of all the classical nontorus pretzel knots. Another application of Delman's technology yields a persistent lamination in the complement of fibred knot 8_{20}. Contrast this with the fact that the suspended stable lamination transverse to the fibration (see i)) is not persistent. See Remark 8.7 v). Very recently Delman observed that his persistent laminations can be chosen to extend to transversely oriented taut foliations on the surgered manifolds.

x) Roberts [99] has shown that if k is an alternating knot, whose alternating projection contains nested Seifert circles, then nontrivial surgery on k yields a manifold with a taut foliation. For other alternating knots "half" of the surgered manifolds have taut foliations. Her foliations are obtained by deforming finite depth foliations which restrict to a foliation on $\partial N(k)$ by longitudes. Actually her technology is applicable to any knot in S^3 (e.g. 8_{20}) provided it bounds an incompressible surface (orientable or not) with a particular local structure, however in these cases she obtains an essential lamination in the surgered manifold (though in 8_{20} she gets taut foliations). Her work combined with that of Delman's, [25] yields the following remarkable result. Nontrivial surgery on a non torus alternating knot produces a manifold with an essential lamination.

xi) Let M be a manifold with $\partial M = T^2$ which fibres over S^1. Roberts [99] has shown that manifolds, obtained by Dehn filling M along ∂-slopes sufficiently close to that of the fibre, contain taut foliations.

xii) Wu [110] has shown that manifolds obtained by nontrivial surgery on non Montesinos arborescent knots have essential laminations.

xiii) Christy [personal communication] has observed that certain spines of certain non Haken manifolds can be deformed to branched surfaces that carry essential laminations. Christy worked this out for standard spines in hyperbolic 3-manifolds with small Matveev [81] complexity. See also Remark 2.2 ii).

xiv) Here is a new construction of taut foliations in 3-manifolds. We work it on the simplest example. Consider the link in S^3 with components K, L shown in Figure 1.1. Let $M = S^3 - \mathring{N}(L)$. Let $P = \partial M = \partial N(L)$. We will show that any manifold N obtained by non trivial surgery on $K \subset M$ has a taut foliation $\mathcal{F}$ such that $\mathcal{F} \mid P$ contains a Reeb component parallel to the longitude. It follows by [49], that N trivially filled along P and hence S^3 nontrivially surgered along K contains a taut foliation. Let S_1, S_2 be the two genus-1 surfaces spanning P consisting of tubes which follow K except as shown in Figure 1.2. The union of S_1 and S_2 separates M into two components A, B, where $K \subset A$. Reimbeddings of the corresponding sutured manifolds are shown in Figure 1.3. Here A is seen from the outside, while B is seen from the inside. An elementary exercise shows that A nontrivially surgered along K contains a taut foliation $\mathcal{F}_A$ such that $\mathcal{F}_A \mid P$ is the suspension of a homeomorphism of I. The sutured manifold $(B, P \cap \partial B)$ has $R_+ = S_1$ and $R_- = S_2$. By explicitly constructing a sutured manifold hierarchy on $(B, P \cap \partial B)$, one concludes tautness of $(B, P \cap \partial B)$ and hence the existence of a taut foliation F_B on B tangent to the S_i's such that $\mathcal{F}_B \mid B \cap P$ is a suspension of a homeomorphism of I. The key observation is that there exists an essential annulus T in B with one

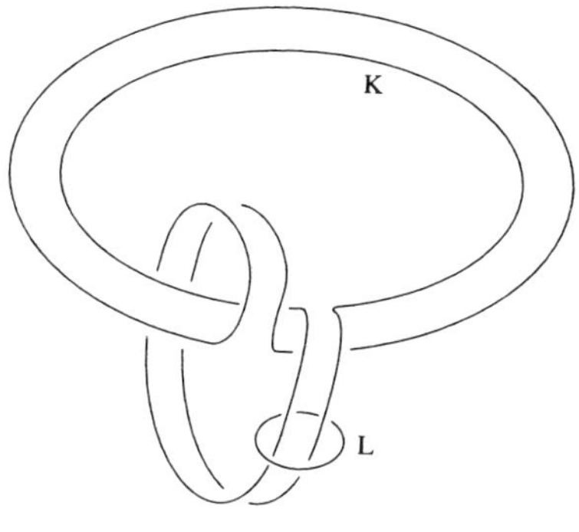

FIGURE 1.1

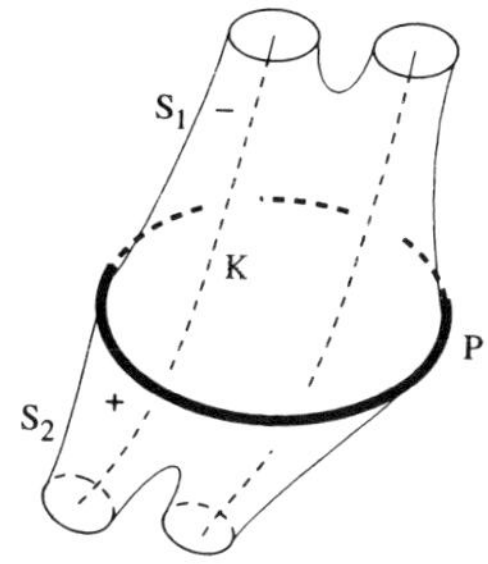

FIGURE 1.2

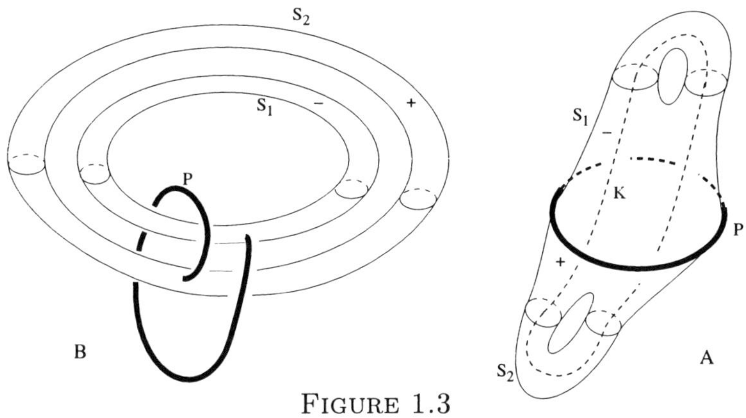

FIGURE 1.3

component in P and the other component in S_1. By [49] the foliation $\mathcal{F} = \mathcal{F}_A \cup \mathcal{F}_B$ can be deformed to have a Reeb component on ∂P parallel to its longitude. Note that Delman, Naimi and Roberts had already known that a manifold obtained by nontrivial surgery on knot 6_1=knot K of Figure 1.1 had a taut foliation.

Call a minimal genus Seifert surface S for a knot k in S^3 *fragile*, if there exists a properly embedded arc t in S such that $T = S - \mathring{N}(t)$ is a compressible surface in $S^3 - \partial T$. The above construction seems to work for a number of knots with fragile Seifert surfaces. (With T being the second term in the sutured manifold hierarchy of "region B" and $L = \partial D^2$ where $D^2 \subset S^3$ is a 2-disc such that $D^2 \cap S = t$.). Does it work for all? One might also consider *homologically fragile* Seifert surfaces, i.e., when T is not thurston norm minimizing for the oriented link ∂T.

Frontier Questions.

FQ 1.1. Let k be the Fintushel Stern knot (see [38]), i.e., the pretzel knot $(-2,3,7)$. Except for slopes $17, 18, 19$, does nontrivial surgery on k yield a manifold with an essential lamination. E.g. $35/2$ surgery.

The Fintushel Stern knot k is fibred and Christy [18] showed that the degeneracy slope of k to be 1(18/1) verifying a 10 year earlier experimental calculation using Penner's thesis [96]. Thus $k(p/q)$ has a genuine lamination if $\mid 18q - p \mid > 1$. Fintushel and Stern calculated that the manifolds $k(17), k(18), k(19)$ have finite π_1. By Roberts [unpublished] $k(\alpha)$ has a taut foliation if $\alpha \in (-\infty, 1)$.

FQ 1.2. Do manifolds obtained by nontrivial surgery on $10_{139}, 10_{142}, 10_{161}, 10_{163}, 10_{165}$ have essential laminations?

By the works of [90], [24] and its methods, [25], [99] and its methods, and [110] all nontrivial surgeries on the other nontorus knots of ≤ 10 crossings have essential laminations.

FQ 1.3. Do manifolds obtained by nontrivial surgery on 9_{49} have taut foliations?

For all the other nontorus knots of ≤ 9 crossings, all nontrivial surgeries yield manifold with taut foliations by [99] and Delman, see Examples 1.1 ix).

FQ 1.4. Except for known phenomena, does non trivial surgery on a 2-bridge link yield a manifold with an essential lamination?

FQ 1.5. Let M be a manifold which fibres over S^1, with pseudo-Anosov monodromy and $\partial M = P$. Let N be the manifold obtained by filling P along the degeneracy slope d. Does N have an essential lamination? What about filling along curves which intersect the degeneracy slope exactly once?

Its well known that one can construct explicit examples of such manifolds by considering torus bundles over S^1. E.g. given a 2×2 matrix A with integer coefficients, determinant 1 and positive eigenvalues $\neq 1$, then A induces a homeomorphism B of T^2 with a fixed point associated to the point $0 \in \mathbb{R}^2$ and hence after a small perturbation a homeomorphism C of $T^2 - \overset{\circ}{D}{}^2$ such that $C \mid \partial D^2 = \text{id}$. Let $M = (T^2 - \overset{\circ}{D}{}^2) \times I / \sim$ where $x \sim C(x)$. In this case $d(\mathcal{F}^s) = 2(x \times S^1)$ where $x \in \partial D^2$ and $\mathcal{F}^s$ is the stable lamination. When the eigenvalues of A are negative and $\neq -1$, then this construction yields a connected degeneracy locus.

Related Problem. Let M be a closed orientable irreducible atoroidal 3-manifold with infinite π_1.

Problem 1.2. Does M have a taut foliation?

Problem 1.3. Does M have an Anosov or pseudo-Anosov flow?

Remark 1.2. Constructions of Anosov flows can be found in [64], [70]. Mosher [58] has shown that if k is a hyperbolic knot, then off of at most two lines in Dehn surgery space all surgeries yield manifolds with pseudo Anosov or Anosov flows. The laminations of Examples 1.1 v) are the stable or unstable laminations of these flows.

Conjecture 1.4. *Some finite sheeted covering of M has an essential lamination.*

Conjecture 1.5. (Waldhausen). *Some finite sheeted covering is Haken.*

Conjecture 1.6. (Thurston). *Some finite sheeted covering of M fibres over S^1.*

Remark 1.3. There are interesting examples [47],[98] of nonfibred manifolds which have coverings which fibre over S^1. A non interesting example is the double covering of two twisted I-bundles glued together. The point of Thurston's conjecture was not necessarily to state a mathematical truth but more to expose our limited knowledge.

2. Branched Surface Theory

Background. See [59] for the basic definitions.

Problem 2.1. When does a branched surface in a 3-manifold fully carry a lamination?

Problem 2.2. Give an algorithm to decide if a branched surface in an irreducible 3-manifold carries an essential lamination.

Problem 2.3. Is there an algorithm to decide if a 3-manifold has an essential lamination?

Problem 2.4. Let M be an atoroidal 3-manifold. Does there exist a triangulation Δ of M such that every essential lamination on M can be isotoped to intersect Δ in Haken normal form?

Problem 2.5. Are there useful branched surfaces which carry nothing?

Remark 2.1. i) Williams [108] pioneered the use of branched surfaces to carry laminations.

ii) Deciding whether a branched surface carries a compact surface is equivalent to finding a solution, nonnegative on each sector, of the finite number of linear branch equations.

iii) In practice Problem 2.2 should follow from 2.1. Suppose that the branched surface B carries λ. Do a preliminary λ-splitting so that there exists no compact surface P with $\partial P \subset \partial_v N(B)$ and $P \subset L$, a leaf of λ. The algorithm of Jaco - Oertel [76] should decide whether or not the complementary regions of $N(B)$ have monogons or compressions of the horizontal-∂. If not and λ contains no spheres or tori (which is the interesting case), then by [59] λ is essential.

iv) In [8] it is shown that an essential lamination λ in the triangulated manifold M can be deformed into a possibly different essential lamination in Haken normal form with respect to the given triangulation. Consequently there exists a finite number (a function of the number of 3-simplicies in a triangulation) of branched surfaces in M, such that if M contains an essential lamination, then one of these branched surfaces carries a (possibly different) essential lamination. This generalizes the corresponding, but stronger theorem of [36] for incompressible surfaces, i.e. every incompressible surface is carried by one of finitely many branched surfaces. Thus by Brittenham a positive solution to Problem 2.2 implies a positive solution to Problem 2.3. In particular carrying out this program for a frontier manifold (See FQ 1.1-1.5) would probably yield either a first negative example to Problem 1.1 or give rise to a new construction of essential laminations.

v) A twisted disc of contact [59] is an obstruction for the branched surface B to fully carry something, for it forces nontrivial holonomy about a homotopically trivial curve. It is easy to construct examples which hide immersed twisted discs of contact.

vi) Naimi has observed that the existence problem of taut foliations on Seifert fibred spaces is equivalent to the question of whether certain simple branched surfaces carry laminations with particular slopes. This points to the subtle nature of Problem 2.1, as does the following

Theorem 2.1. (Lee Mosher) [84]. *There is no algorithm to decide if a general branched surface abstractly carries a lamination.*

Proof. Associated to a finite collection of $\mathbb{R}^2$-prototiles (or dominos) one readily constructs a compact branched surface such that the prototiles generate a tiling of $\mathbb{R}^2$ if and only if the associated branched surface carries a lamination. Since by [3] there exists no algorithm to decide whether or not a finite collection of $\mathbb{R}^2$-prototiles generates a tiling of $\mathbb{R}^2$, Mosher's theorem follows. □

Remark 2.2. i) See [85] for the general definition of *carries* which I refer to here as *abstractly carries.* The usual definition implicitly implies that the branched surface has a fibred neighborhood and hence embeds in a 3-manifold. Christy [17] gives the first and simplest example of a branched surface which does not embed in any 3-manifold.

ii) Any compact branched surface B smoothly embeds in some closed 4-manifold.

iii) An analogue of Theorem 2.1 to Problem 2.1 is the fact that the word problem is unsolvable for 4-manifold fundamental groups while the word problem is unknown for 3-manifold groups.

iv) Problem 2.5 is another way of asking whether branched surfaces are interesting only because they carry laminations, or whether they can have a life of their own. There may be an interface with Matveev's theory of standard spines [81]. Christy [18] gives a necessary and sufficient condition for deciding if a standard spine can be deformed into a branched surface.

3. Compression Problem

Background 3.1. A closed 2-sided surface S in the 3-manifold N can be compressed, by the loop theorem, to a union of 2-spheres and π_1-injective surfaces. If $f : M \to N$, M is irreducible, and S is incompressible in N then, by Stallings, f is homotopic to g where $g^{-1}(S)$ is incompressible in M. The challenge is to find laminations versions of these results.

Definition 3.2. The lamination $\lambda \subset M$ is *transversely essential* if

i) All curves in leaves of λ with nontrivial holonomy are homotopically nontrivial in M.

ii) λ is carried by a branched surface B with efficient vertical fibres , i.e. if τ is an arc contained in a vertical fibre of $N(B)$, then τ cannot be homotoped relative to its endpoints into the leaf containing $\partial\tau$.

Remark 3.3. i) Since the Reeb stability theorem asserts that curves with nontrivial holonomy are π_1-injective in their leaves, condition i) is weaker than the statement, "leaves of λ are π_1-injective in M". However i) is needed to make the notion of compressing a lamination make sense. Suppose L is a leaf of λ and $\gamma \subset L$ bounds a disc D transverse to λ, with $\mathring{D}$ embedded in M. If D is embedded, then by i) D can be enlarged to a disc E, with $\partial E \subset M - \lambda$ and $\lambda \mid E$ a lamination by circles. (Recall that we are assuming that λ is nowhere dense.) The disc E gives rise to a whole compact set of compressions. If D is not embedded, then ∂D lies on a 1-sided leaf L. In that case, split L (into its double covering) to reduce to the embedded case.

ii) If $f : M \to N$, λ is essential in N and transverse to f, then $f^{-1}(\lambda) = \mu$ is a transversely essential lamination. A straightforward application of Stallings' lemma shows that a compression of μ which does not involve splitting of leaves can be realized by a homotopy of f.

Examples 3.4. *(compression cycle)* In contrast to the situation of compact surfaces, a transversely essential lamination may require an infinite number of compressions to transform itself to an essential lamination. Here is the simplest example, viewed in the sutured manifold (M,γ) where $M = T \times I$, $T = T^2 - \overset{\circ}{D}{}^2$ and sutures$(\gamma) = s(\gamma) = \partial T \times \{0,1\}$. That is view ∂M as a manifold with corners, i.e., ∂M is decomposed into 5 smooth surfaces which transversely meet along their boundary circles. These surfaces are $A(\gamma) = N(\partial T \times \{0,1\}), T \times 0 - \overset{\circ}{A}(\gamma), T \times 1 - \overset{\circ}{A}(\gamma)$ and $\partial T \times I - \overset{\circ}{A}(\gamma)$. See Figure 3.1. or [39, p.449]. Figure 3.1 also shows a branched surface consisting of $T \times 0, T \times 1$ together with an annulus, far away from $\partial T \times I$, connecting them. Construct a lamination λ_0 of M, transverse to $A(\gamma)$, tangent to $R(\gamma) = \partial M - \overset{\circ}{A}(\gamma)$, carried by the branched surface of Figure 3.1 consisting of the following 4 leaves. $T \times 0, T \times 1, \partial T \times I$ and S. (We suppress the fact that e.g. $T \times 0$ really means $T \times 0 - \overset{\circ}{A}(\gamma)$). Here the "core" of S is an annulus almost vertical in $T \times I$ and the ends of S respectively spiral about $T \times 0$ and $T \times 1$. So each end of S looks like "half" of an infinite cyclic covering of $T \times 1$ or $T \times 0$ and in particular has a countable number of boundary components on each component of $A(\gamma)$. Compress λ_0 along the embedded disc of Figure 3.2a to obtain a new lamination λ_1. λ_1 is isotopic to the lamination consisting of the leaves of λ_0, together with a leaf parallel to $\partial T \times I$. After any finite number of compressions of λ_0 the resulting lamination will still be compressible. However after an "infinite" number of compressions one obtains a limit lamination, which is essential, consisting of $T \times \{0,1\}$ and a countable number of leaves parallel to $\partial T \times I$. In the limit the leaves $T \times \{0,1\}$ "split" into $T \times \{0,1\} \cup R$, where R is parallel to $\partial T \times I$.

Example 3.4 is the simplest example of what is called a *compression cycle*, in part because the compressions are carried by an embedded *compression lamination* transverse to λ. In fact by deforming the 2-complex of Figure 3.3a to that of Figure 3.2b one obtains a branched surface B which carries the compression lamination. In Figure 3.3a one

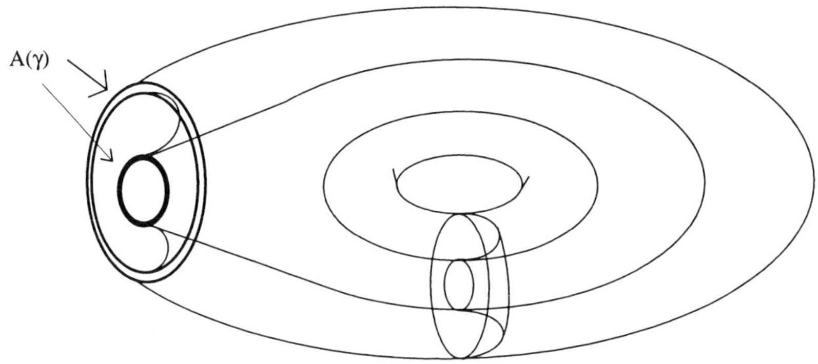

FIGURE 3.1

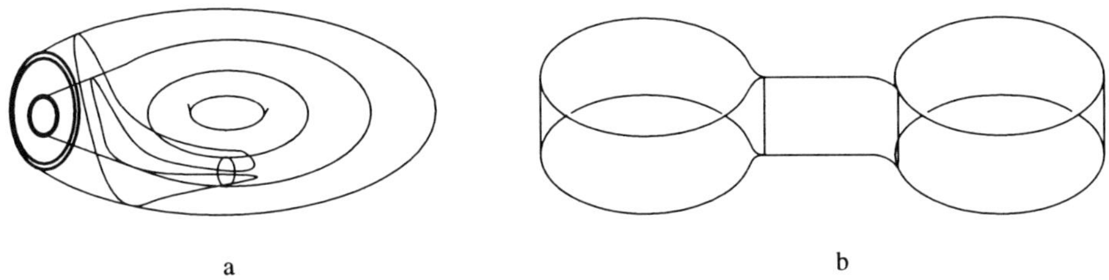

FIGURE 3.2

sees $\lambda \cap B$. We leave the embedding of $B \subset M$ as an exercise to the reader.

Examples 3.5. *(compression complex)* Here is probably the simplest example of a lamination requiring an infinite number of compressions, which do not follow one after the other in the organized fashion of Example 3.4. Let E be the disc with 2 holes. Let (M, γ) be the sutured manifold with $M = E \times I$ and $s(\gamma) = \partial E \times \{0, 1\}$. The laminations constructed below will contain $R_-(\gamma)$ as a leaf, will be tangent to $R_+(\gamma)$ and will be transverse to $A(\gamma) = N(\partial E \times \{0, 1\})$. Here $R_-(\gamma) = E \times \{0, 1\}$ and $R_+(\gamma) = \partial E \times I$. Again I abuse notation, for the surfaces $R(\gamma)$ actually have been cut down in size to make room for $A(\gamma)$. Embed the laminated 1-complex Y of Figure 3.3a into M as in Figure 3.4. The lamination on Y extends "horizontally" to a lamination λ_0 on (M, γ), although along two arcs one has to bend it near, then infinitely spiral around $R_-(\gamma)$ to achieve tangency. See the left hand side of Figure 3.5. λ_0 has 4 leaves, three of which are the components of $R_-(\gamma) \cup R_+(\gamma)$. The fourth is qualitatively very similar to a leaf of the lamination of Example 3.4. We leave it to the reader to analyze its compressions. Next consider the branched lamination B' of Y shown in Figure 3.3b. B' carries a 1-dimensional lamination μ' which has no isolated leaves and is such that any arc transverse to a sector hits μ' in a Cantor set,

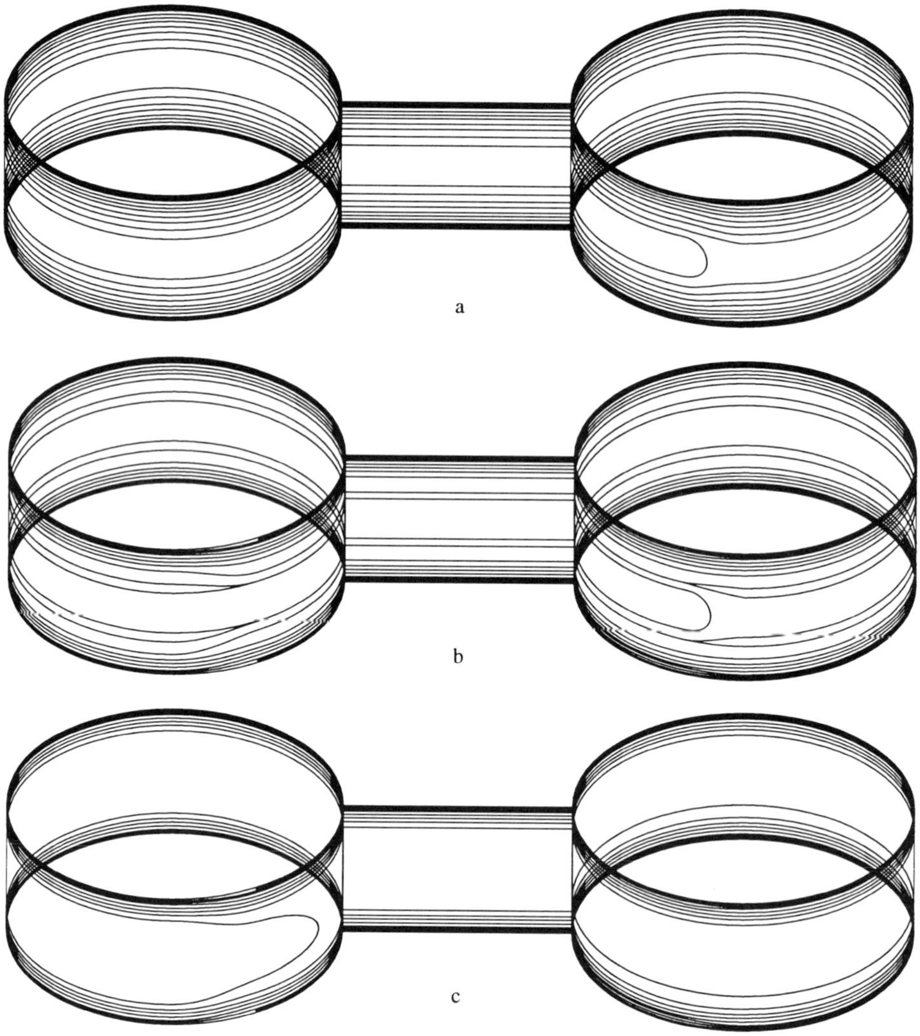

FIGURE 3.3

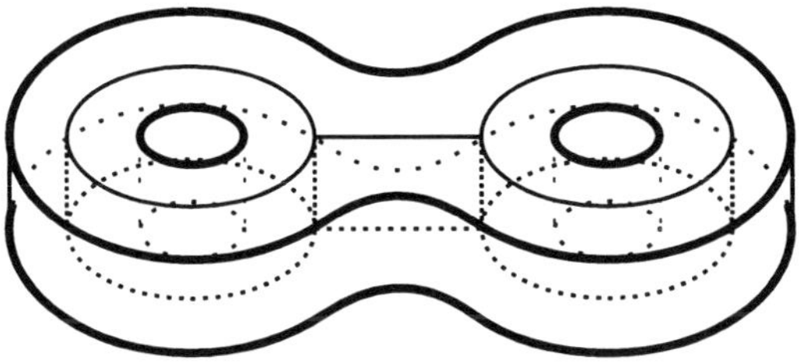

FIGURE 3.4

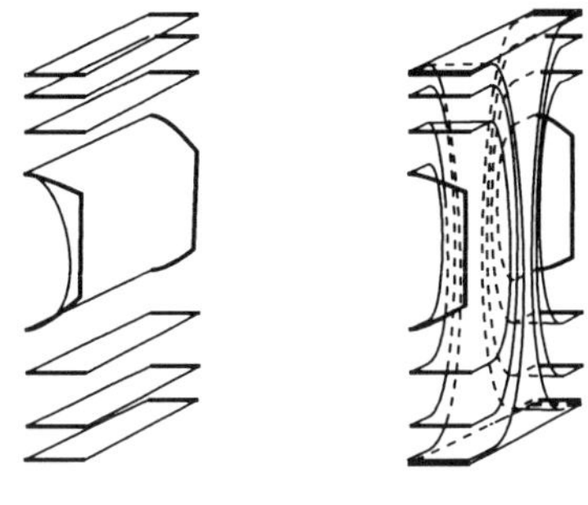

FIGURE 3.5

(rather than just a point as in Figure 3.3a). Nevertheless μ' extends to a lamination μ_0 on $(E \times I, \gamma)$. After a finite number of compressions of μ_0, one obtains the lamination μ_1, which intersects Y as in Figure 3.3c. The point of this example is that by choosing the holonomy of μ_0' correctly the corresponding μ_0 will compress in a very irregular fashion and one might get, after finite time, a lamination which intersects Y as in Figure 3.3b, except the "Reeb" direction is reversed. Nevertheless one can "compress" μ_0 via the single geometric operation of Figure 3.5. In the limit the leaf $R_-(\gamma)$ has been split into $R_-(\gamma) \cup R_+$ where R_+ is parallel to $R_+(\gamma)$.

Example 3.5 is an example of a *compression complex*, i.e. the restriction of a lamination μ to a sutured manifold (M, γ) which requires an infinite number of compressions to fully compress. The limit lamination is obtained more or less by excising μ from $\overset{\circ}{M}$, and reforming near $R_+(\gamma)$ with the leaves of $R_-(\gamma)$ being split into $R_-(\gamma) \cup R_+$ where R_+ is parallel to $R_+(\gamma)$. In particular, up to isotopy, $\overset{\circ}{M}$ has been totally evacuated of lamination.

Problem 3.1. If λ is compressible in M does one of the following occur?

1) After a finite number of compressions λ is essential.

2) After a finite number of compressions λ is essential modulo compression complexes.

3) After a finite number of compressions λ compresses down to a union of S^2's .

Problem 3.2. If $f : M \to N$ is a map of nonzero degree where M is irreducible, closed orientable and N contains an essential lamination λ, does one of conclusions 1), 2) hold for $f^{-1}(\lambda)$? Is f homotopic to g such that $g^{-1}(\lambda) = \mu$ is essential in M?

Problem 3.3. What if f is a homotopy equivalence?

Remark 3.6. i) In the case that f is a homotopy equivalence, then a positive answer to Problem 3.1 implies that f is homotopic to a homeomorphism. Let λ_0 be essential in N and μ_0 be $f^{-1}(\lambda_0)$ compressed as in 1,2). Let h be the homotopy inverse and let λ_1 be $h^{-1}(\mu_0)$ compressed as in 1,2). In a similar manner proceed to construct $\lambda_2, \lambda_3, \cdots$. Eventually μ_i is obtained from $f^{-1}(\lambda_i)$ via a finite number of compressions which involve no splittings of leaves. For if either compressing a compression complex or splitting a 1-sided leaf was required, then the number of non I-bundle closed complementary regions of λ_{i+1} is greater than that of λ_i. This contradicts the fact that the number of such regions is bounded above [57]. On the other hand if μ_i is obtained from $f^{-1}(\lambda_i)$ via a finite number of compressions, then Stallings' lemma implies that f can be homotoped to g so that $g^{-1}(\lambda_i) = \mu_i$. Now if λ is transversely orientable (resp. not transversely orientable), then [9] (resp. [53], an extension of the argument of [9]) shows that f is homotopic to a homeomorphism.

ii) Morgan and Shalen [87] showed that if λ is a measured lamination dual to a nontrivial action of $\pi_1(M)$ on an $\mathbb{R}$-tree, then M contains an incompressible measured lamination. This result is a variant of the compression problem, in that their limit lamination is not obtained from a finite geometric process as above, rather they do an infinite number of compressions, blink their eyes, and suddenly new limit measures induce an incompressible measured lamination.

iii) A very interesting special case is when the transversely essential lamination μ is transversely orientable, e.g. the case of pullbacks of taut foliations. The case of transversely orientable laminations may be more amenable to complexity arguments. E.g. the closed complementary regions can be obtained from a product manifold via the addition of $n < \infty$ 1 and 2-handles, see [63]. The complexity n is not increased and may decrease upon compressing.

iv) An exercise in the use of the sphere theorem and [59] shows that if $f : M \to N$ is a homotopy equivalence and N contains an essential lamination, then M is covered by $\mathbb{R}^3$.

v) It may be both enlightening and profitable to attempt to find a counterexample to Problem 3.1 in the case that λ is not transversely orientable.

4. Theory of Order Trees

The notion of *order tree* (Defintion 6.9 [59]) simultaneously generalizes the notion of $\mathbb{R}$-tree and (not necessarily Hausdorff) simply connected 1-manifold. However order trees $\mathcal{T}$ are topologized by declaring

U open if for every segment σ of $\mathcal{T}$, $U \cap \sigma$ is open in σ, so the order tree topology on an $\mathbb{R}$-tree may be finer than the metric topology. In what follows all order trees under discussion are assumed to be $\mathbb{R}$-order trees, i.e., 2nd countable order trees whose segments are order isomorphic to compact intervals of $\mathbb{R}$.

Remark 4.1. i) By [59] if λ is an essential lamination in the closed orientable 3-manifold M, such that λ has no isolated leaves and $\tilde{\lambda}$ is the lift of λ to the universal covering of M, then $\mathcal{T}(\lambda)$ the *space of leaves* of $\tilde{\lambda}$ is an $\mathbb{R}$-order tree. I.e. $\mathcal{T}$ is the quotient space of $\mathbb{R}^3$, where closed complementary regions and non boundary leaves (i.e. leaves not on the boundary of a closed complementary region) are identified to points. Recall that any lamination can be transformed into one without isolated leaves just by replacing the isolated leaves by I-bundles of leaves.

ii) The same argument shows that if λ has no isolated leaves and is transversely essential, then $\mathcal{T}(\lambda)$ the space of leaves of $\tilde{\lambda}$ is an order tree.

iii) Haefliger and Reeb [68] showed that there is a 1-1 correspondence between {simply connected 1-manifolds} and {spaces of leaves of foliations of $\mathbb{R}^2$ by lines}/ $\sim$, where the equivalence relation $\sim$ accounts for operations such as turning a "Reeb $I \times \mathbb{R}$" upside down.

iv) Palmeira [95] used [68] to show that if $\mathcal{F}$ is a foliation of $\mathbb{R}^n$ by codimension-1 planes, then $\mathcal{F}$ is topologically conjugate to a (foliation on $\mathbb{R}^2$) $\times\ \mathbb{R}^{n-2}$.

v) [57] generalized iii), iv) as follows. There exists a 1-1 correspondence between {$\mathbb{R}$-order trees} and {laminations of $\mathbb{R}^2$ by proper lines without isolated leaves}/$\sim$. If λ is an essential lamination on M, then $\tilde{\lambda}$ is topologically conjugate to (lamination on $\mathbb{R}^2$ by lines)$\times\mathbb{R}$.

Problem 4.1. Develop a theory of group actions on order trees.

Problem 4.2. What 3-manifold groups act nontrivially on order trees? (A nontrivial action does not have a global fixed point.) What about fundamental groups of hyperbolic 3-manifolds, perhaps after passing to a subgroup of finite index?

Problem 4.3. If the closed oriented irreducible 3-manifold M acts nontrivially on an order tree does M possess an essential lamination?

Problem 4.4. If $\pi_1(M)$ acts nontrivially on an order tree is $\tilde{M} = \mathbb{R}^3$?

Problem 4.5. If M is atoroidal and $\pi_1(M)$ acts minimally (i.e., every orbit is dense) on an order tree $\mathcal{T}$, then is $\pi_1(M)$ Gromov negatively curved?

Special Cases. Problems 4.1-4.5 are also very interesting when restricted to the following special cases.

a) $\mathcal{T} = \mathbb{R}$

b) $\mathcal{T}$ is Hausdorff (I.e $\mathcal{T}$ is *tight* See Definition 4.3, Remark 4.4.)

c) $\mathcal{T}$ is a simply connected 1-manifold

d) $\mathcal{T}$ is transversely orientable (i.e., $\mathcal{T}$ is covered by coherently oriented segments)

Problem 4.6. If λ is essential in M and N is obtained by Dehn surgery on a curve in a leaf of λ or on a curve in a complementary region of λ, what is the effect on $\mathcal{T}(\lambda)$.

Remark 4.2. i) Morgan and Shalen [87] proved that if $\pi_1(M)$ acts nontrivially on an $\mathbb{R}$-tree via isometries, then M has an incompressible measured lamination, and hence a π_1-injective surface.

ii) A positive solution to Problem 3.1 implies one for Problem 4.2 for dual to an action of $\pi_1(M)$ on an order tree $\mathcal{T}$, there is a transversely essential lamination in M. In fact given an equivariant triangulation $\tilde{\Sigma}$ of $\tilde{M}$, map vertices of $\tilde{\Sigma}$ into $\mathcal{T}$ equivariantly. Equivariantly extend this map to the 1 and 2 simplices using iii) and iv) of the definition of order tree. This induces a lamination on the 2-skeleton of Σ which restricts to a lamination by circles on $\partial\tau$ for each 3-simplex τ of Σ. This gives rise to an equivariant branched surface on $\tilde{M}$ which carries $\tilde{\lambda}$ with the property that each leaf of $\tilde{\lambda}$ passes through a given sector at most once. Therefore $\tilde{\lambda}$ has no holonomy and contains no monogons. It follows that $\tilde{\lambda}$ covers a transversely essential lamination λ. A finite and complete compression of λ would imply that the group action had a global fixed point. Conversely as noted in Remark 4.1 ii) the space of leaves of the lift of a transversely essential lamination is an order tree.

iii) Mark Brittenham points out that non Haken Seifert fibred spaces make a first test case for Problem 4.3. For it is now completely understood (see Remarks 1.1 i)) exactly which non Haken Seifert fibred spaces have essential laminations, and such laminations are dual to actions of their fundamental groups on $\mathbb{R}$. Also it might not be impossible to classify actions of such groups on $\mathbb{R}$.

iv) Very recently Fenley [35] has found examples of Anosov flows on hyperbolic 3-manifolds such that $\mathcal{T}(\mathcal{F}^s) \neq \mathbb{R}$, where $\mathcal{F}^s$ is the stable lamination. Previously Bonatti and Langevin [4] found examples of Anosov flows in graph manifolds (unions of Seifert fibred spaces glued together along boundary tori) such that $\mathcal{T}(\mathcal{F}^s) \neq \mathbb{R}$.

Definition 4.3. The essential lamination λ is *tight* if $\mathcal{T}(\lambda)$ is Hausdorff.

Remark 4.4. λ is tight if and only if every embedded arc is homotopic rel endpoints to an efficient arc.

Problem 4.7. If M has an essential lamination λ, does M have a tight lamination? Special cases are when λ is genuine or when λ is a taut foliation.

Remark 4.5. i) Since a Hausdorff simply connected 1-manifold is $\mathbb{R}$, a taut foliation λ is tight if and only if $\mathcal{T}(\lambda) = \mathbb{R}$.

ii) Problem 4.7 is one facet of a homotopy version of the following.

Problem 4.8. Let V be a solid torus in the closed orientable 3-manifold M such that $M - \mathring{V}$ is irreducible. If M has an essential lamination λ, does it have an essential lamination μ such that $\mu \mid M - \mathring{V}$ is a lamination by incompressible and boundary incompressible leaves?

Problem 4.9. If M has a taut foliation, does M have a taut foliation with smooth transverse structure?

Remark 4.6. I suspect not. This last problem is related to the question of whether there exists a smooth simply connected 1-manifold on which $\pi_1(M)$ acts smoothly.

5. Minimal Surface Theory

Problem 5.1. If λ is an essential lamination in the 3-manifold M with a fixed Riemannian metric, can λ be deformed to μ so that each leaf of μ is least area in its isotopy class? I.e. if L is a leaf of μ, and C, K are compact embedded surfaces in M such that $C \subset L$ and K is homotopic to C rel ∂K, then $Area(C) \leq Area(K)$.

Remark 5.1. 0) This problem is left vague through the undefined nature of the word deformed. For example one must be able to deal with cylindrical components, i.e. sublaminations of the form (2 dimensional Reeb component)$\times S^1$. Besides isotopy perhaps it should allow for splittings of leaves or collapsing I-bundles of M bounded by leaves of λ to leaves of μ, or restriction to sublaminations. In particular collapsing along non properly embedded I-bundles, i.e., doing an un-Denjoy operation should be allowed.

i) This is known for incompressible surfaces by [86]. The stronger statement, that an embedded, 2-sided, incompressible surface least area in its isotopy class is least area in its homotopy class is due to [37]. The corresponding statement for 1-sided incompressible surfaces is a well known open problem in minimal surface theory.

ii) If $\mathcal{F}$ is a taut C^2 foliation, then by [102] there exists a Riemannian metric for M such that compact subsets of leaves are area minimizing in their relative homology class.

iii) A positive solution to Problem 5.1 would be a major technical development in the theory of essential laminations. In particular the Meeks Yau roundoff trick would yield a positive solution to the following purely topological

Problem 5.2. If λ', μ' are essential laminations in M, then λ' and μ' can be deformed to λ, μ so that each leaf of λ is either a leaf of μ or transverse to μ. Furthermore there is no product region of M bounded by a compact subset C of a leaf of λ together with a compact subset K of a leaf of μ.

Remark 5.2. The known mathematics of minimal surface theory is so vast that there exist multivolume survey books. Mercifully the semi-expository paper of Hass and Scott [73] presents much of what the topologist really needs to know in just a few pages. See [115] for a minimal surfaces proof of Novikov's theorem and the details of Thurston's proof of Sullivan's theorem.

Project 5.3. Through out this paper we have assumed that leaves of laminations are smoothly immersed in the ambient manifold. A major gap in the literature is a proof of the following long "known" regularity result which asserts that this is always true, up to topological conjugacy.

Theorem 5.4. *If $\mathcal{F}$ is a topological foliation or lamination on the 3-manifold M, then $\mathcal{F}$ is topologically conjugate to a foliation or lamination such that each leaf is smoothly immersed in M.*

6. Ideal Geometry

Let λ be an essential lamination in the hyperbolic 3-manifold M. Let $\tilde{\lambda}$ be the lift of λ to its universal covering $\mathbb{H}^3$.

Problem 6.1. If $L = \overset{\circ}{D}{}^2$ is a leaf of $\tilde{\lambda}$, does the embedding of $\overset{\circ}{D}{}^2$ in $\mathbb{H}^3$ extend continuously to a map $f : D^2 \to \mathbb{B}^3 = \mathbb{H}^3 \cup S^2_\infty$?

Problem 6.2. Is L quasi-isometrically embedded? (I.e. is the restriction of the metric of $\mathbb{H}^3$ to L equal, up to uniformly bounded distortion, to the path metric coming from the induced Riemannian metric?)

Problem 6.3. If λ is a taut foliation does there exist a flow ϕ transverse to λ such that the flow lines of ϕ are uniformly quasigeodesic in $\mathbb{H}^3$. (I.e., is the restriction of the $\mathbb{H}^3$ metric to flow lines equal, up to uniformly bounded distortion, to the path metric coming from the induced Riemannian metric?)

Problem 6.4. In hyperbolic 3-manifolds, for what Anosov or pseudo-Anosov flows are the flow lines uniformly quasi-geodesic?

Problem 6.5. (Fenley). Let $\mathcal{F}$ be a foliation of the closed orientable 3-manifold M such that $\pi_1(M)$ is Gromov-negatively curved. Is $\mathcal{T}(\mathcal{F}) = \mathbb{R}$ if and only if each leaf has the entire S^2_∞ as its limit set? If not characterize when the latter condition holds.

Remark 6.1. i) For the case of surface bundles over S^1, the answers to Problems 6.1-6.3 are yes, no, yes, thanks to the celebrated work of Cannon Thurston [22]. In fact in Problem 6.1 $f \mid \partial D^2$ is a space filling curve of S^2_∞. Furthermore the leaves of the transverse laminations are quasi-isometrically embedded. The solution to Problem 6.1 has been extended by Fenley [30] to the case of most depth-1 foliations (which are not perturbed fibrations). He shows that the depth-1 leaves lift to non quasi-isometrically embedded planes which limit on Sierpinski curves, i.e. closed sets whose complements consist of a countable number of open discs. It had been previously known by Thurston that the depth-0 leaves lift to quasiisometrically embedded planes which limit to Jordan curves.

ii) Alberto Candel [11] generalizing work of Cantwell-Conlon has shown that if λ is an essential lamination in an atoroidal 3-manifold M, then M has a Riemannian metric whose restriction to the leaves of λ is a metric of constant negative curvature. This implies that (with respect to any metric) the leaves of $\tilde{\lambda}$ have natural compactifications to D^2. The first statement is a special case of a more general theorem of Candel.

iii) If $\mathcal{F}$ is a foliation, then Fenley [31] has shown that the answer to Problem 6.2 is no, more precisely there is at least one leaf of $\mathcal{F}$ whose lift is not quasiisometrically embedded. Furthermore if $\mathcal{T}(\mathcal{F}) = \mathbb{R}$, then the lift of each leaf limits on all of S^2_∞, thereby establishing one direction of Problem 6.5. It is not known whether the lifts extend continuously.

iv) If $\mathcal{T}(\mathcal{F}^s) = \mathbb{R}$, where $\mathcal{F}^s$ is the stable foliation of an Anosov flow in a hyperbolic 3-manifold, then by Fenley [32] the flow lines are not uniformly quasi-isometrically embedded. Fenley conjectures that for Anosov flows the answer to Problem 6.4 is always no.

v)Fenley and Mosher [112] prove that for pseudo Anosov flows transverse to finite depth foliations the answer to Problem 6.4 is always yes. Partial results are obtained in [83].

vi) Fenley actually proves his results [31]-[32] in the more general setting that $\pi_1(M)$ is Gromov negatively curved.

vii) Zeghib [111] has shown that if M has negative sectional curvature, then the leaves of a lamination cannot be totally geodesic, unless the leaves are compact and finite in number.

Perhaps these problems can be viewed as warmups to the following outstanding well known problems.

Problem 6.6. Let λ be an essential lamination in the closed atoroidal manifold M. Does $\tilde{M}$ naturally compactify to $\mathbb{B}^3$ so that each leaf $\overset{\circ}{D}{}^2$ of $\tilde{\lambda}$ extends to a continuous map of D^2 to $\mathbb{B}^3$.

Problem 6.7. Are atoroidal manifolds with essential laminations Gromov negatively curved?

Problem 6.8. (Hyperbolization conjecture). Do atoroidal manifolds with essential laminations have hyperbolic structures?

Remark 6.2. i) Its not yet known how to use the essential laminations hypothesis, beyond what is known [104] for Haken manifolds. One should keep in mind that that by [22] S^2_∞ is obtained as a quotient map of ∂D^2. The quotient information is encoded purely in terms of the invariant stable and unstable laminations on $\overset{\circ}{D}{}^2$ arising from the pseudo-Anosov maps. Also the limit points of the pseudo-Anosov flow lines are dense in S^2_∞. At least in the case of "good" pseudo-Anosov flows, (e.g. ones obtained by starting from a transverse flow to a finite depth foliation and doing a reasonable Dehn surgery along a closed orbit) can we not take the hint and use the flow lines to help find the compactification? In the case of depth-1 foliations on hyperbolic manifolds can we see the $\mathbb{B}^3$-compactification purely in terms of the topology of the various foliations and laminations?

ii) Atoroidal 3-manifolds with genuine laminations reek of negative curvature, at least near the *seam*, i.e. the non foliation locus.

7. Algebraic Topology

This section surveys what is known for the following classical 3-manifold problems, for closed orientable manifolds with essential laminations. More results can be found in Kirby's problem set [79].

Problem 7.1. (Homotopy equivalence implies homeomorphism). Let $f : M \to N$ be a homotopy equivalence between a closed orientable irreducible M^3 and a manifold N with essential lamination. Is f homotopic to a homeomorphism? What if $M = N$? What if M also has an essential lamination?

Problem 7.2. (Homotopy implies isotopy for homeomorphisms). Let M have an essential lamination. If $f : M \to M$ is a homeomorphism homotopic to id_M, is f isotopic to id_M?

Problem 7.3. (Finiteness of mapping class group). If M is an atoroidal manifold with an essential lamination is Homeo(M) /Homeo$_0$(M) finite?

Problem 7.4. (Missing boundary problem). Let M have an essential lamination. If G is a finitely generated subgroup of $\pi_1(M)$ is $M_G = \overset{\circ}{X}$, where X is a compact 3-manifold and M_G is the covering of M with $\pi_1(M_G) = G$. I.e., is the covering determined more or less just by its abstract π_1?

Remark 7.1. i) One approach to problem Problem 7.1 is discussed in Remarks 3.6 i). The case of $M = N$ in Problem 7.1 asks if manifolds with essential laminations have exotic self maps. (By combining the results of [92], [5], and [69] one concludes that there exists a homotopy equivalence $f : L(8,1) \to L(8,1)$ not homotopy equivalent to a homeomorphism.) Waldhausen [105] answers Problem 7.1 positively if N is irreducible and contains an embedded π_1-injective surface.

ii) Consider the map ϕ:Homeo(M)/Homeo$_0$(M) $\to$ Out($\pi_1(M)$)= Aut($\pi_1(M)$)/Inn($\pi_1(M)$). Since M is a $K(\pi, 1)$ a positive solution to Problem 7.1 for the case $N = M$ is equivalent to ϕ being surjective. A positive solution to Problem 7.2 is equivalent to ϕ being injective.

iii) Again [105] answers 7.2 positively if M has an embedded π_1 injective surface. [56] gives a positive solution to Problem 7.2 in the case that M contains a genuine lamination λ with a $\overset{\circ}{D}{}^2 \times S^1$ complementary region, whose closure is not an I-bundle. E.g. most surgeries on fibred hyperbolic knots. See Examples 1.1 i).

iv) Problem 7.3 is true for Haken manifolds by Johannson [74] and for manifolds with genuine laminations by [114].

v) If $G = 1$, then Problem 7.4 is true (i.e., $M_1 = \mathbb{R}^3$) for Haken manifolds by [105], for manifolds with Reebless foliations by [95], for manifolds with essential laminations by [59]. If $G = \mathbb{Z}$, Problem 7.4 is true for manifolds with essential laminations by [54]. See [79] for what's known in other cases, most notably [6].

Remark 7.2. Candel [12] has shown that an atoroidal manifold M with a tight lamination has a rank two free subgroup in its fundamental group provided that $\mathcal{T} \neq \mathbb{R}$. If $\mathcal{T} = \mathbb{R}$, then it has a free rank two semigroup.

I know of no results regarding other classical fundamental group problems. E.g. Do 3-manifolds with essential laminations have residually finite, coHopfian or Hopfian fundamental groups? Is the word problem solvable?

Problem 7.5. (Thurston). If M is closed, orientable, atoroidal and $\psi \in H^2(M)$ is such that $x^*(\psi)=1$, is ψ the Euler class of some transversely orientable taut foliation on M?

Remark 7.3. See [103, p.129]. Here x^* is the dual thurston norm. By [39], Problem 7.5 is true for the vertices of the unit x^*-sphere of $H^2(M)$. In fact if S is any thurston norm minimizing surface which is a rational multiple of a point t on the unit x-sphere (on $H_2(M)$) where lies in the interior of a face dual to a vertex ψ of the x^*-sphere, then any taut foliation which has S as a leaf realizes ψ as its Euler class. Such a foliation exists by [39].

For a thurston norm minimizing surface S which projects to a higher codimension face of the x-sphere, is there a relation between the various Euler classes ψ of dual norm 1 which satisfy $\psi(S) = -\chi(S)$ and the various finite depth foliations (corresponding to different sutured manifold hierarchies) which realize S as a compact leaf.

8. Knot Theory

Conjecture 8.1. (Strong property P). *No manifold obtained by nontrivial surgery on a knot k in S^3 contains a fake 3-cell.*

Problem 8.2. (Achilles knots). For what knots k in S^3 does non trivial surgery always yield an irreducible 3-manifold with infinite π_1?

Remark 8.1. i) Strong property P would be a consequence of positive proofs of the property P and weak cabling conjectures. See Kirby's problem set for a long list of partial results towards the property P conjecture (which asserts that one cannot obtain a fake 3-sphere by surgery on a knot in S^3) and weak cabling conjecture (which asserts that surgery on a hyperbolic knot yields an irreducible 3-manifold). By Gordon - Luecke [62] one cannot obtain S^3 by nontrivial surgery on a nontrivial knot in S^3.

ii) The Achilles knots are by definition the positive solutions to Problem 8.2. The terminology Achilles knot is a natural since there is only one surgery which "kills" the complement. The Fintushel Stern knot, the (-2,3,7) pretzel is a non Achilles knot. Actually (-2,3,7) can be viewed as one of an infinite set $\mathcal{B}$ of 1-bridge braids in solid tori discovered and classified by Berge [2]. Each $k \in \mathcal{B}$, has the property that $k \subset D^2 \times S^1 = V$ and a nontrivial surgery on k yields a $D^2 \times S^1$. When V is trivially embedded in S^3, then nontrivial surgery on k, yields a lens space and so these *Berge* knots are non Achilles. Berge also proved that (-2,3,7) is the only knot in V such that distinct non trivial surgeries yield a $D^2 \times S^1$ and that there are exactly two such surgeries. Compare [48].

iii) Although the theory of surgery on knots has received great attention these last 30 years, as of 1990 the collection of proven Achilles knots was quite sparse, even among knots of ≤ 10 crossings, although Weeks' [106]-[107] thesis and SnapPea strongly suggested that all the prime nontorus knots of < 10 crossings were Achilles. By [59] if each nontrivial surgery on k has an essential lamination, then k is Achillies. Thus by Examples 1.1 viii)-xii) Delman, Naimi, Roberts and Wu showed that many knots were Achilles, in particular the nontorus alternating knots [25]. The knots listed in FQ1.2 are also the prime non torus knots of ≤ 10 crossings not known to be Achilles.

Problem 8.3. Compute $depth(k)$ =min$\{n \mid$ there exists a depth-n foliation $\mathcal{F}$ on $S^3 - \overset{\circ}{N}(k)$ such that $\mathcal{F} \mid \partial N(k)$ is a foliation by circles$\}$.

Remark 8.2. By Stallings $depth(k) = 0$, i.e., k is fibred if and only if $\pi_1(S^3 - \overset{\circ}{N}(k))$ has a finitely generated commutator subgroup. By Seifert a necessary condition is $1/2(\deg \Delta(k)) =$ genus k, Δ being the Alexander polynomial. Geometric techniques for detecting if $depth(k) = 0$ can be found in [46]. By [44] the knots of ≤ 10 crossings and the alternating knots have $depth(k) \leq 1$. Conlon and Cantwell [13] give examples for each n of satellite knots with depth$(k) = n$. Very recently Kobayashi [80] gave the first example of a hyperbolic knot k with depth(k)=2. See [16] for a classification of certain depth-1 foliations on some knot complements. One can restate Problem 8.3 by not requiring that $\mathcal{F} \mid \partial N(k)$ be a foliation by circles.

Problem 8.4. Let $k(0)$ denote the manifold obtained by 0-frame (i.e. longitudinal) surgery on k. If $k(0)$ has a taut depth-n foliation, is $depth(k) \leq n$?

Remark 8.3. i) In [41] it is shown that k is a fibred knot if and only if $k(0)$ fibres over S^1. This problem is the natural generalization of that work to depth-n knots.

ii) There is something mysterious about this problem and what it says about S^3.

Problem 8.5. (Smooth Reeb Placement Problem). Let k be a knot in S^3. Does there exist a smooth foliation $\mathcal{F}$ of S^3 having a unique Reeb component with core k?

Remark 8.4. Suspect not. If you either drop the smoothness condition (which refers to the transverse structure) or know genus$(k) > 1$, then the answer is yes [39]. Cantwell and Conlon [14] have observed that the degree of smoothness of a finite depth foliation reflects topological information of the sutured manifold hierarchy.

Definition 8.5. If λ is an essential lamination of $S^3 - \mathring{N}(k)$ then (if possible) define the *degeneracy slope of* $\lambda = d(k, \lambda) = n(p/q)$ where $n(p/q)$ denotes $n \geq 1$, p/q curves in meridian/longitude coordinates, if there exist n pairwise disjoint, properly embedded non parallel annuli in a closed complementary region of λ such that one component lies in a leaf of λ and the other a p/q curve on $\partial N(k)$. Furthermore n is the maximal number of such annuli. Since trivial surgery on k yields S^3, which has no essential laminations, it follows as in Example 1.1 i) that if $q \neq 0$, then $\mid q \mid = 1$ and $n = 1$.

Remark 8.6. By definition $d(k, \lambda) = n(p/q)$ is an invariant of the lamination but the multiplicity n is not an invariant of the knot because Delman constructed a lamination λ of 8_{20} with $d(k, \lambda) = 2(1, 0)$, while $d(k, \mathcal{F}^s) = 1(1/0)$, where F^s is the suspended stable lamination of the fibration. See Remark 8.7 v).

Problem 8.6. Compute the possible values of $d(k, \lambda)$, where λ is essential in $S^3 - \mathring{N}(k)$. In particular when λ is transverse to either a fibration or more generally a finite depth foliation. Is the slope (i.e. without multiplicity) an invariant of the knot k?

Conjecture 8.7. *If λ is an essential lamination in $S^3 - \mathring{N}(k)$ where $S^3 - \mathring{N}(k)$ is atoroidal and $d(k, \lambda) = 1(p/1)$, then $\mid p \mid \geq 2$.*

Remark 8.7. i) Here is another view of $d(k)$ in the case of k fibred. If ϕ is a flow transverse to the fibration $\mathcal{F}$ such that $\phi \mid \partial N(k)$ flows along meridians, then ϕ induces an automorphism of the fibre S, called the *monodromy*. The monodromy is well defined up to isotopy of S, rel ∂S. One can suppress the condition that $\phi \mid \partial N(K)$ is the meridianal flow to obtain the traditional definition of monodromy, which is only well defined up to isotopy of S. Now choose ϕ so that this "weak" monodromy is the normal representative, e.g. pseudo Anosov, periodic, or reducible. In that case $\phi \mid \partial N(k)$ will contain some closed orbit J. Up to multiplicity J is the degeneracy slope. Thus the degeneracy slope measures the difference between the meridinal flow on $\partial N(k)$ and the normal representative flow on $\partial N(k)$.

ii) With this viewpoint the notion of degeneracy slope $d(k)$ generalizes to all fibred knots. Thus for a (p,q)-torus knot k, $d(k) = 1(pq/1)$. The manifold $k(pq/1)$ is a connected sum of lens spaces, $k(\alpha)$ is a lens space if $|< \alpha, pq/1 >| = 1$ and $k(\alpha)$ has Heegaard genus 2 otherwise. Thus the distance of the surgery slope from the degeneracy slope is a measure of the degeneracy of the surgered manifold.

iii) Using [43, Corollary 1.4] and the point of view of i), one can show that if the fibre F of the fibred knot k decomposes as a Murasugi sum, and $d(k) = d(k, \mathcal{F}^s) = 1(p/1)$, then $| p |> 2$. Furthermore taking a hint from Hugh Morton one can construct an atoroidal fibred knot k with $d(k) = 1(-2/1)$. For example, first construct, the (-1,2) cable about knot 8_{21} to obtain a fibred satellite knot k with $d(k) = 1(-2,1)$. Next do a Stallings twist (i.e. a $+1$ surgery along an unknotted untwisted curve in a fibre) about a sufficiently complicated curve in a fibre to obtain a hyperbolic knot k'. This curve should be sufficiently uncomplicated so that $d(k') = 1(-2,1)$. (The fibre for such a k' was hung from my office ceiling for several years. Hyperbolicity was established experimentally using SnaPea [107].) This implies that there exists an atoroidal fibred knot which does not decompose as a Murasugi sum, yielding a counterexample to Conjecture 8.1 of [46]

iv) Penner's 1982 thesis [96] (which computes the action of the mapping class group on measured lamination space), implemented by Penner on computer, experimentally showed in 1982 that $d((-2,3,7)) = 1(18/1)$ and $d(10_{145}) = 1(6/1)$. See [18] for rigorous proofs. For non-torus fibred 2-bridge knots [55] computes $d(k) = n(1,0)$ where $n =$ 2(changes in sign of the continued fraction expansion with even coefficients).

v) The computation of $d(8_{20})$ was first verified computationally (with Kazez) using Penner's program. (In this case once we knew the answer it was effortless to do by hand) In fact 8_{20} gave the first example of a fibred knot whose monodromy is pseudo Anosov, but whose *capped off monodromy* is not, in this case reducible. The capped off monodromy is the monodromy of the associated fibration of the 0-frame surgered manifold, terminology due to Casson-Gordon.

Theorem 8.8. *If* $d(k,\lambda) = q(1,0)$, *then* $1 \leq q \leq 4(genus(k)) - 2$, *otherwise* $d(k) = 1(q/1)$ *where* $| q | \leq 4(genus(k)) - 2$. *If* k *is fibred,* *and* $\lambda = \mathcal{F}^s$ *is the stable lamination transverse to the fibration* $\mathcal{F}$, *then* $2 \leq |q|$.

Proof. Let S be a minimal genus surface for k. After a preliminary isotopy of S, S is transverse to λ and the set of non I-bundle closed complementary regions of $\lambda \mid S$ is a union of surfaces of non positive Euler characteristic. (E.g. a 3-pronged disc has Euler characteristic -1/2). Let $R \subset S$ be the closed complementary region containing ∂S. If $\{A_i\}$ is a maximal set of annuli defining $d(k)$, then after an isotopy, $R \cap (\cup A_i)$ contains no two arcs parallel in R with exactly one endpoint on ∂S. The Poincaré - Hopf index formula implies that $q \leq 4(\text{genus}(S)) - 2$ and so the upper bound follows.

Here is a hint for the lower bound. First put k in thin position and then put the fibration $\mathcal{F}$ in normal form in S^3, this more or less means that $S^3 = S^2 \times I/\sim$ where $S^2 \times 0$ and $S^2 \times 1$ are contracted to points and furthermore the knot complement fibration is transverse, except along saddle tangencies, to the horizontal fibration $\mathcal{H}$ of S^3 by S^2's, at least in a substantial part of S^3. Let X denote the oriented locus of tangency between $\mathcal{F}$ and $\mathcal{H}$, where the orientation is inherited from the transverse orientation of $\mathcal{F}$. The transverse orientation of $\mathcal{F}$ also induces an orientation on the boundary cusps of the transverse lamination $\mathcal{F}^s$. These boundary cusps can be made to closely approximate X, both geometrically and with orientation. For Euler characteristic reasons in an appropriate level 2-sphere $Q = S^2 \times s$, one will find $|C \cap Q| = |k \cap Q| = m$ and $|X \cap Q| = m - 2$ where C is one of the cusps. This means that C passes near the same saddle tangency of $\mathcal{F}$ and $\mathcal{H}$ twice which means that the cusp C crosses the fibre at least twice and therefore that $|q| \geq 2$. □

References

1. J. W. Alexander, *On the Subdivision of 3-Space by a Polyhedron*, Proc. Nat. Acad. Sci. **10** (1924) 6-8.
2. J. Berge, *The knots in* $D^2 \times S^1$ *with Nontrivial Dehn Surgeries yielding* $D^2 \times S^1$, Top. and App. **38** (1991) 1-19.
3. R. Berger, *The Undecidability of the Domino Problem*, Mem. AMS **66** (1966).
4. C. Bonati and R. Langevin, *Un example de flot d'Anosov Transitif Transverse a un Tore et non Conjugue a une Suspension*, Preprint.
5. F. Bonahon, *Diffeotopies des Espaces Lenticulaires*, Topology **22** (1983) 305-314.
6. _____, *Bouts des varietes Hyperbolics de Dimension Trois*, Ann. Math. **124** (1986) 71-158.

7. M. Brittenham, *Essential Laminations in Seifert Fibred Spaces*, Topology **32** (1993) 61-85.
8. _____, *Essential Laminations and Haken Normal Form I*, Pacific J. **168** (1995) 217-234.
9. _____, *Essential Laminations and Deformations of Homotopy Equivalences: From Essential Pullback to Homeomorphism*, Top. and App. **53** (1994) 317-324.
10. _____, *Small Seifert-fibred spaces and Dehn surgery on 2- bridge knots*, Preprint.
11. A. Candel, *Uniformization of Surface Laminations*, Ann. Ecole Norm. Sup. **26** (1993) 489-516.
12. _____, *Laminations with Transverse Geometric Structure*, Preprint.
13. L. Conlon and J. Cantwell, *Depth of Knots*, Top. and App. **42** (1991) 277-289.
14. _____, *Topological Obstructions to Smoothing Proper Foliations*, Cont. Math., to appear.
15. _____, *Every Surface is a Leaf*, Topology **26** (1987) 265-285.
16. _____, *Surgery and Foliations of Knot Complements*, J. Knot Theory and Ram. **2** (1993) 369-397.
17. J. Christy, *Immersing Branched Surfaces in Dimension-3*, Proc. AMS **115** (1992) 853-861.
18. _____, *Standard Spines and Branched Surfaces*, Preprint.
19. A. Casson and D. Jungreis, *Seifert Fibered Spaces and Convergence Groups*, Invent. Math. **118** (1994) 441-456.
20. E. Claus, 1990 U. Texas, Austin thesis.
21. C. Camacho and A. Neto, translated by S. Goodman, *Geometry of Foliations*, Birkhauser, 1986.
22. J. Cannon and W. P. Thurston, *Group Invariant Peano Curves*, Preprint.
23. C. Delman, *Essential Laminations and Dehn Surgery on 2-Bridge Knot Complements*, Top. and App. **63** (1995) 201-221.
24. _____, in preparation.
25. C. Delman and R. Roberts, *Alternating knots have property P*, in preparation.
26. D. Epstein, *Periodic Flows on 3-Manifolds*, Ann. Math. **95** (1975) 66-82.
27. _____, *Pointwise Periodic Homeomorphisms*, Proc. London Math. Soc **42** (1981) 415-460.
28. D. Epstein, K. Millett, and D. Tischler, *Leaves without Holonomy*, J. London Math. Soc. **16** (1977) 548-552.
29. D. Eisenbud, W. Neumann, and U. Hirsch, *Transverse Foliations of Seifert Bundles and self Homeomorphisms of the Circle*, Comm. Math. Helv. **56** (1981) 638-660.
30. S. Fenley, *Asymptotic Properties of Depth 1 Foliations in Hyperbolic 3-Manifolds*, J. Differential Geom **36** (1992) 269- 313.
31. _____, *Quasi-isometric Foliations*, Topology **31** (1992) 667-676.
32. _____, *Quasigeodesic Anosov Flows and Homotopic Properties of Flow*

Lines, J. Differential Geom. **41** (1995) 479-516.
33. _____, *On End Periodic Homoemorphisms and 3-Manifolds*, Math. Zeit. to appear.
34. _____, *Anosov flows in 3-Manifolds*, Ann. Math. **139** (1994) 79-115.
35. _____, *The Structure of Branching in Anosov Flows of 3-Manifolds*, Preprint.
36. W. Floyd and U. Oertel, *Incompressible Surfaces via Branched Surfaces*, Topology **23** 117-125.
37. M. Freedman, J. Hass and P. Scott, *Least Area Incompressible Surfaces in 3-Manifolds*, Invent. Math. **71** (1983) 609-642.
38. R. Fintushel and R. Stern, *Constructing Lens Spaces by Surgery on Knots*, Math. Zeit. **175** (1980) 33-51.
39. D. Gabai , *Foliations and the Topology of 3-Manifolds*, J. Differential Geom. **18** (1983) 479–536 .
40. _____ , *Foliations and the Topology of 3-Manifolds. II*, J. Differential Geom. **26** (1987) 461-478.
41. _____ , *Foliations and the Topology of 3-Manifolds. III*, J. Differential Geom. **26** (1987) 479-536 .
42. _____, *The Murasugi Sum is a Natural Geometric Operation. I* Cont. Math. **20** (1983) 131-145.
43. _____, *The Murasugi Sum is a Natural Geometric Operation. II*, Cont. Math. **44** (1985) 93-100.
44. _____ , *Foliations and Genera of Links*, Topology **23** (1984) 381-394.
45. _____, *Genera of the Arborescent Links*, AMS Memoirs **339** (1986).
46. _____, *Detecting Fibred Links in* S^3, Comm. Math. Helv. **61** (1986) 519-555.
47. _____, *On 3-Manifolds Finitely Covered by Surface Bundles*, London Math. Soc. Lec. Notes **112** (1987) 145-167.
48. _____, *1-Bridge Braids in Solid Tori*, Top. and App. **37** (1990) 221-235.
49. _____, *Taut Foliations of 3-Manifolds and Suspensions of* S^1, Ann. Inst. Fourier, Grenoble **42** (1992) 193-208.
50. _____, *Foliations and 3-Manifolds*, Proc. ICM Kyoto-1990 **1** (1991) 609–619 .
51. _____ , *Convergence Groups are Fuchsian Groups*, Ann. Math. **136** (1992) 447-510.
52. _____, *On the Geometric and Topological Rigidity of Hyperbolic 3-Manifolds*, J. AMS, to appear.
53. _____, *An extension of a theorem of Brittenham*, Preprint.
54. D. Gabai and W. H. Kazez, *3-Manifolds with Essential Laminations are Covered by Solid Tori*, J. London Math. Soc. **47** (1993) 557-576.
55. _____, *Pseudo Anosov Maps and Surgery on Fibred 2-Bridge Knots*, Top. and App. **37** (1990) 221-235.
56. _____, *Homotopy, Isotopy and Genuine Laminations of 3-Manifolds*, These proceedings.
57. _____, *Order trees and lamination of the plane*, in preparation.

58. D. Gabai and L. Mosher , in preparation.
59. D. Gabai and U. Oertel, *Essential Laminations in 3-Manifolds*, Ann. of Math. **(2) 130** (1989) 41–73.
60. E. Ghys, *Topologie des Feuilles Generiques*, Ecole Normale Ann of Math. **141** (1995) 387-422.
61. C. Godbillon, *Feuilletages*, Prog. in Math. **93** (1991).
62. C. Gordon and J. Luecke, *Knots are determined by their Complements*, JAMS **2** (1989) 371-415.
63. H. Goda, *Heegaard splitting for Sutured Manifolds and Murasugi Sum*, Osaka J. Math. **29** (1992) 21-40.
64. S. Goodman, *Dehn Surgery and Anosov Flows*, Springer Lecture Notes in Math. **1007** (1983) 300-307.
65. A. Haefliger , *Varietes feuilletes*, Ann. Scuola Norm. Sup. Pisa. **16** (1962) 367-397.
66. J. Hass, *Minimal Surfaces in Foliated Manifolds*, Comment. Math. Helv. **61** (1986) 1-32.
67. J. Hempel, *3-Manifolds*, Ann. of Math. Stud. **86** (1976).
68. A. Haefliger and G. Reeb, *Varietes (non Separees) a une Dimension et Structures Feuilletees du Plan*, L'Ensign. Math. **3** (1957) 107-125.
69. C Hodgson and H. Rubinstein, *Involutions and Isotopies of Lens Spaces*, Lec. Notes. Math., Springer **1144** (1983) 60-96.
70. M. Handel and W. P. Thurston, *Anosov Flows on new 3-Manifolds*, Invent. Math. **59** (1980) 95-103.
71. A. Hatcher, *Examples of Essential Laminations*, Ann. Inst. Fourier, Grenoble **42** (1992) 313-325.
72. G. Hector and U Hirsch, *Introduction to the Geometry of Foliations*, Aspects of Math. **E1** (1981).
73. J. Hass and P. Scott, *The existence of Least Area Surfaces in 3-Manifolds*, TAMS **310** (1988) 87-114.
74. K. Johannson, *Homotopy Equivalences of 3-Manifolds with Boundary*, Lect. Notes Math. Springer **761** (1979) 1-303.
75. W. Jaco, Lectures on 3-Manifold Topology, CBMS Vol. 43, 1980.
76. W. Jaco and U. Oertel, *An Algorithm to Decide if a Manifold is Haken*, Topology **23** (1984) 195-209.
77. M. Jankins and W. Neumann, *Homomorphisms of Fuchsian Groups to PSL(2,R)*, Math. Annalen, Comm. Math. Helv. **60** (1985) 480-495.
78. M. Jankins and W. Neumann, *Rotation Numbers of Products of Surface Homeomorphisms*, Math. Annalen **271** (1985) 381- 400.
79. R. Kirby, *Update to the 1976 Stanford Problem Set*, these proceedings.
80. T. Kobayashi, *Example of a Hyperbolic Knot which does not admit a Depth-1 Foliation*, Preprint.
81. S. Matveev, *A Certain Method for the Construction of 3-Manifolds*, Moscow U. Math. Bull. **30** (1975) 7-14, MR 52 # 4293.
82. J. Milnor, *On the Existence of a Connection with Curvature Zero*, Comm. Math. Helv. **32** (1957) 215-223.

83. L. Mosher, *Examples of Quasi-Geodesic flows on Hyperbolic 3-Manifolds*, Topology 90, (ed. Apanasov, et. al.), de Gruyter, 1992, 227-241.
84. _____, personal communication.
85. L. Mosher and U. Oertel, *Spaces which are not Negatively Curved*, Preprint.
86. W. H. Meeks III, L. Simon and S. T. Yau, *Embedded Minimal Surfaces, Exotic Spheres, and Manifolds with Positive Ricci Curvature*, Ann. of Math **(2) 91** (1982) 621-659.
87. J. Morgan and P. Shalen, *Degenerations of Hyperbolic Structures. II*, Ann. of Math. **127** (1988) 403-456.
88. _____, *Degenerations of Hyperbolic Structures. III*, Ann. of Math. **127** (1988) 457-519.
89. S. P. Novikov, *Topology of Foliations* , Trans. Moscow Math. Soc., 1965 268-304.
90. R. Naimi, *Constructing Essential Laminations in some 3-Manifolds*, Pasific J., to appear.
91. _____, *Foliations of Seifert Fibred Spaces*, Comm. Math. Helv, to appear.
92. P. Olum, *Mappings of Manifolds and the notion of Degree*, Ann. of Math. **58** (1953) 453-480.
93. U. Oertel, *Closed Incompressible Surfaces in Complements of Star Links*, Pacific J. **111** (1984) 209-230.
94. U. Oertel, *Measured Laminations in 3-Manifolds*, TAMS **305** (1988) 531-573 .
95. C. F. B. Palmeira, *Open Manifolds Foliated by Planes*, Ann. Math. **107** (1978) 109-131.
96. R. Penner, *The Action of the Mapping Class Group on Curves in Surfaces*, L'Enseign. Math. **30** (1984) 39-55.
97. G. Reeb, *Sur Certaines Proprietes Topologiques des Varietes Feuilletees*, Actualities Sci. Indust. **1183** (1952) 91-158.
98. A. Reid, *A Non-Haken Hyperbolic 3-Manifold Covered by a Surface Bundle*, MSRI Preprint 084-92.
99. R. Roberts, *Constructing Taut Foliations*, Comm. Math. Helv, to appear.
100. _____, *Taut foliations in punctured surface bundles*, in preparation.
101. J. Stallings, *On Fibering Certain 3-Manifolds*, Topology of 3-Manifolds - Proc. 1961 Ga. Top. Conf., Prentice Hall, (1962) 95-100.
102. D. Sullivan, *A Homological Characterization of Foliations Consisting of Minimal Surfaces*, Comm. Math. Helv. **54** (1979) 218-223.
103. W. P. Thurston , *A Norm For the Homology of 3-Manifolds* , AMS Memoirs **339** (1986) 99-130.
104. _____, *Three-Dimensional Manifolds, Kleinian groups and Hyperbolic geometry*, Proc. Symp. Pure Math. **39** (1983) 87-111.
105. F. Waldhausen, *On Irreducible 3-Manifolds which are Sufficiently Large*, Annals of Math. **87** (1968) 56–88 .
106. J. Weeks, *Hyperbolic structures on three-manifolds*, 1985 Princeton Univ. Ph.D. Thesis.
107. _____, *SnapPea: a computer program for creating and studying hyperbolic*

3-manifolds, available by anonymous ftp from geom.umn.edu.
108. R. Williams, *Expanding Attractors*, Pubs. IHES **43** (1973) 473-487.
109. J. Wood, *Bundles with Totally Disconnected Structure Group*, Comment. Math. Helv. **46** (1971) 271-273.
110. Y. Q. Wu, *Dehn Surgery on Arborescent Knots*, Preprint.
111. Zeghib, *Laminations et Hypersurfaces Geodesiques Des Varietes Hyperboliques*, Ann. Scient. Ec Norm Sup. **24** (1991) 171-188.
112. S. Fenley and Mosher, *Quasi-geometric Flows in Hyperbolic 3-manifolds*, Preprint.
113. D. Gabai, *Malleable lamination in 3-manifolds*, in preparation.
114. D. Gabai and W. H. Kazez, *Finiteness of the mapping class group for 3-manifolds with genuine laminations*, Preprint.
115. J. Hass, *Minimal surfaces in foliated 3-manifolds*, Comm. Math. Helv. **61** (1986) 1-32.

California Institute of Technology, Pasadena, CA 91125 USA
E-mail address: Gabai@ math.caltech.edu

AMS/IP Studies in Advanced Mathematics
Volume **2**, 1997 (Part 2)

PROBLEMS IN LOW-DIMENSIONAL TOPOLOGY

EDITED BY ROB KIRBY

CONTENTS

INTRODUCTION

In April, 1977 when my first problem list [38, Kirby, 1978] was finished, a good topologist could reasonably hope to understand the main topics in all of low dimensional topology. But at that time Bill Thurston was already starting to greatly influence the study of 2- and 3-manifolds through the introduction of geometry, especially hyperbolic. Four years later in September, 1981, Mike Freedman turned a subject, topological 4-manifolds, in which we expected no progress for years, into a subject in which it seemed we knew everything. A few months later in spring 1982, Simon Donaldson brought gauge theory to 4-manifolds with the first of a remarkable string of theorems showing that smooth 4-manifolds which might not exist or might not be diffeomorphic, in fact, didn't and weren't. Exotic $\mathbb{R}^4$'s, the strangest of smooth manifolds, followed. And then in late spring 1984, Vaughan Jones brought us the Jones polynomial and later Witten a host of other topological quantum field theories (TQFT's). Physics has had for at least two decades a remarkable record for guiding mathematicians to remarkable mathematics (Seiberg–Witten gauge theory, new in October, 1994, is the latest example).

Lest one think that progress was only made using non-topological techniques, note that Freedman's work, and other results like knot complements determining knots (Gordon-Luecke) or the Seifert fibered space conjecture (Mess, Scott, Gabai, Casson & Jungreis) were all or mostly classical topology.

So editing a problem list in 1994 is a very different task than in 1977. It would not have been possible for this editor without an enormous amount of help from others. For no particular reason, I did not keep track at first of who provided help with the Updates of the lists from 1977 and 1982 [39, Kirby, 1984], so there are no names attached to the Updates. However Geoff Mess alone must have provided me with half the Updates in dimensions 2 and 3, as well as others. I also received much help from Joel Hass, Cameron Gordon, Dieter Kotschick, Walter Neumann, Peter Teichner, Larry Taylor, Jonathan Hillman, Peter Kronheimer, Marty Scharlemann, Ron Stern, Andrew Casson, Francis Bonahon, Paulo Ney de Souza, Hyam Rubinstein, Lee Rudolph, Robert Myers, Bob Gompf, Selman Akbulut, Chuck Livingston, Tom Mrowka, Mike Freedman, Pat Gilmer, Michel Boileau, Peter Scott, Abby Thompson, Steve Bleiler, Curt McMullen, Raymond Lickorish, Tsuyoshi Kobayashi, Yasha Eliashberg, Yukio Matsumoto, Danny Ruberman, Žarko Bižaca, Wolfgang Metzler, Jim Milgram, Dave Gabai, Darryl McCullough, and many others who contributed to Updates involving their own work. One can imagine an editor who reads and understands hundreds of papers so as to personally write and vouch for each Update; that is not the case here. Rather I am indebted to many erudite and hardworking friends.

The new Problems often have a name or names attached; occasionally the name is the originator of a conjecture or question, but most of the time is the person who helped me write the problem. I'd like to thank also a handful of people who reviewed parts of the penultimate draft: Mess, Gordon, Neumann, Kotschick, Akbulut and Gompf.

One might expect, with a problem list of this size, that the list is all inclusive. Wrong. Of course I have made attempts to cover obvious areas, but I never wished to take on the task of covering everything. For example, laminations are already beautifully covered by Dave Gabai in another problem list in these Proceedings. In the 1977 list, I particularly tried to get problems involving related subjects, but this time, that task was too daunting and no great effort was made. There are not as many problems involving contact structures, graph theory, dynamics, for example, as there could have been.

Will Kazez suggested this task in June 1992, no doubt hoping that I would be done shortly after the August, 1993, Georgia conference. But not much was accomplished before the Georgia conference at which many of the problems were proposed. More were added at conferences in Warwick (August, 1993), Oberwolfach (September 1993), the Park City gauge theory conference (July, 1994), Huia, New Zealand (December, 1994), Princeton (January, 1995), Max Planck Institute, Bonn (May, 1995), and Gokova, Turkey (May 1995); and many problems turned up by e-mail or through personal contacts.

Another half year would have elapsed if I had not had the help of Paulo Ney de Souza for Updates, computer drawn figures, wise advise on editing, and (with Faye Yeager) high tech typing and editing. He is primarily responsible for the huge task of organizing the bibliography and making it complete and accurate. Significant further help in LaTeXing was provided by Larry Taylor, and in producing figures by Silvio Levy, Leo Tenenblat, and Jonas Gomes.

The format of the 1977 list has, in a Procrustean way, been continued. There are still five chapters, one each for the four dimensions plus miscellaneous. The old problems keep their numbers, except that the 1982 list of 4-manifold problems with a few 2-in-4 knot problems, have had the N dropped, e.g. Problem N4.45 became Problem 4.45. Then the new Problems continue the numbering, so, for example, in Chapter 1, Problems 1.1 through 1.51 are from 1977, Problems 1.52 through 1.57 are from 1982, and Problems 1.58 through 1.105 are new. This means that, for example, problems on knot groups can appear in three different sections rather than together. But there is liberal cross referencing, as in the table of contents for each chapter.

The old Problems are stated without changes, except for completed references and corrections of errors.

The texts of the Problems contain abreviated references, typically just the author, journal, and year, which should usually be enough for the reader to guess the paper rather than pause to consult the bibliography where the full citation occurs.

There are indices of conjectures, and an index of mathematical terms including symbols, knots and manifolds. Finally, there is a list of old problem lists.

This Problem List is available in electronic format at the URL:

`http://math.berkeley.edu/`

Berkeley, California

1 Jan 1996

Chapter 1

Knot Theory

- Problems 1.1–1.51 (1977), 1.52–1.57 (1982), 1.58–1.105 (new).
- $S^1 \hookrightarrow S^3$, 1.1–1.47, 1.52, 1.53, 1.58–1.101.
- $S^2 \hookrightarrow S^4$, 1.48–1.51, 1.54–1.57, 1.103–1.105.
- Braids, 1.7, 1.8, 1.84, 1.100.
- Knot groups, 1.9–1.14, 1.57, 1.85, 1.86.
- Properties P, R, 1.15–1.18, 1.82.
- Branched covers, 1.21–1.29, 1.74.
- Concordance and slice knots, 1.19, 1.30–1.47, 1.52, 1.53, 1.93–1.97.
- Various genera, 1.1, 1.20, 1.40–1.42, 1.83.
- Crossing, unknotting and tunnel numbers, 1.63–1.73.
- Hyperbolic knots, 1.75–1.77.
- Dehn surgery, 1.15–1.18, 1.77–1.82.
- Jones polynomial, etc., 1.87–1.92.
- Contact, complex structures, 1.98–1.102.

Introduction

Definitions and notation: most definitions are given in the problems, but here are some that are widely used.

A *diagram* for a knot K in S^3 is, intuitively, how we draw a knot on the blackboard; that is, it is a generic projection onto a plane (but preserving over and undercrossings) of a representative in the isotopy class of submanifolds in S^3 which is K.

Then the *crossing number* $c(K)$ of a knot K in S^3 is the minimal number of crossings in a diagram for K, the minimum being taken over all possible diagrams of K.

The *unknotting number* $u(K)$ of a knot K in S^3 is the minimum, taken over all diagrams of K, of the number of crossings which must be changed to obtain a diagram of the unknot.

The *tunnel number* $t(K)$ of a knot K in S^3 is the minimal number of arcs which must be added to the knot, forming a graph with three edges at a vertex, so that the complement in S^3 (of an open regular neighborhood of the graph) is a handlebody. The boundary of this handlebody is then a minimal Heegaard splitting of the knot complement. This graph is the simplest graph, formed by adding arcs to the knot, which (allowing edges to slide over edges) can be moved into a plane, yet contains the knot. This construction amounts to the same thing as boring holes in the complement of K, whence the name tunnel.

Two knots $f_i : S^n \hookrightarrow S^{n+2}$, $i = 0, 1$, are *concordant* if the f_i extend to an imbedding $F : S^n \times I \hookrightarrow S^{n+2} \times I$ (this is occasionally called *cobordism*, but that term should be reserved for the case in which any oriented N^{n+1} with boundary $S^n \cup -S^n$ is allowed in place of $S^n \times I$).

A knot is *slice* if it is concordant to the unknot. A knot $K = f(S^1) \subset S^3$ is called *ribbon* if f extends to an immersion $f : B^2 \to S^3$ whose singularities are always of the form:

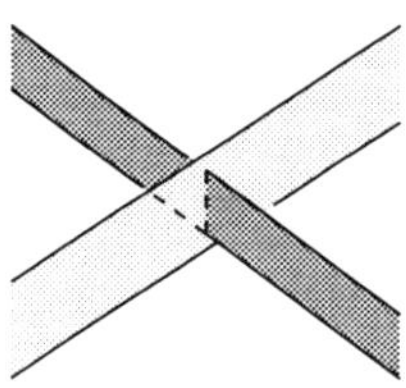

For knots K in S^3, the *genus* of K is the minimal genus over all oriented Seifert surfaces for K; the *free genus* is the minimal genus over all oriented Seifert surfaces for K whose complements have free fundamental group; the *4-ball genus* is the minimal genus over all oriented surfaces in B^4 with boundary K; the *r-shake genus* is the minimal genus over all closed surfaces, which are smoothly imbedded in B^4 with a 2-handle attached along K with framing r ($r = 0$ is the most interesting case), and which represent the generator of H_2.

If K is oriented, then the *reverse* of K is K with the opposite orientation, the *obverse* of K is the reflection of K in a plane, and the *inverse* of K is the concordance inverse of K; note that the composition of two of these gives the third. K is called *reversible, obvertible, or invertible* if there is an orientation preserving homeomorphism $h : S^3 \to S^3$ taking K to its reverse, obverse, or inverse, respectively.

An unoriented knot K is *amphicheiral* if it is isotopic to its obverse, which is equivalent to the existence of an orientation reversing homeomorphism $h : S^3 \to S^3$ such that $h(K) = K$. If K is oriented and h preserves the orientation of K, then K is *positively amphicheiral*; if the orientation is reversed, then K is *negatively amphicheiral* (the latter holds only if K is isotopic to its inverse).

In the last two paragraphs, if h can be chose to be an involution, then the adjective *strongly* should be adjoined to the definition.

Let J be a knot in a solid torus $T = S^1 \times B^2$; J represents an integer in $H_1(T; \mathbb{Z})$ called the *winding number.* Imbed T in S^3 as a tubular neighborhood of a knot J' using the 0-framing of J'; then the image of J is a knot K called the *satellite* of J' using the *pattern* J; also, J' is a *companion* of K.

Let $f : S^1 \times B^2 \hookrightarrow S^3$ be a trivialization of the normal disk bundle of $K = f(S^1 \times 0)$, for which $f(S^1 \times (1,0))$ lies on a Seifert surface for K (equivalently, $f(S^1 \times (1,0))$ represents $0 \in H_1(S^3 - K; \mathbb{Z})$). This trivialization is called the *0-framing* of K, and it defines the longitude λ $(= f(S^1 \times (1,0)))$ (the meridian μ is just $f(point \times \partial B^2)$). Framing n is obtained from $n \in \pi_1(SO(2))$ (in this case, $f(S^1 \times (1,0))$ should wind n times around K as in a right-handed screw). If a 2-handle is added to B^4 along K with framing n, then the boundary is the result of *n-surgery* on S^3 along K. *(p,q)-Dehn surgery* is more general; one removes $f(S^1 \times B^2)$ and glues it back in so that *point* $\times B^2$ is glued to $p\mu + q\lambda$ (the n-surgery above corresponds to (n,1)-Dehn surgery).

Problem 1.1 (Lickorish) Conjecture: *Given a knot K, any band connected sum with an unknot is still a knot. This follows from the*

Conjecture: $genus(K) + genus(L) \leq genus(K \#_b L)$.

Remarks: The group H of $K\#_b$(unknot) has a quotient which is obtained from the group G of K by adding one new generator and one new relation in which the exponent sum of the new generator is 1. It follows from ([373, Gerstenhaber & Rothaus, 1962, Proc. Nat. Acad. Sci. U.S.A.] (also see Problem 5.7)), that if G has a nonabelian quotient Q which imbeds in a compact connected Lie group, then H has a quotient which contains Q, and hence is nonabelian. The first conjecture is therefore true for knots K whose group G has this property, in particular those with $\Delta(t) \neq 1$ and those with G group (Problem 3.33).

Update: The conjectured inequality is true; equality holds iff there is a minimal genus surface for $K \#_b L$ which is the union of a minimal genus surface for K, one for L, and the band b. This was proved independently and simultaneously by Gabai [356, 1987b, Topology] and Scharlemann [970, 1989, J. Differential Geom.].

Problem 1.2 (A) (T. Matumoto) *Suppose the band connected sum of a trivial link (of two components) is the trivial knot. Is the band isotopic to the trivial band?*

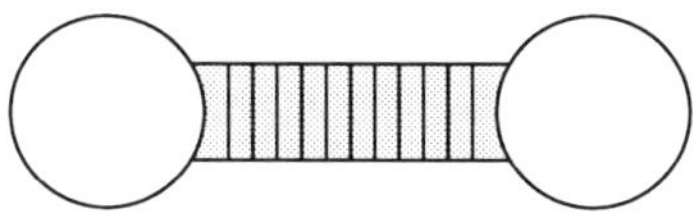

(B) (H. Morton) *Suppose we attach a half-twisted band to the unknot and get the unknot. Is the band isotopic to the trivial band?*

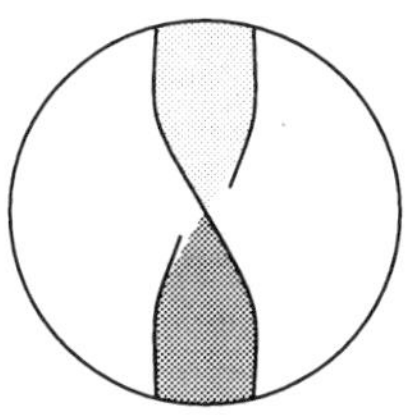

Remarks: Note that these problems are relevant to the question of whether an imbedded T^2 (or $\mathbb{RP}^2$) in $\mathbb{R}^4$ with 4 (or 3) critical points is unknotted (see Problem 4.30).

Update:

(A) Yes [968,Scharlemann,1985a,Invent. Math.].

(B) Yes [109,Bleiler & Scharlemann,1988,Topology].

Problem 1.3 (Gordon) *Is there a knot K such that if a crossing is changed as below, one gets the unknot in one case and the unlink in the other?*

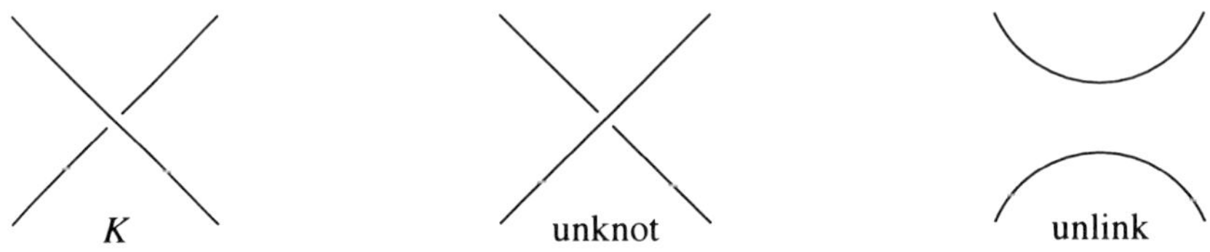

Remarks: Such a knot K bounds a smooth imbedded 2-ball D in B^4 whose complement is a homotopy circle since $(B^4, D) \times I$ is unknotted. There are examples of the latter, e.g., doubles of slice knots [407,Gordon & Sumners,1975,Math. Ann.]. The answer is no if the answer to Problem 1.2(A) is yes.

Update: There is no such knot because the answer to Problem 1.2(A) is yes.

Problem 1.4 (Milnor) *Is the unknotting number of the (p,q)-torus knot, $(p,q) = 1$, equal to $(p-1)(q-1)/2$?*

Remarks: The unknotting number is the minimum number of crossings which must be changed to get the unknot. For the algebraic geometric background for this problem, see [767,Milnor,1968b].

The number $(p-1)(q-1)/2$ is half the degree of the Alexander polynomial of the (p,q)-torus knot. However, it is not even true in general that the unknotting number is $\geq \frac{1}{2}\deg(\Delta(t))$, for the knot 8_{20}

(see Figure 1.4.1) has unknotting number 1 and $\Delta(t) = 1 - 2t + 3t^2 - 2t^3 + t^4$.

Figure 1.4.1. 8_{20}

Update: The equality holds [623,Kronheimer & Mrowka,1993,Topology]. In fact, they prove more: if a knot $K = V \cap S^3$ where V is a complex curve in $\mathbb{C}^2$ with an isolated singularity at the origin, and if V_ε is the Milnor fiber, then the unknotting number of K is equal to the genus of V_ε. A lower bound for the unknotting number of an arbitrary knot K is its 4-ball genus; if K bounds a complex surface V, then the 4-ball genus of K is equal to the genus of V, [960,Rudolph,1993,Bull. Amer. Math. Soc.].

Problem 1.5 (Van Buskirk) Conjecture: *K is amphicheiral iff K is invariant under reflection through the origin.*

Remarks: True for knots with ≤ 10 crossings. Note that a knot K is invariant under reflection through a plane (hence amphicheiral) iff $K = J\# - J$ for some knot J. For such knots, the conjecture is true.

Update: The conjecture is false [447,Hartley,1980,Math. Zeit.]. Hartley gives an example of an orientation reversing diffeomorphism $h : (S^3, K) \to (S^3, K)$, preserving the orientation of K (thus K is positively amphicheiral), and shows there is no orientation reversing involution of S^3 keeping K invariant as a set (K is not strongly amphicheiral).

Problem 1.6 (Montesinos) Conjecture: *Each invertible knot is strongly invertible.*

Remarks: A knot K is *invertible* if there exists an orientation preserving homeomorphism μ of S^3 which takes K to itself but reverses the orientation of K. It is *strongly invertible* if μ is also an involution; in this case, μ is equivalent to a rotation about an axis [1090,Waldhausen, 1969,Topology]. (Added in proof. False, W. Whitten.)

Update: The conjecture is false; examples were given in [447,Hartley, 1980,Math. Zeit.] and [1114,Whitten,1981,Pacific J. Math.]. In fact Whitten proves that a knot K is strongly invertible iff each double of K is strongly invertible; since any double of a knot is invertible [994, Shubert,1953,Acta Math.], begin with a non-invertible knot and any of its doubles are counterexamples.

Boileau [114,1985] gives a complete answer to Montesinos' Conjecture; namely that an invertible knot K is strongly invertible iff K is not a satellite with winding number zero (e.g. a double).

Note that *reversible* is the modern name for what is called *invertible* in this old problem; probably the word *invertible* should be reserved for a knot which is isomorphic to its concordance inverse (see the introduction to this Chapter).

Problem 1.7 (Stallings) *Characterize those braids β of n strands whose closed braid $\hat{\beta}$ is the unknot. Specifically, for σ_i the generators of the braid group, let σ_{ij}, $i < j$, be defined by*

$$\sigma_{ij} = (\sigma_i\sigma_{i+1}\ldots\sigma_{j-2})\sigma_{j-1}(\sigma_i\sigma_{i+1}\ldots\sigma_{j-2})^{-1}.$$

The braid σ_{ij} crosses the ith and j^{th} strands in front of all those strands in between.

Conjecture: *$\hat{\beta}$ is the unknot iff $\hat{\beta}$ is conjugate to a product of $n-1$ elements of the form σ_{ij} or σ_{ij}^{-1}.*

Update: The conjecture is false, for it would imply that there are only finitely many non-conjugate n-braids which close to the unknot; however, Morton [805,1978,Topology] showed that the braids $\sigma_1\sigma_2^{2i+1}\sigma_3\sigma_2^{-2i}$ give an infinite family of non-conjugate 4-braids which close to the unknot.

Furthermore, the braids in the conjecture are precisely those n-braids whose closure spans a disk meeting the braid axis in exactly n points [807,Morton,1985].

Problem 1.8 (Stallings) Suppose β is a word in the generators $\sigma_1, \ldots, \sigma_{n-1}$ and their inverses in the braid group B_n. If the length of β is minimal over all words representing the same element of B_n, call β *minimal.*

Conjecture: *If the last letter of a minimal word β is σ_i^ε, then the word $\beta\sigma_i$ is again minimal ($\varepsilon = \pm 1$).*

Update: Still open in general.

Problem 1.9 (Fox & Birman) Let G be the knot group of a non-trivial knot K and let $\mu \in G$ be represented by a meridian. Let $N(\mu^2)$ be the normal closure of μ^2 in G.

Conjecture: *$G/N(\mu^2)$ is never abelian, i.e., is never $\mathbb{Z}/2\mathbb{Z}$.*

Remarks: Let H be the kernel of the obvious homomorphism $G \to \mathbb{Z}/2\mathbb{Z}$ and note that the normal closure of μ^2 in H is still $N(\mu^2)$. Then $H/N(\mu^2)$ is of index 2 in $G/N(\mu^2)$ so the conjecture is that $H/N(\mu^2)$ is never trivial. But H is the knot group of K in the double branched cover M^3; if $H/N(\mu^2)$ is trivial, then M^3 is a homotopy 3-sphere which must be fake since by Waldhausen there is no involution on S^3 with knotted fixed point set.

Update: The conjecture is true, using the remarks and the fact that the proof of the Smith Conjecture (see Problem 3.38) shows that the double branched cover of K is never a homotopy 3-sphere.

Problem 1.10 (L. Moser) *Is there a geometric characterization of knots whose groups have one relator?*

Remarks: The groups of 2-bridge knots are presented on 2 generators and one relator where the generators are meridians. The groups of torus knots also are presented on 2 generators with one relator but the generators are not meridians.

Update: No characterization is yet known, but it is conjectured to be tunnel number one knots. Bleiler [103,1994,Proc. Amer. Math. Soc.] has proven this conjecture for cable knots, and Bleiler & A. Jones ([107,

1995b] and [106,1995a]) have verified most of the steps to prove it for satellite knots. Note that this conjecture is a special case of: when does the minimal number of generators for π_1 of a 3-manifold equal the minimal genus of a Heegaard splitting? Further note that Boileau & Zieschang [127,1984,Invent. Math.] exhibit a Seifert fibered space with two generators but genus three (this phenomenon has not yet been seen in a 3-manifold with non-trivial boundary).

Problem 1.11 (Cappell & Shaneson) *Is every knot, whose group is generated by 2 meridians, actually a 2-bridge knot? Same for n meridians and n-bridge knots.*

Remarks: (Bailey) If yes, then Fox's *bushel basket* of homotopy 3-spheres [324,1962b] contains only S^3 (see [161,Burde,1971,Canad. J. Math]).

Update: Yes for knots and links when $n = 2$, if the Orbifold Geometrization Conjecture is true (see Problem 3.46).

Problem 1.12 (J. Simon) *Let $G_K = \pi_1(S^3 - K)$.* **Conjectures***: If there is a non-trivial epimorphism $\phi : G_L \twoheadrightarrow G_K$, then*

(A) *$n(G_L) > n(G_K)$ where $n(G)$ is the minimum number of (meridian?) generators; (Added in proof, 1977: False, e.g. the group of the torus knot $(3p, 2), p$ odd, maps onto the trefoil knot group [450, Hartley & Murasugi, 1978, Canad. J. Math].)*

(B) *$genus(L) \geq genus(K)$.*

Remarks: (B) is known if genus $K = \frac{1}{2}\deg\Delta_K(t)$ or if $\phi(l_L) = (l_K)^n$, $l =$ longitude.

(C) *Given K, there exists a number N_K such that any chain of epimorphisms of knot groups $G_K \twoheadrightarrow G_{L_1} \twoheadrightarrow G_{L_2} \twoheadrightarrow \ldots \twoheadrightarrow G_{L_n}$ with $n \geq N_K$ contains an isomorphism.*

Remarks: This implies knot groups are Hopfian (Problem 3.33). Knot groups seem like the right place to start, but the conjecture could be made for compact 3-manifold groups.

(D) *Given K, there exist only finitely many knot groups G for which there is an epimorphism $G_K \twoheadrightarrow G$.*

Update: In (C), note that knot groups are residually finite, hence Hopfian (see the Update to Problem 3.33). Also note Problem 3.100 which is a related problem for closed 3-manifolds.

Problem 1.13 (J. Simon) Conjecture: *Let K be a (p,q)-cable about a nontrivial knot K_0 such that $|p| = 1$ or 2, and let L be a knot such that $\pi_1(S^3 - K) \cong \pi_1(S^3 - L)$; then $S^3 - K$ and $S^3 - L$ are homeomorphic.*

Remarks: The conjecture covers the only remaining case where it is not yet known whether the group of a prime knot in S^3 determines the complement up to homeomorphism. If we draw the (p,q)-torus knot on $\partial(S^1 \times B^2)$ (the knot represents $q \in H_1(S^1 \times B^2)$) and we assume $|q| \neq (1,0)$ and tie $S^1 \times B^2$ into a knot K_0 so that $S^1 \times (1,0)$ is homologically trivial in $S^3 - K_0$, then the resulting knot is the (p,q)-cable about K_0.

The knot is composite or cable iff the complement of K, $C^3(K)$, admits a proper imbedding of an annulus A that is essential in the sense that (i) $\pi_1(A) \to \pi_1(C^3(K))$ is monic, and (ii) A cannot be pushed into $\partial C^3(K)$ by a homotopy fixing ∂A; then K is composite if a boundary curve of A generates $H_1(C^3(K))$ and is cable otherwise [999,Simon, 1973,Ann. of Math.].

Suppose $\pi_1(C^3(K)) \cong \pi_1(C^3(L))$. If $C^3(K)$ has no essential annulus, then $C^3(K)$ is homeomorphic to $C^3(L)$ [303,Feustel,1976,Trans. Amer. Math. Soc.; Theorem 10]. If in addition K has Property P (see Problem 1.15), then K and L are equivalent knots. Assume now that $C^3(K)$ has an essential annulus A. If K is composite, then L is also composite and their prime factors are equivalent, e.g., granny and square knots [304,Feustel & Whitten,1978,Canad. J. Math].

Suppose K is cable. Then K is a torus if $\pi_1(C^3(K))$ has a nontrivial center [162,Burde & Zieschang,1966,Math. Ann.] or if $C^3(K)$ admits no essential imbedding of a torus [Feustel, ibid.]; in this case K is equivalent to L. Hence assume K_0 is nontrivial. If $|p| \geq 3$ or K_0 has Property P, then K is equivalent to L [Feustel & Whitten, ibid.]; otherwise L is a $(\pm p, \pm q)$-cable about a knot L_0 such that $C^3(K_0) \cong C^3(L_0)$, and we arrive at the conjecture.

Note that the papers [Feustel & Whitten, ibid.] and [1000,Simon, 1975,Proc. Amer. Math. Soc.] show that the conjectures, **all knots**

have Property P for S^3 and **complements of prime knots are determined by their groups**, are nearly equivalent.

Update: The conjecture is true [1115,Whitten,1986,Bull. Amer. Math. Soc.], [405,Gordon & Luecke,1989,J. Amer. Math. Soc.], [1028,Swarup, 1986,Bull. Amer. Math. Soc.].

Problem 1.14 (J. Simon) *Characterize those knots K in S^3 for which the commutator subgroup G' of $\pi_1(S^3 - K)$ has infinite weight (is not normally generated by a finite number of elements).*

Conjecture*: K has infinite weight if K has a companion of winding number zero.*

Remarks: The conjecture is true for the untwisted double of any knot.

Construct a knot K by putting a knot J in a solid torus T and tying T in a knot J'; then J' is called a companion of K and J is a satellite of K. K represents an integer in $H_1(T;\mathbb{Z})$ which is called the winding number of J'.

If G' has finite weight, then the Alexander polynomial of K is monic.

Update: The conjecture is false [671,Livingston,1987,Proc. Amer. Math. Soc.].

The modern terminology for satellites is this, using the above notation: J is called a *pattern*, meaning it describes a satellite for any knot that T is tied into; J' is still *a companion* (K can have many companions); K is a *satellite* of J' with pattern J.

Problem 1.15 *Does every nontrivial knot K have Property P; that is, does Dehn surgery on K always give a nonsimply connected manifold?*

Remarks: Knots with Property P (introduced by Bing & Martin, [89, Bing & Martin,1971,Trans. Amer. Math. Soc.]) include:

- *torus knots* (Seifert), [471,Hempel,1964,Proc. Amer. Math. Soc.], [Bing & Martin, ibid.], [398,Gonzalez-Acuña,1970,Bol. Soc. Mat. Mexicana];

- *twist knots* [Bing & Martin, ibid.], [Gonzalez-Acuña, ibid.], [927, Riley,1974,Quart. J. Math. Oxford Ser. (2)];
- *composite knots* [850,Noga,1967,Math. Zeit.], [Bing & Martin. ibid.], [Gonzalez-Acuña, ibid.];
- *doubled knots* [Bing & Martin, ibid.], [Gonzalez-Acuña, ibid.];
- *weakly splittable knots* [224,Connor,1969]; most *cable knots* [997, Simon,1970], [Gonzalez-Acuña, ibid.];
- some *pretzel knots* [Simon, ibid.], [Riley, ibid.];
- some 2-*bridge knots* [Riley, ibid.], [722,Mayland, Jr.,1977];
- and other knots [827,Nakagawa,1975,Publ. Res. Inst. Math. Sci.], [845,Neuzil,1975,Trans. Amer. Math. Soc.], [998,Simon, 1971,Trans. Amer. Math. Soc.], [665,Litherland,1979,Proc. London Math. Soc.].

Update: Gordon & Luecke [405,1989,J. Amer. Math. Soc.] prove that Dehn surgery on a non-trivial knot never gives S^3. But a homotopy 3-sphere is still possible. But Dehn surgery on the following classes of knots always gives non-trivial fundamental group:

- *satellite knots* [357,Gabai,1989,Topology].
- 2-*bridge knots* [1032,Takahashi,1981] by considering representations into $PGL(2,\mathbb{C})$.
- 2-*bridge knots* Delman [241,1995,Topology Appl.] and Naimi [826, 1995] independently constructed essential laminations and by [364, Gabai & Oertel,1989,Ann. of Math.] these have infinite fundamental group.
- *classical pretzel knots* and more generally *Montesinos knots* with all *odd denominators* [241,Delman,1995,Topology Appl.].
- *nonflat alternating knots* [929,Roberts,1995].

Problem 1.16 *Does every nontrivial knot K have Property R, that is, does surgery on K with framing 0 always give a manifold other than $S^1 \times S^2$, as expected?*

Remarks: If surgery gives $S^1 \times S^2$, then K has trivial Alexander polynomial, is prime, is not a doubled knot [811,Moser,1974,Pacific J. Math.], is a slice of an unknotted S^2 in S^4 [580,Kirby & Melvin,1978, Invent. Math.], and does not have a genus one unknotted Seifert surface [637,Lambert,1977,Proc. Amer. Math. Soc.].

Does 0-surgery on K give a manifold which is not even a homotopy $S^1 \times S^2$? This is equivalent, using infinite cyclic covers, to the question: If G is a knot group and $l \in [G,G]$ is the longitude, is $[G,G]$ not normally generated by l? (Note that the normal closure of l in G coincides with the normal closure of l in $[G,G]$, since l commutes with a meridian, which generates $G/[G,G] = \mathbb{Z}$.) Also, see Problem 5.7.

A homology 3-sphere contains a knot K whose 0-surgery produces $S^1 \times S^2$ iff the homology 3-sphere is the boundary of a 4-manifold made with a 0-handle, a 1-handle and a 2-handle. For these homology 3-spheres the appropriate conjecture is that 0-surgery on K alone produces $S^1 \times S^2$.

Update: Gabai [355,1987a,J. Differential Geom.] proved that every knot has Property R. Moreover he proved that 0-surgery on K gives an irreducible 3-manifold M for which $genus(K)$ equals the minimum genus of surfaces in M representing the homology class of a capped off Seifert surface for K (see also Problem 1.41B). Thus M is not even homotopy equivalent to $S^1 \times S^2$.

The last question in the Remarks appears to be wide open.

Problem 1.17 (R. Edwards, after F. Laudenbach & Poénaru) *Suppose K is a nontrivial knot in S^3 with longitude l, tubular neighborhood N, and group $G = \pi_1(S^3 - K)$. It is an appealing* **conjecture**, *made by Poénaru (and others?), that*

(A) *(algebraic version): $l \in G$ cannot be a product of conjugates of itself, with zero exponent sum; that is, there do not exist $a_1, \dots, a_{2n} \in G$ such that $l = a_1^{-1} l a_1 a_2^{-1} l^{-1} a_2 \dots a_{2n}^{-1} l^{-1} a_{2n}$ (the exponents of l alternate here only for convenience).*

(B) *(geometric version): There cannot be an imbedding of a sphere-with-$(2n+1)$-holes F^2 in $S^3 - intN$, with the $2n+1$ boundary* *components of F imbedded onto parallel copies of l in ∂N so that their algebraic sum in ∂N is l.*

Remarks:

1. Clearly (A) $\Rightarrow$ (B). Conversely, (B) $\Rightarrow$ (A), for given that (A) fails, there is a map$f : (F^2, \partial F^2) \to (S^3 - \text{int}N, \partial N)$, imbedding ∂F^2 onto $2n+1$ copies of l with algebraic sum l in ∂N. Let M_K denote the 3-manifold obtained by doing 0-framed surgery on S^3 along K. Apply the Sphere Theorem to the capped-off map f taking S^2 into M_K, to get an imbedded, nonseparating S^2 in M_K. Such an S^2 must represent a generator of $H_2(M_K) = \mathbb{Z}$, and so it provides, after puncturing, the desired geometric surface F^2 of (B).

2. If l normally generates $[G,G]$ (cf. previous problem), then (A) is false, as one sees by writing $l \in [[G,G],[G,G]]$ as $l = \prod_j [c_j, d_j]$, $c_j, d_j \in [G,G]$, and then writing each c_j and d_j as a product of conjugates of l. In other words, if the Poénaru Conjecture holds for a knot K, then K has homotopy Property R.

3. This question is related to the previous one, for if M_K is homotopy-equivalent to $S^1 \times S^2$ (M_K as above), then K bounds in S^3 the punctured sphere F^2 as in (B). (Is the converse true? Cf. the final remarks of Problem 5.7.) In the study of 4-manifolds, this question arises from the following question: Suppose W^4 is a Mazur-like contractible 4-manifold constructed by attaching a 2-handle to $S^1 \times B^3$ along a degree 1 curve Γ in $S^1 \times \partial B^3$, and suppose $\partial W^4 = S^3$. Is Γ necessarily unknotted in $S^1 \times S^2$?

4. F. Laudenbach has shown that the conjecture is true for $n = 1$.

Update: Gabai's work (see Problem 1.16 Update) implies that (B) and hence (A) are true, because 0-surgery on K is irreducible.

Problem 1.18 (J. Martin) (A) *Suppose T_1 and T_2 are solid tori with $T_2 \subset \text{int}\, T_1$, such that the wrapping number of T_2 in T_1 is nonzero, and the winding number of T_2 in T_1 is zero. Can T_2 be removed and sewn back differently (Dehn surgery) so that the result is still a solid torus?*

Remarks: Winding number zero means the homomorphism $H_1(T_2) \to H_1(T_1)$ is zero; nonzero wrapping number means that T_2 does not lie in a 3-ball in T_1. An example of such a T_2 would provide a knot without Property P (Problem 1.15).

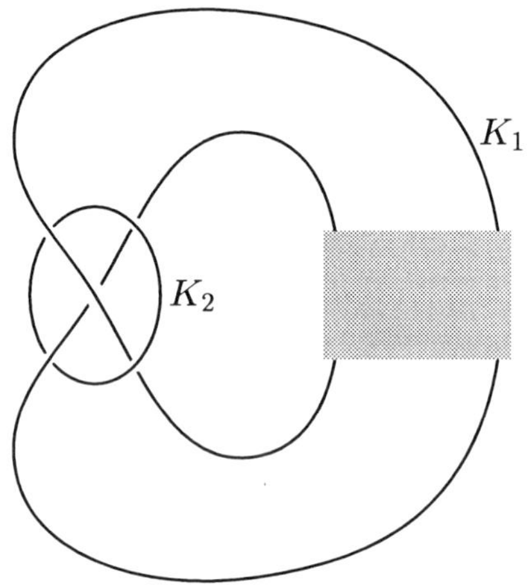

(For let T_1' be the result of Dehn surgery on T_2. Then winding number zero implies that the diffeomorphism between T_1 and T_1' preserves the meridian. Imbed T_1 in a knotted fashion in S^3 so that T_2 is knotted in S^3. Then Dehn surgery on T_2 gives S^3.)

(B) *Let K_1 and K_2 be unknots in S^3 as drawn above, and assume that K_1 and K_2 are geometrically linked but algebraically unlinked. Can an example be found so that if the overcrossing in K_1 is changed to an undercrossing, then K_1 remains unknotted?*

Remarks: An example for (B) provides one for (A). For +1-surgery on K_2 changes the crossing in K_1. If K_1 is unknotted in both cases, then choose $T_1 = S^3 - K_1$ and $T_2 = K_2 \times B^2$, satisfying (A).

(C) *Let T be the solid torus and imagine T standardly imbedded in each of S^3 and $S^1 \times S^2$. Is there a simple closed curve J in T such that*

1. *J does not bound a disk in T, and*
2. *J bounds a disk in each of S^3 and $S^1 \times S^2$?*

Remarks: If so, there is a knot failing Property P or Property R. For perform +1-surgery on J in T and get either (i) a solid torus, or (ii) a cube with a knotted hole (= knot complement). If (i), proceed as in (A). If (ii) then $S^1 \times S^2$ is the union of a solid torus and a cube with a knotted hole; that is, it is the result of 0-surgery on the knot making the knotted hole.

Update: The example asked for in (A) does not exist because surgery on a knot cannot give S^3 (Property P for S^3, see Problem 1.15); the example asked for in (B) does not exist because it would give an example for (A) (for an earlier solution, see [108,Bleiler & Scharlemann,1986, Math. Ann.]); finally no example exists satisfying (C) because of (A) or Property R (see Problem 1.16).

Problem 1.19 (Akbulut & Kirby) Conjecture: *If 0-framed surgeries on two knots give the same 3-manifold, then the knots are concordant.*

Remarks: This is true if one knot is the unknot ([580,Kirby & Melvin, 1978,Invent. Math.], see Problem 1.16). If homotopy 4-spheres are spheres, then it is true if one knot is slice. In general all known concordance invariants of the two knots are the same; this is true even if we assume only that the 0-surgeries give homology bordant 3-manifolds.

Update: Still open. Note that 0-framed surgery on the unknot is the only case giving $S^1 \times S^2$ [355,Gabai,1987a,J. Differential Geom.].

Problem 1.20 (Giffen & Siebenmann) A Seifert surface F for a knot K in S^3 is called free if $\pi_1(S^3 - F)$ is free (equivalently, $S^3 - F$ is an open handlebody). The construction in [323,Fox,1962a] yields such a free Seifert surface.

(A) *What is the smallest genus among free Seifert surfaces of K? Call this the free genus. Relate the free genus to other invariants of knots. Note that the free genus seems to be arbitrarily large for genus 1 knots; consider untwisted doubles.*

(B) *Which knots bound an incompressible free Seifert surface? Fibered knots do. Some knots do not [687,Lyon,1972,Proc. Amer. Math. Soc.].*

(C) *Trotter gives examples of nonunique free incompressible Seifert surfaces [1064, 1975]. Are these examples non-unique (up to isotopy?) when pushed into B^4?*

Update: (A) The free genus can be arbitrarily large for twisted doubles, [794,Moriah,1987,Proc. Amer. Math. Soc.], and for untwisted doubles [672,Livingston,1988,Proc. Amer. Math. Soc.].

(B) No information.

(C) No information.

Problem 1.21 (Gordon) Let L be a nontrivial link in S^3 with two unknotted components C_1, C_2, and linking number zero. Take the k-fold branched cover over C_1.

Conjecture: *The k lifts of C_2 form a nontrivial link.*

Remarks: If it is trivial, then there are many counterexamples to the Smith conjecture (Problem 3.38) for $\mathbb{Z}/k\mathbb{Z}$ actions [401,Gordon,1977, Quart. J. Math. Oxford Ser. (2)].

Update: True, because the Smith conjecture is true; but it follows more easily from the equivariant sphere theorem [743,Meeks, III & Yau,1980, Ann. of Math.].

Problem 1.22 (Montesinos) *Find a set of moves on links of S^3 so that two links have the same 2-fold branched covering space iff it is possible to pass from one link to the other using this set of moves.*

Update: Still open.

Problem 1.23 (Montesinos) *If a knot $K \subset S^3$ is amphicheiral, then the 2-fold covering space branched over K is symmetric (has an orientation reversing diffeomorphism). Is the converse true?* (Conjecture: *No.)*

Update: Still open. A related question [118,Boileau, Gonzalez-Acuña, & Montesinos,1987,Math. Ann.] was answered by Callahan [167,1994] who constructed examples of knots which admit no non-trivial symmetries, but have Dehn surgeries which yield 3-manifolds which are 2-fold branched coverings of S^3 and hence have a symmetry of order two.

Problem 1.24 (Fox & Perko) *Does every simple 4-fold branched cover of a knot K have precisely three distinct branch curves?*

Remarks: A simple 4-fold branched cover corresponds to a representation of $\pi_1(S^3 - K)$ onto S_4 with meridians going to elements of order 2. Every orientable, closed 3-manifold is a 3-fold (irregular) branched cover of S^3 over some K, and that representation of $\pi_1(S^3 - K)$ onto $D_3 = S_3$ lifts to a representation onto S_4 which is simple or takes meridians to elements of order 4 [871,Perko,1975].

Update: The answer is yes [272,Edmonds & Livingston,1984,Topology Appl.].

Problem 1.25 (Cappell & Shaneson) Let M_α be an irregular P-fold dihedral cover of a knot α. Let $\alpha_0, \alpha_1, \ldots, \alpha_r$, $r = (p-1)/2$, be the branching curves in M_α where α_0 has branching index 1. Let $v_{i,0} = l(\alpha_i, \alpha_0)$.

Prove that $v_{i,0} \equiv 2 \pmod 4$ *if M_α is a $\mathbb{Z}/2\mathbb{Z}$-homology sphere.*

Remarks: For 2-bridge knots, $v_{i,0} \equiv 2 \pmod 4$ [872,Perko,1976, Invent. Math.]. Cappell & Shaneson have shown that $\sum_{i=1}^{r} v_{i,0} \equiv (p-1) \pmod 4$ for a $\mathbb{Z}/2\mathbb{Z}$-homology sphere. Consequently, $v_{i,0} \equiv 2 \pmod 4$ if $p \equiv 3 \pmod 4$ and if 2 generates $\mathbb{Z}_p^*/\{\pm 1\}$. For $p = 3$, the formula reduces to $v_{1,0} \equiv 2 \pmod 4$ [180,Cappell & Shaneson, 1975,Bull. Amer. Math. Soc.].

Update: A proof that $v_{1,0} \equiv 2 \pmod 4$ is given in [183,Cappell & Shaneson,1984].

Problem 1.26 (Murasugi) Suppose the first homology group of the 2-fold cyclic branched cover of a knot $\alpha \subset S^3$ is $\mathbb{Z}/p\mathbb{Z}$ (hence $p = |\Delta_\alpha(-1)|$), and let M_α be the irregular p-fold dihedral cover of α.

Conjecture: *If M_α is a $\mathbb{Z}$-homology sphere, then* $\sum_{i=1}^{r} v_{i,0} \equiv \sigma(\alpha) \pmod 8$ *where $\sigma(\alpha)$ is the signature of α and $v_{i,0}$ is defined above.*

Remarks: The conjecture holds with *equality* for 2-bridge knots ([450, Hartley & Murasugi, 1978, Canad. J. Math], also see [872, Perko, 1976, Invent. Math.]).

Update: The conjecture is false [446, Hartley, 1977]. However $\sum_{i=1}^{(p-1)/2} v_{i,0} \equiv p-1 \pmod 4$ [183, Cappell & Shaneson, 1984], and $\sigma(\alpha) = |\Delta_\alpha(-1)| - 1 \pmod 4$ [819, Murasugi, 1965, Trans. Amer. Math. Soc.]; since $|\Delta_\alpha(-1)| = p$, it follows that the conjecture holds modulo 4.

Problem 1.27 (Goldsmith) *Do there exist distinct prime knots K and K' in S^3 all of whose cyclic branched covers are homeomorphic?*

Update: No, because Kojima [600, 1986] has proved that for a prime knot K there is an integer $N_K > 0$ such that two prime knots, K and K', are homeomorphic if their n-fold branched covers are the same for $n > \max\{N_K, N_{K'}\}$. However see Problem 1.74.

Problem 1.28 (Goldsmith) Let $M^3 \xrightarrow{\pi} S^3$ be an n-fold cyclic branched cover of S^3 along a knot K. Let A be an unknot in $S^3 - K$. If K is a closed braid about A, then $\pi^{-1}(A)$ is a fibered knot or link in M^3.

Question: *Is the converse true?*

Update: The answer is yes if n is a prime power, or more generally, if M is a rational homology sphere [271, Edmonds & Livingston, 1983, Comment. Math. Helv.].

Problem 1.29 (Cappell & Shaneson) *Is every closed, oriented 3-manifold the dihedral branched covering space of a ribbon knot?*

Update: Still open.

Problem 1.30 (Cappell & Shaneson) *Are the classical PL and TOP knot concordance groups the same?*

Remarks: Clearly $C^{PL} \to C^{TOP}$ is onto. This question may be easier than the Hauptvermutung for $B^2 \times \mathbb{R}^2$.

Update: No by the answer to Problems 1.36 and 1.37. In fact, the kernel should be infinitely generated.

Problem 1.31 (Y. Matsumoto) Let $\mathcal{H}_A$ = {all knots in all homology 3-spheres which bound PL acyclic 4-manifolds, modulo homology bordism of pairs}.

Is the natural map $C^{PL} \to \mathcal{H}_A$ an isomorphism?

Update: Akbulut's example proving the Zeeman conjecture [20,Akbulut,1991b,Topology] gives a knot K in a homology 3-sphere Σ which is not concordant to a knot in S^3 via a concordance that lies in a particular homology bordism from Σ to S^3; however K is concordant to the unknot in S^3 via a different bordism (the two homology bordisms are two different Mazur manifolds minus their 0-handles). So the problem is still open.

Problem 1.32 (Gordon) *Does the classical knot concordance group contain any nontrivial elements of finite order other than 2?*

Update: Many more elements are known of infinite order, but still none of finite order other than two.

Problem 1.33 *If K is a slice knot, is K a ribbon knot? (See Problem 4.22.)*

Update: No progress.

Problem 1.34 (Casson) *Find an algorithm for determining whether a knot is slice or ribbon .*

Remarks: There is an algorithm for the (harder?) problem of determining whether a knot is unknotted [428,Haken,1961,Acta Math.].

Update: No progress.

Problem 1.35 (Cappell & Shaneson) *The μ-invariant formula of Cappell & Shaneson [180, 1975, Bull. Amer. Math. Soc.] detects non-ribbon knots. Does it detect non-slice knots as well? Relate this to the Casson–Gordon invariant.*

Update: No progress.

Problem 1.36 (L. Taylor) *If a knot has Alexander polynomial equal to one, is it a slice knot?*

Remarks: Any such knot is algebraically slice, i.e., there is a basis so that the Seifert matrix has the form $\left(\begin{smallmatrix} 0 & A \\ B & C \end{smallmatrix}\right)$ (Taylor). These knots have no metacyclic covers, so the Casson–Gordon method does not apply.

Here is a possible generalization to links. A boundary link is a link L whose components bound disjoint imbedded surfaces $F_1, \ldots, F_r$ in S^3. Note that if a knot has a Seifert matrix of the form $\left(\begin{smallmatrix} 0 & 0 \\ B & C \end{smallmatrix}\right)$, then its Alexander polynomial is one. Hence, define a *good* boundary link to be one for which there is a summand $A_i \subset H_1(F_i; \mathbb{Z})$ such that $2 \dim A_i = \dim H_1(F_i; \mathbb{Z})$ and the intersection of every element of A_i and every other element of $H_1(F_j; \mathbb{Z})$ is zero for all i, j.

Question. Is a good boundary link slice?

Note that a slice link is not necessarily a boundary link [1008, Smythe, 1966].

Update: All Alexander polynomial one knots are topologically slice. In fact, if K is a knot in a homology sphere which bounds a smooth contractible 4-manifold X^4, then K has Alexander polynomial one iff K bounds a locally flat 2-ball D in X with $\pi_1(X - D) = \mathbb{Z}$, [336, Freedman & Quinn, 1990; page 210].

On the other hand, all the pretzel knots in Problem 1.37 are not smoothly slice. Many knots can be shown not smoothly slice using various gauge theoretic techniques.

Whether good boundary link are topologically slice is still open.

Problem 1.37 (Casson) (A) *The knot*

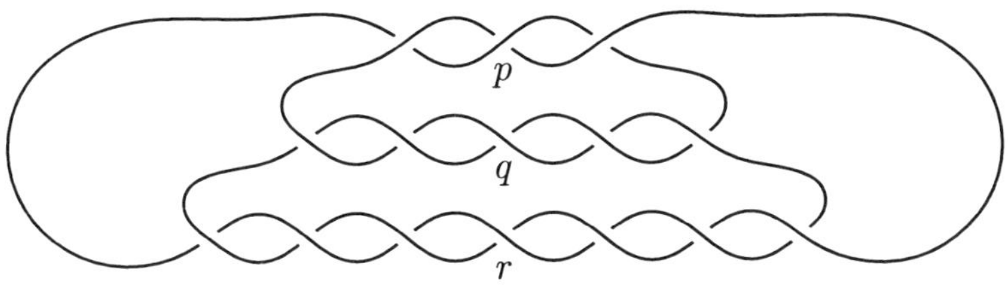

Figure 1.37.1. (p,q,r)-pretzel knot

($p = -3,\ q = 5,\ r = 7$ in illustration) has Alexander polynomial 1 if p, q, r are odd and $qr + rp + pq = -1$. Is it slice?

(B) *The double branched covering of this knot is the Brieskorn homology sphere $\Sigma(|p|, |q|, |r|)$. Does it bound a homology ball?*

Remarks: An affirmative answer to (A) implies that $\Sigma(p, q, r)$ bounds a $\mathbb{Z}/2\mathbb{Z}$-homology ball.

If the Brieskorn sphere $\Sigma(|2bc + 1|, |2a(b - d) + 1|, |2d(c - a) + 1|)$ bounds a homology ball for some numbers a, b, c, d with $ad - bc = 1$, then the homology class (a, b) in some homology $S^2 \times S^2$ is representable by an imbedded S^2. For example, if $\Sigma(3, 5, 7)$ bounds, then $(2, 3)$ is representable. (See Problem 1.42.)

Update: Fintushel & Stern have shown (Theorem 10.7 in [310,1985, Ann. of Math.]) that none of the $\Sigma(|p|, |q|, |r|)$ bound $\mathbb{Z}/2\mathbb{Z}$-acyclic 4-manifolds, so none of the pretzel knots in (A) are slice. In fact they show that the $\Sigma(|p|, |q|, |r|)$ have infinite order in Θ_3^H, the homology bordism classes of homology 3-spheres. In [312,Fintushel & Stern,1990, J. London Math. Soc.] they show that the collection of $\Sigma(2, 3, 6k - 1)$ gives infinitely many, free, generators of Θ_3^H, and for k even one gets such generators for the kernel of the Rohlin homomorphism $\Theta_3^H \to \mathbb{Z}/2\mathbb{Z}$.

Problem 1.38 Conjecture *(unlikely): The untwisted double of a knot is slice $\Leftrightarrow$ the knot is slice.*

Update: The odds on the (smooth version of the) conjecture are improving since many untwisted doubles are not slice, [960,Rudolph,1993,

Bull. Amer. Math. Soc.] or [961,Rudolph,1995,Invent. Math.].

Some untwisted doubles of non-slice *links* are also not slice. In particular, iterated untwisted positive doubles of the Whitehead link are never (smoothly) slice (Rudolph); this implies that one of the simplest Casson handles is not smoothly standard (see the following Problem 1.39).

The topological version of the conjecture fails easily because the untwisted double of any knot K has Alexander polynomial one and is therefore topologically slice (see Problem 1.36), whereas any K which is not algebraically slice is not topologically slice.

Problem 1.39 (Casson) Drawn below is the Whitehead link and an untwisted double of the Whitehead link.

This construction can be iterated by replacing  by 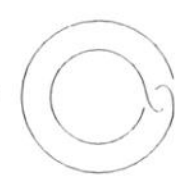 or 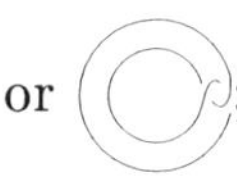; call the nth iterate W_n.

Is any W_n null-concordant?

Remarks: If so, then there exists a proper homotopy $S^3 \times \mathbb{R}$ which is not $S^3 \times \mathbb{R}$. If not, there exists an end of a 4-manifold, $Q \cong^P S^2 \times S^2 - pt$, which is fake (A. Casson). It is also interesting to know if W_n is null-concordant in some contractible 4-manifold.

The manifold Q can be described as follows. Let $X \subset S^1 \times \text{int} B^2$ be the continuum of Whitehead [1110,Whitehead,1962], defined as $X = \cap_{i>0} \alpha^i (S^1 \times B^2)$, where $\alpha : S^1 \times B^2 \to S^1 \times B^2$ is the imbedding shown below (α should have 0-twisting about its core):

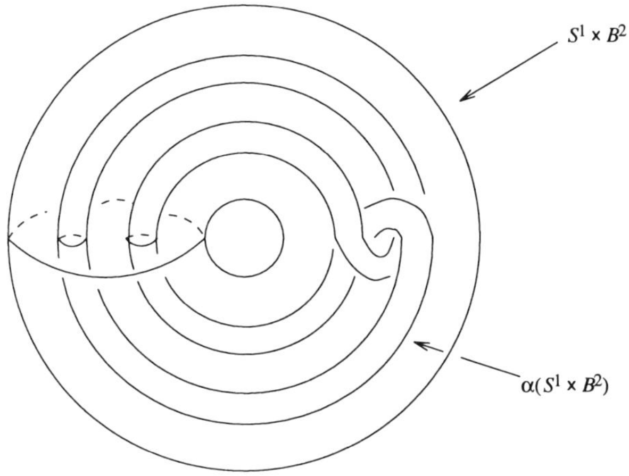

Regarding $S^2 \times S^2$ as the union of a 0-handle B_0^4, two 2-handles $B_1^2 \times D^2$, $B_2^2 \times D^2$ and a 4-handle B_3^4, define a compact subset $C = B_3^4 \cup cX_1 \cup cX_2$, where each X_i is a Whitehead continuum constructed in $\text{int}B_i^2 \times \partial D^2 \subset \partial B_3^4$, the *cocore* attaching tube of the ith 2-handle, and where X_i is coned off to the center of the 2-handle $B_i^2 \times D^2$. Then $Q = S^2 \times S^2 - C$.

The open subset $B_1^2 \times \text{int}D^2 - cX_i$ of the open 2-handle $B_i^2 \times \text{int}D^2$ is called a *flexible* 2-handle by Casson; this is the simplest such. (The others are constructed by replacing X by a Whitehead-like continuum obtained by ramifying at each stage the original Whitehead imbedding α.) Any such flexible 2-handle is proper homotopy equivalent, rel∂, to the standard open 2-handle $B^2 \times \text{int}D^2$. One fundamental question is whether this can ever be a diffeomorphism. The manifold Q can be regarded as the union of an open 0-handle $\text{int}B_0^4$ and two flexible 2-handles.

The compact subset $C = S^2 \times S^2 - Q$ is cell-like (i.e., homotopic to a point in any arbitrarily small neighborhood) and satisfies the cellularity criterion (i.e., Q is 1-connected at ∞; see [737, McMillan, Jr., 1964, Ann. of Math.]). However, C is smoothly cellular ($\equiv$ the intersection of a nested sequence of smooth 4-balls) $\Leftrightarrow$ some W_n is null-concordant.

Update: If W_n is constructed using only right-handed (or left-handed) clasps, then it is not smoothly null concordant [100, Bižaca, 1995, Proc. Amer. Math. Soc.].

Many flexible handles (now called Casson handles) are exotic, but it is not known if all of them are. Note that exotic means that the Casson handle does not have a smoothly imbedded B^2 spanning the $S^1 \times 0$ in the boundary; it is possible to construct an open 2-handle, which does have a spanning B^2 but is not smoothly standard, by end-connected summing with an exotic $\mathbb{R}^4$ (Gompf). A Casson handle is known to be exotic if it contains a branch all of whose kinks (or clasps) are right handed [Bižaca, ibid.].

For a good description of Casson handles, see [422,Guillou & Marin, 1986a].

Problem 1.40 (A) *Let K be a torus knot. Is the genus of K equal to the 4-ball genus of K?*

Remarks: The 4-ball genus is the minimal genus of a bounding surface in B^4. The genera are equal for $(2,q)$-torus knots. The genus of the (p,q)-torus knot is $(p-1)(q-1)/2$. Since $n \in H_2(\mathbb{CP}^2;\mathbb{Z}) = \mathbb{Z}$ is represented by an imbedded 2-sphere which is smooth except for a cone on the $(n-1,n)$-torus knot, the Remark in Problem 4.36 gives a lower bound for the 4-ball genus of the $(n-1,n)$-torus knot.

(B) *L. Rudolph has shown that every link in S^3 is isotopic in S^3 to one which bounds a nonsingular complex analytic curve in $B^4 \subset \mathbb{C}^2$. What can be said about the minimal genus of such a curve?*

Update:

(A) Yes [623,Kronheimer & Mrowka,1993,Topology]. Also, see the Update for Problem 1.4.

(B) The statement attributed to Rudolph is erroneous. In fact, Rudolph shows in [956,1983,Topology] that every quasipositive closed braid is isotopic in S^3 to one which bounds a non-singular complex analytic curve in $B^4 \subset \mathbb{C}^2$ (what Rudolph calls a *transverse* $\mathbb{C}$-link). The converse is open (see Problem 1.100).

There are many knots which are not quasipositive braids. The first was the figure-8 knot[959,Rudolph,1990,Math. Proc. Cambridge Philos. Soc.] using the HOMFLY polynomial; many more were constructed that way, and more still using [Kronheimer & Mrowka, ibid.].

Problem 1.41 (A) (Akbulut & Kirby) Let M^4_K be constructed by adding a 2-handle to B^4 along a knot K with the 0 framing. Define the 0-shake genus of K to be the minimal genus of a smooth, imbedded surface in M_K representing the generator of $H_2(M_K)$.

Does the 0-shake genus equal the 4-ball genus of K? Probably not.

Remarks: The r-shake genus of K (obvious definition) can be less than the 4-ball genus, for $r \neq 0$ [15,Akbulut,1977,Math. Proc. Cambridge Philos. Soc.].

(B) (Giffen) *What is the minimal genus of a smooth, imbedded surface in ∂M_K representing the generator of $H_2(\partial M_K)$?*

Conjecture: *It equals the genus of K.*

Remarks: (L. Taylor) The conjecture is true for fibered knots (even in homology spheres); however, the conjecture fails in general for knots in homology spheres. Perhaps it would be easier to show that the Seifert surface of K union the 2-handle is incompressible in ∂M_K. If so, or if the conjecture holds, then every knot has Property R (Problem 1.16).

Update:

(A) Still open.

(B) The conjecture is true [355,Gabai,1987a,J. Differential Geom.]. In fact, Gabai shows that the Gromov norm and the singular Thurston norm of the generator of $H_2(\partial M_K)$ are both linear functions of genus K.

Problem 1.42 (Y. Matsumoto) *Does the following link in S^3 bound a smooth punctured sphere in B^4? If so, $(2,3) \in H_2(S^2 \times S^2; \mathbb{Z})$ is represented by a smooth S^2. Can it be represented by a torus?*

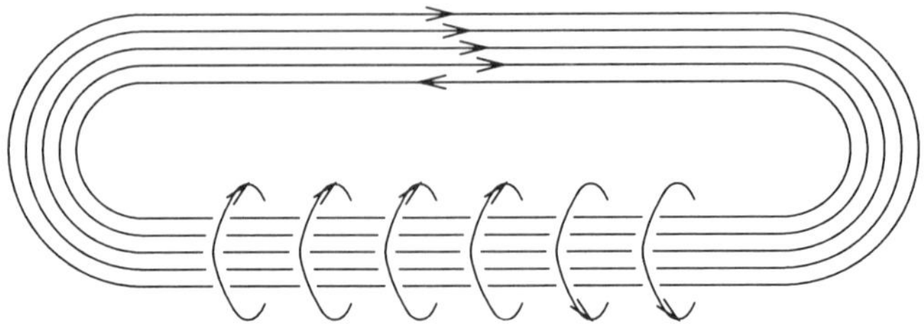

Remarks: This is the simplest unsolved case one encounters in trying to represent $(2,3) \in H_2(S^2 \times S^2; \mathbb{Z})$ by a smooth imbedded S^2. This can be done iff such a link as above, with $2+2k$ circles in one group and $3+2l$ circles in the other group oriented to give $(2,3)$, bounds a smooth, imbedded punctured S^2 in B^4. The above is the case $k = 2$, $l = 1$. The simpler cases are ruled out by the Murasugi–Tristram inequality

$$|\sigma_p| + |\eta_p| \leq \mu - 1$$

where σ_p is the p-signature of the link, η_p is the nullity ($\equiv$ dim minus rank of the associated form) and μ is the number of components of the link. For $p = 2$ we get $|6| + |2+2l| \leq |5+2k+2l-1|$, which rules out the cases $k \leq l$, l arbitrary. For $p = 3$ we get $|5| + |1+2k| \leq |5+2k+2l-1|$, ruling out k arbitrary, $l = 0$.

Update: The class $(m,n) \in H_2(S^2 \times S^2; \mathbb{Z})$ can be represented by a smoothly imbedded S^2 if $|m| \leq 1$ or $|n| \leq 1$. Otherwise it cannot be [628,Kuga,1984,Topology]. (m,n) can be represented by a locally flat, imbedded S^2 whenever m and n are relatively prime.

Apparently the question of representing $(2,3)$ by a smooth T^2 is still open.

Problem 1.43 (Scharlemann) *Are there knots $f : S^1 \to S^3$ such that for any locally flat concordance $F : S^1 \times I \to S^3 \times I$ the map $\pi_1(S^3 - f(S^1)) \to \pi_1(S^3 \times I - F(S^1 \times I))$ is injective?*

Conjecture: *This is true for torus knots.*

Remarks: This is true for torus knots if F must be a fibered concordance.

Update: The conjecture is true [187,Casson & Gordon,1983,Invent. Math.].

Problem 1.44 (Kauffman) *Does link concordance imply link homotopy? (Added in proof April 1, 1977: Yes, Giffen and (independently) Goldsmith.)*

Update: Yes, [379,Giffen,1979,Math. Scand.] and [388,Goldsmith, 1979,Comment. Math. Helv.].

The following definitions are used in the next three problems. Given a knot K in S^3, an *algebraically-one strand* is a way of imbedding an unknotted $T = S^1 \times B^2$ in S^3, containing K, so that K and $p \times \partial B^2$, $p \in S^1$, link algebraically once. Thus K goes algebraically once around T. There is a similar definition for algebraically-l strand.

A (k, l)-twist on K is obtained by taking some algebraically-l strand, i.e., some T, and twisting T k full times in a right-handed direction around $S^1 \times 0$.

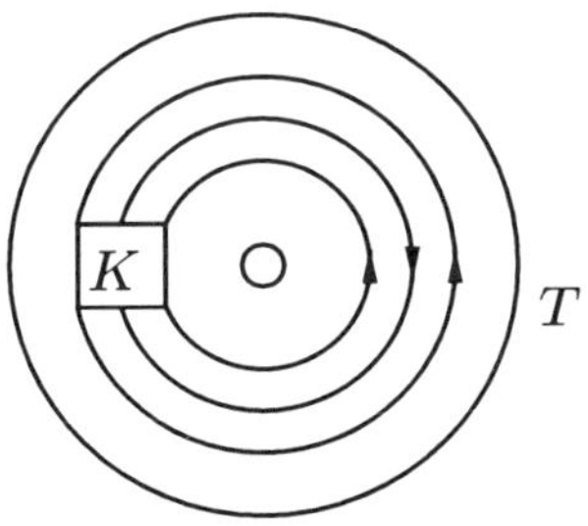

Problem 1.45 (Akbulut) Conjecture: *There exist a knot K and an algebraically-one strand such that no matter what knot is tied in the strand (in T), the new knot is not slice in any homotopy 4-ball (with $\partial = S^3$).*

Remarks: If the conjecture is true, there exists a knot in the boundary of a contractible 4-manifold which does not bound an imbedded PL 2 ball (see Problem 4.21). Specifically, suppose K and T satisfy the conjecture. Then if we add a 2-handle to K (with any framing) in $\partial(S^1 \times B^3)$, we get a contractible manifold W and $S^1 \times p$, $p \in \partial B^3$, does not bound a PL 2-ball. For, if there is a PL 2-ball, with singularity equal to a cone on a knot J, then K with J tied in it is slice in a homotopy 4-ball.

Update: No progress, but the application of the conjecture to the existence of a knot has been shown (see Update to Problem 4.21 and [20, Akbulut, 1991b, Topology]).

Problem 1.46 (Akbulut & Kirby) Conjecture: *Given a knot K with Arf invariant zero, there is a $(\pm 1, \pm 1)$-twist changing K into:*

(A) *An algebraically slice knot (Seifert matrix concordant to zero). Very likely true.*

(B) *A slice or ribbon knot. Perhaps.*

(C) *The unknot. Surely false.*

There is a better chance for the conjecture that K is concordant to a knot K' which, after a $(\pm 1, \pm 1)$-twist, satisfies (A) or (B) or (C).

Update:

(A) True (Casson).

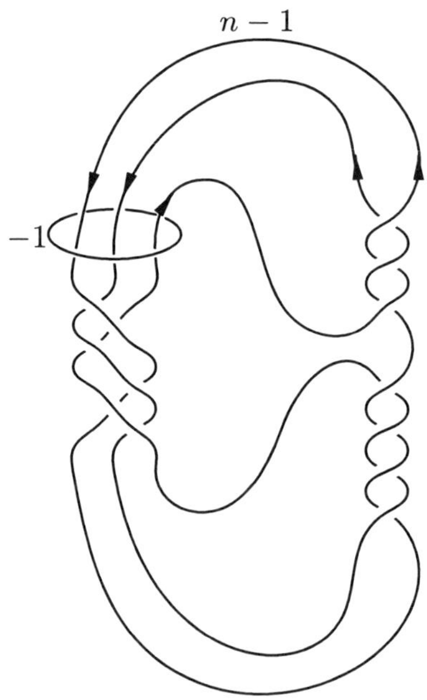

Figure 1.46.1.

(B) and hence **(C)** is false (even if a concordance to K' is allowed) (Akbulut). For, suppose to the contrary that (B) is true. Consider the $(2,7)$-torus knot K with framing n, and the 3-manifold M defined by surgery on K. Giving K a $(\pm 1, \pm 1)$ twist amounts to blowing up a ± 1 unknot U with $lk(U, K) = \pm 1$ (see Figure 1.46.1 for an example with geometric linking 3), which changes the framing on K to $n \pm 1$. This two component link describes a 4-manifold S^4 with $\partial W = M$ and intersection form $\left(\begin{smallmatrix} \pm 1 & \pm 1 \\ \pm 1 & n\pm 1 \end{smallmatrix}\right)$. If we pick n so that $n \pm 1 = 0$, then because K is *now* slice, there is a smoothly imbedded S^2 in W^4 with self intersection zero; surger it to see that M bounds an acyclic 4-manifold X^4. But M also is ∓ 1 surgery on K which bounds a 4-manifold with definite intersection form Γ_{16} (if $n = +1$) or $2E_8 \oplus \left(\begin{smallmatrix} 0 & 1 \\ 1 & 0 \end{smallmatrix}\right)$ (if $n = -1$); this contradicts the results in [250,Donaldson,1987b,J. Differential Geom.].

Problem 1.47 (Casson) *Given any Arf invariant zero knot K, is it possible to change K to the unknot by a series of $(1, \pm 1)$-twists on K?*

Remarks: (Casson) Yes, if K is ribbon or slice (since we can connect sum with a ribbon knot and get a ribbon knot). Any K can be reduced to a connected sum of granny knots. If the granny knot could be changed to the unknot with 17 $(1, \pm 1)$-twists, then there exists an even, signature 16, $\beta_2 = 18$ closed 4-manifold (in $17(-\mathbb{CP}^2)\#2\mathbb{CP}^2$, try to represent $(1, 1, \ldots, 1, 3, 3)$ by a smooth imbedded S^2; in $\#2\mathbb{CP}^2$, $(3, 3)$ is represented by a PL sphere with a singularity equal to a cone on the granny knot, and the 17 $-\mathbb{CP}^2$'s allow 17 $(1, \pm 1)$–twists to unknot the granny knot).

Update: No progress, except that the 4-manifold above does not exist so one cannot change the granny knot to the unknot as asked above. This problem seems to be outmoded.

Problem 1.48 (J. Levine) *A general question is what groups π are fundamental groups of the complement of some knotted S^2 in S^4?*

Recall that a group π has weight 1 if it is normally generated by one element, and deficiency one if it has a presentation $\{x_1, \ldots, x_n, t : \mathbf{R}_1, \ldots, \mathbf{R}_n\}$ with one more generator than relation.

(A) *Given π such that $H_1(\pi) = \mathbb{Z}$, π has weight one and deficiency one,* *then π is the group of an S^2 in a homotopy 4-sphere [571, Kervaire, 1965, Bull. Soc. Math. France]. Which of these are realizable by knots in S^4? π is realizable if the induced presentation of the trivial group defined by setting $t = 1$ is trivializable by Andrews–Curtis moves.*

(B) *Let $\Lambda = \mathbb{Z}[t, t^{-1}]$ and let the Λ-module A of an $S^2 \to S^4$ be π'/π'' with the induced action of $\pi/\pi' = \mathbb{Z}$. Which Λ-modules are realizable? If A is $\mathbb{Z}$-torsion free (implied by deficiency one) the answer is known since there are enough deficiency one π to get all such A's.*

Update: Questions 1 and 2 have not been answered, but the status of fundamental groups of knotted 2-spheres in S^4 as of 1988 is given in [485, Hillman, 1989] with an update to 1993 in [486, Hillman, 1994; Chap. X].

Problem 1.49 (Lomonaco) *Does there exist a smooth, prime S^2 in S^4 such that the deficiency of the fundamental group of its complement is < 0?*

Remarks: The deficiency is always ≤ 1 because the complement is a homology S^1. E. Artin [46, 1925, Abh. Math. Sem. Univ. Hamburg] constructed knots of deficiency one by spinning, and Fox constructed a knot of deficiency 0 in [323, 1962a]. J. Levine and C. Giffen have recently found many nonprime knots of arbitrarily large negative deficiency, by connect-summing arbitrarily many knots of deficiency zero.

Update: Still open.

Problem 1.50 (Gordon) *Can a branched cyclic cover of a (locally flat) knot $S^n \to S^{n+2}$ ever be a $K(\pi, 1)$, for $n \geq 2$?*

Update: No progress.

Problem 1.51 (Kawauchi) Suppose a knotted 2-sphere Σ^2 in S^4 is of Dehn's type (i.e., the homomorphism $\pi_2(\partial N) \to \pi_2(S^4 - \Sigma^2)$ is trivial where N is a tubular neighborhood of Σ^2).

Questions: *Is Σ^2 algebraically unknotted (i.e., is $S^4 - \Sigma^2 \simeq S^1$)? Geometrically unknotted?*

Remarks: The infinite cyclic cover of $S^4 - \Sigma^2$ is acyclic; if in addition $\pi_3(S^4 - \Sigma^2) = 0$ is satisfied, then Σ^2 is algebraically unknotted [567, Kawauchi, 1986, Osaka J. Math.]. If there exists an algebraically knotted Σ^2 of Dehn's type, it would have infinitely many ends.

Update: A knotted n-sphere of Dehn's type must be algebraically unknotted by a result of Swarup [1025, 1975, J. Pure Appl. Algebra], together with the fact that finitely presentable groups are accessible [258, Dunwoody, 1985, Invent. Math.], (for a simpler proof, see [1029, Swarup, 1993]). It follows that the knot is unknotted, except that when $n = 2$, it unknots topologically; whether it unknots smoothly is still unknown.

Problem 1.52 (Kauffman) Conjecture: *If K is a slice knot in S^3 and F^2 is an orientable Seifert surface for K, then there exists a simple closed curve α in F such that*

- *α is null (meaning that the linking number of α with a parallel in F is zero and $0 \neq \alpha \in H_1(F;\mathbb{Z})$),*
- *the Arf invariant of α is zero.*

If true, can one then find a null α which is slice?

Update: Still open if slice means smoothly slice. However, Freedman [330, 1984] showed that any knot with Alexander polynomial one has a topological slice disk which is locally flat. So the untwisted double of the figure-8 knot (or any other knot with non-trivial Arf invariant) provides a counterexample in the topological category.

Also, if K has a topological slice disk which is locally flat, $genus(F) = 1$ and $|\Delta_K(-1)| \neq 1$, then there is a null α such that some interesting bordism invariants (depending on $|\Delta_K(-1)|$) of α must vanish [384, Gilmer & Livingston, 1992, Math. Proc. Cambridge Philos. Soc.] and [382, Gilmer, 1993, Comment. Math. Helv.].

Problem 1.53 *Does mutation preserve the concordance type of a knot in S^3? (Mutation is the operation of a knot which removes a tangle, twists it $180°$, and glues it back in).*

Update: No [568,Kearton,1989,Proc. Amer. Math. Soc.]. But if the knot is oriented and the mutation is the one (of three types) which preserves orientation, then the problem is still open.

Problem 1.54 (Hillman) *When is the result of surgery on a knot in S^4 aspherical?*

Remarks: The knot group must be an orientable Poincaré duality group of formal dimension four [482,Hillman,1980,Houston J. Math.], but is this condition sufficient?

Update: The closed 4-manifold $M(K)$ obtained by surgery on a 2-knot K is aspherical if and only if the knot group πK is a PD_4^+-group (an orientable Poincaré duality group of dimension 4) and the image of the orientation class of $M(K)$ in $H_4(\pi K;\mathbb{Z})$ is nonzero. (See Theorem II.5 of [486,Hillman,1994].) The latter condition holds if $\pi K' \neq \pi K''$ (i.e., if the infinite cyclic covering space of $M(K)$ is *not* homologically S^3), by the Corollary to Theorem III.8 of [485,Hillman,1989].

The localization argument shows that if πK has a large enough torsion free abelian normal subgroup then $M(K)$ is aspherical. (See Theorems III.3 and III.4 of [ibid.]). This argument has since been extended to elementary amenable normal subgroups with restricted torsion. (See Chapter X of [486,Hillman,1994].) Using L^2-cohomology, Eckmann has shown that if πK is amenable then $M(K)$ is aspherical if and only if πK has one end and $H^2(\pi K;\mathbb{Z}[\pi K]) = 0$ [266,Eckmann,1993].

Problem 1.55 (A) *If a smooth 2-sphere K in S^4 has group $\pi_1(S^4 - K) = \mathbb{Z}$ (this implies that $S^4 - K \simeq S^1$), is it smoothly unknotted?*

Remarks: K is unknotted in the topological category [329,Freedman,1982,J. Differential Geom.]. Also, see Problem 4.41.

(B) *Let L be a link in S^4 with unknotted components and let $\pi_1(S^4 - L)$ be free on a set of meridians. Is L trivial (topologically or smoothly)?*

Remarks: G. A. Swarup [1026,1977,J. Pure Appl. Algebra] has shown that the exterior $S^4 - L$ has the right homotopy type *rel* boundary. However, the homotopy type rel boundary of a knot exterior does not in general determine the homeomorphism type of the knot exterior [878,Plotnick,1983,Math. Zeit.] .

(C) *When is a 2-link splittable? In particular, is it sufficient that the group be a free product with each factor normally generated by a meridian?*

Update:

(A) No progress.

(B) If L is a 2-link whose link groups is freely generated by meridians, then there is a topological concordance from L to the unlink whose exterior is an s-cobordism rel boundary [487,Hillman,1995].

(C) No progress.

Problem 1.56 *Are all 2-links slice?*

Remarks: All 2-knots are slice [571,Kervaire,1965,Bull. Soc. Math. France]. All boundary links are slice ([Kervaire, ibid.], [425,Gutierrez, 1973,Bull. Amer. Math. Soc.], [182,Cappell & Shaneson,1980,Comment. Math. Helv.]), so the problem is to show that every 2-link is concordant to a boundary link. Note that L is a boundary link iff there exists a homomorphism $\varphi : \pi_1(S^4 - L) \to F_\mu$ (= free group on number of components) taking meridians to generators [Gutierrez, ibid.]. More generally, it is sufficient to find $\varphi : \pi_1(S^4 - L) \to P$ where the normal closure of image φ is P, P is a higher dimensional μ-component link group, and $H_3(P; \mathbb{Z}/2\mathbb{Z}) \cong H_4(P; \mathbb{Z}) = 0$, [211,Cochran,1984d,Trans. Amer. Math. Soc.].

An easier problem is: does the $\mathbb{Z}/2\mathbb{Z}$ invariant of Sato–Levine [963, Sato,1984,Topology Appl.] vanish for all 2-links? It vanishes for certain classes of 2-component links, e.g. when one of the components is unknotted [210,Cochran,1984c,Topology Appl.].

Update: This problem is still wide open, even for any even-dimensional link. Some invariants have been found, but they have all turned out to be zero. Le Dimet [643,1988,Mém. Soc. Math. de France (N.S.)] has shown the problem to be equivalent to a hard problem in homotopy theory.

The Sato–Levine invariant above has been shown to always be zero [861,Orr,1987,Comment. Math. Helv.].

Problem 1.57 (A) *Is the center of a 2-knot group finitely generated?*

Remarks: The only known centers are $\mathbb{Z}$, $\mathbb{Z} \oplus \mathbb{Z}/2\mathbb{Z}$, and $\mathbb{Z} \oplus \mathbb{Z}$, and they are realized by twist spun trefoil knots, [483,Hillman, 1981].

(B) *Is the center of the group of a 2-link with more than one component trivial?*

Remarks: The argument of Hausmann and Kervaire may be readily modified to show that any finitely generated abelian group is the center of the group of some μ-component n-link for each $\mu \geq 1$, $n \geq 2$. In the classical case, $n = 1$, the center must be $0, \mathbb{Z}$ or $\mathbb{Z} \oplus \mathbb{Z}$.

Update: No progress has been made except there is now a 2-knot whose center is $\mathbb{Z}/2\mathbb{Z}$ [1134,Yoshikawa,1982,Bull. Austral. Math. Soc.].

NEW PROBLEMS

Problem 1.58 (X.-S. Lin) *Suppose that oriented knots K_+ and K_- differ at exactly one crossing at which K_+ is positive and K_- is negative. If $K_+ = K_-$, does it follow that K_+ equals*

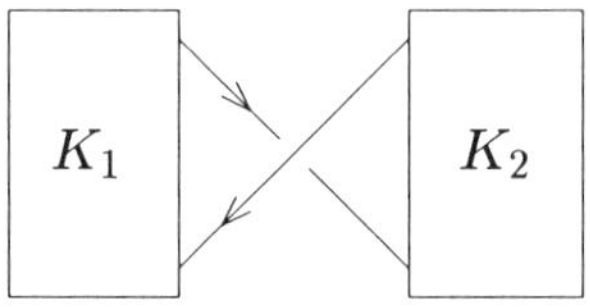

where either K_1 or K_2 could be the unknot?

Remarks: This is known if $K_\pm$ is the unknot (this follows from the technique in the proof of Theorem 1.4 in [972,Scharlemann & Thompson,1989,Comment. Math. Helv.]). If $K_+ = K_-$ by an orientation reversing homeomorphism, then the answer is no (for example, consider the pretzel knots $K\pm = (3, \pm 1, -3)$; one is the mirror image of the other, so they are *equal* by an orientation reversing homeomorphism (see the old Problem 1.37(A) for a picture of the (p, q, r)-pretzel knot)).

Problem 1.59 (Przytycki) **(i)** The local change in an oriented link diagram which replaces  by k positive half-twists 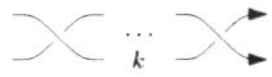 is called a t_k-move.

(ii) For k even, the local change replacing  by  is called a $\bar{t}_k$-move.

(iii) For an unoriented diagram, replacing  by k right-handed half-twists  , is called a k-move.

We say that two oriented links L_1 and L_2 are t_i (respectively $\bar{t}_j$ or $t_i, \bar{t}_j$)-*equivalent* if there is a sequence of $t_i^{\pm 1}$ (respectively $\bar{t}_j^{\pm 1}$ or $t_i^{\pm 1}, \bar{t}_j^{\pm 1}$)-moves and isotopies which converts L_1 to L_2.

We say that two unoriented links L_1 and L_2 are $k - equivalent$ if there is a sequence of $k^{\pm 1}$-moves and isotopies which converts L_1 to L_2.

k-moves were probably considered explicitly for the first time by Kinoshita [575,Kinoshita,1957,Osaka Math. J.].

(1) Conjecture: (Montesinos & Nakanishi) *Any link is* 3-*equivalent to a trivial link.*

Remarks: Nakanishi first considered the conjecture in 1981. Earlier Montesinos analyzed 3-moves in relation to 3-fold dihedral branch coverings and asked a related but different question. Conjecture **(1)** easily holds for algebraic links (in the Conway sense) [225,Conway,1969]. Settling the conjecture for a link with braid index at most 5 (and bridge index at most 3) is reduced to a finite number of cases because Coxeter [228,Coxeter,1957] showed that the quotient of the braid group $B_n / < \sigma_1^3 >$ is finite for $n \leq 5$.

According to Nakanishi (1994) the smallest known obstruction to Conjecture **(1)** is the 2-parallel of the Borromean rings (notice that it is a 6-string braid), Figure 1.59.1

Figure 1.59.1.

(2) **Conjecture:** *Any 2-tangle can be reduced, using 3-moves, to one of the four 2-tangles of Figure 1.59.2, where additional trivial components are allowed in the tangles, [901, Przytycki, 1994a].*

S_+ S_- S_0 S_∞

Figure 1.59.2.

Remarks: Conjecture **(2)** holds for algebraic 2-tangles.

(3) **Conjecture:**

(a) **(Nakanishi 79)** *Any knot is 4-equivalent to the unknot.*

(b) **(Kawauchi 85)** *If two links are link-homotopic then they are 4-equivalent.*

Remarks: Conjecture **3(a)** holds for algebraic (in the Conway sense) knots and 3-braid knots. The smallest known unsolved case of Conjecture **3(a)** is a 2-cable of the trefoil knot (see Figure 1.59.3), which is a 4-braid knot (Nakanishi 1994). See [830, Nakanishi, 1984, Sūrikaisekikenkyūsho Kōkyūroku], [833, Nakanishi & Suzuki, 1987, Osaka J. Math.], [809, Morton, 1988], [894, Przytycki, 1988].

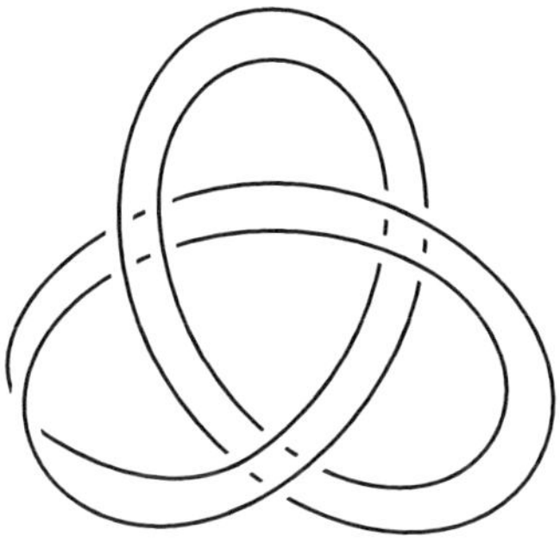

Figure 1.59.3.

(4) **Conjecture: (Przytycki 86)** *Any oriented link is $t_3, \bar{t}_4$ equivalent to a trivial link.*

Remarks: Conjecture **(4)** holds for algebraic (in the Conway sense) links and 3-braid links (for 3-bridge links and links with the braid index no more than 5 the conjecture should follow from the Coxeter theorem (see Remark to Conjecture **(1)**)). Conjecture **(4)** also holds for links with matched diagrams [896,Przytycki, 1990,Math. Proc. Cambridge Philos. Soc.] (see Problem 1.60). Also see [809,Morton,1988], [894,Przytycki,1988].

(5) **Conjecture: (Przytycki 87)**. *Any oriented link is $t_3, \bar{t}_6$ equivalent to a trivial link.*

Remarks: Conjecture **(5)** holds for closed 3-braids, [900,Przytycki,1993].

Consider the following (Figure 1.59.4) (n, k)-move which changes n horizontal twists to k vertical twists.

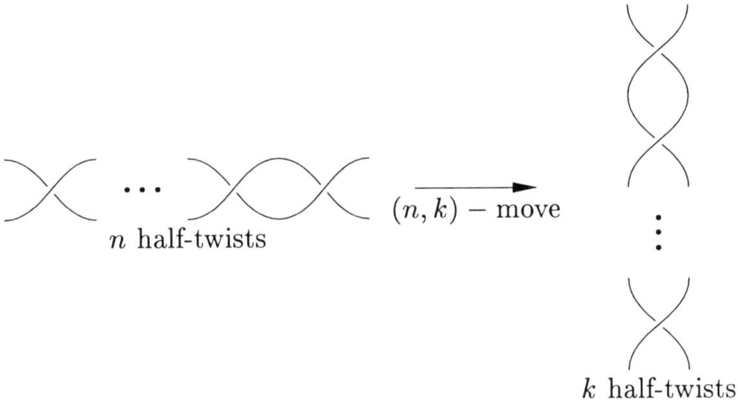

Figure 1.59.4.

(6) **Conjecture: (Nakanishi, Harikae 92)** *Any link can be reduced to an unlink by (2,2) moves.*

Remarks: Conjecture **(6)** holds for algebraic links (in the Conway sense) [445,Harikae & Uchida,1993], [831,Nakanishi,1992,Sūrikaisekikenkyūsho Kōkyūroku], [832,Nakanishi,1994,J. Knot Theory Ramifications], [901,Przytycki,1994a].

(7) **Question: (Przytycki 95)** *Can any link be reduced by (2,3) moves to an unlink?*

Remarks: The answer is yes for algebraic links (in the Conway sense).

Remarks to parts 1-7: A k-move on a link preserves $H_1(M_L^{(2)};\mathbb{Z}/k\mathbb{Z})$, where $M_L^{(n)}$ is the n-fold branched cover of S^3 with branching set the link (thus different trivial links are not k-equivalent); similarly, a (k,n)-move preserves $H_1(M_L^{(2)};\mathbb{Z}/(kn+1)\mathbb{Z})$ and t_3 and $\bar{t}_4$ moves preserve $H_1(M_L^{(3)};\mathbb{Z}/2\mathbb{Z})$ [894,Przytycki,1988], [901,Przytycki,1994a].

If $t^k = (-1)^k$ and $t \neq -1$, then the t_k move changes the Jones polynomial by $\pm i^k$; if $t^{2k} = 1$ and $t \neq -1$, then the Jones polynomial is preserved by a $\bar{t}_{2k}$ move; $(2,2)$ moves preserve the Kauffman polynomial of unoriented framed links up to a factor ± 1 at $(a,x) = (1, 2cos(\frac{2\pi}{5}))$

[894,Przytycki,1988]. Any $2k+1$ move is a combination of two $(2,k)$-moves (for example a 5-move is a combination of $(2,2)$-moves (see Figure 1.59.5), but not vice versa [445,Harikae & Uchida,1993], [901,Przytycki, 1994a]).

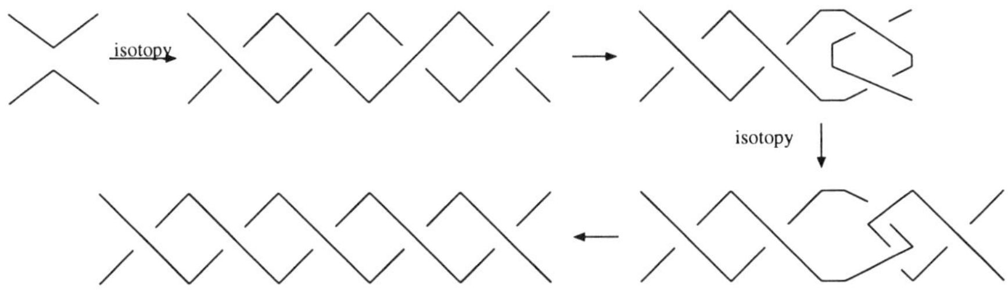

Figure 1.59.5.

Problem 1.60 (Przytycki) We say that an oriented diagram D is a *matched diagram* if one can pair up the crossings in D so that each pair looks like one of those in Figure 1.60.1 (notice the antiparallel orientation of a clasp).

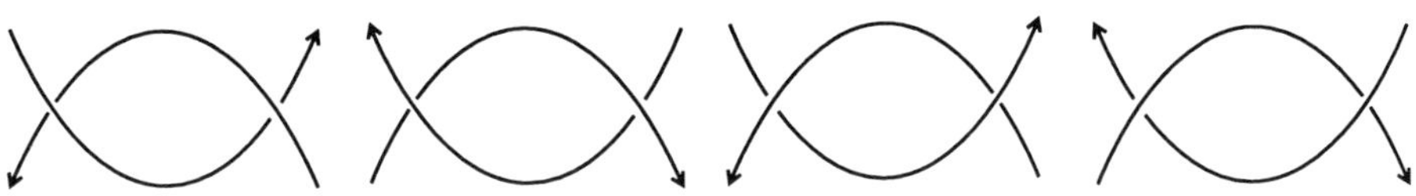

Figure 1.60.1.

(1) Conjecture: (Przytycki 87)

(a) There are oriented knots without a matched diagram.

(b) Any oriented link is t_3-equivalent to a link with matched diagram.

Remarks: Matched diagram were first considered in [890,Przytycka & Przytycki,1987], and their introduction was motivated by the proof [539,Jaeger,1988,Proc. Amer. Math. Soc.] that computation of the skein (Homflypt) polynomial is NP-hard. A similar concept was considered before by Conway who constructed knots which probably do not possess a matched diagram (see [40, Anstee, Przytycki, & Rolfsen,1989,Topology Appl.], [890,Przytycka & Przytycki,1987], [892,Przytycka & Przytycki,1993]).

Consider the map Ψ from matched oriented diagrams to unoriented diagrams obtained by reducing each clasp to a single crossing.

(2) **Conjecture: (Przytycka & Przytycki 93)** *Consider oriented links up to t_3 moves and up to change of orientation of each split component of the link; then Ψ descends to a bijection (onto unoriented links).*

Remarks: This conjecture is stronger than Conjecture 1(b). References: [892,Przytycka & Przytycki,1993], [900,Przytycki,1993].

Problem 1.61 *Find an invariant that distinguishes an oriented knot from its reverse (same knot with opposite orientation). Is a random knot irreversible?*

Remarks: It is possible to distinguish knots from their reverses by showing that there is no automorphism of the knot group which inverts the meridian and the longitude, as it would have to if there was an isotopy from a knot to its reverse [1063,Trotter,1964,Topology]. Even more, there are knots which are not concordant to their reverses ([670, Livingston,1983,Quart. J. Math. Oxford Ser. (2)] using a refinement of the Casson–Gordon techniques in [381,Gilmer,1983,Quart. J. Math. Oxford Ser. (2)]). Vassiliev invariants which are derived from Lie algebras will not distinguish the reverse of a knot. It is now known [1079, Vogel,1995] that there are weight systems for Vassiliev invariants which are not derived from the classical Lie algebras, but these do not yet distinguish knots any better (see Problem 1.89).

Although there is no single invariant for distinguishing reverses, in practice it seems that there are algebraic means available to deal with any specific example (see e.g. [448,Hartley,1983,Topology]). Furthermore, there are geometric methods: the SnapPea program will usually decide if a knot complement is hyperbolic, and if it is, a symmetries program will decide if there is an automorphism of the knot group which reverses the longitude; if SnapPea doesn't work there are slower algorithms based on the fact that the knot complement is Haken.

Problem 1.62 (Jones & Przytycki) A *Lissajous knot* K is a knot in $\mathbb{R}^3$ given by the parametric equations

$$x = cos(\eta_x t + \phi_x)$$

$$y = cos(\eta_y t + \phi_y)$$

$$z = cos(\eta_z t)$$

for integers η_x, η_y, η_z.

Question: *Which knots are Lissajous?*

Remarks: The crossing number of a Lissajous knot projected onto the xy-plane is $2\eta_x\eta_y - \eta_x - \eta_y$. The Jones polynomial J satisfies $J_{Lissajous}(i) = 1$ which implies that the Arf invariant of the knot is 0. Also, the Alexander polynomial of a Lissajous knot is a square modulo 2 (note that it follows from this that the Arf invariant is zero, and that, for example, the 8_5 knot is not Lissajous).

If the integers η_x, η_y, η_z, are all odd, then K is strongly positively amphicheiral; it follows that the Alexander polynomial is a square [449, Hartley & Kawauchi,1979,Math. Ann.], and more generally that the Alexander module over the ring $\mathbb{Q}[t^{\pm 1}]$ is a double (i.e. $A \oplus A$) [ibid.]. Furthermore, strongly positively amphicheiral implies that K is algebraically slice [673,Long,1983].

If one of the η's is even, then K has period 2 [111,Bogle, Hearst, Jones, & Stoilov,1994,J. Knot Theory Ramifications]. Also (Przytycki), there is an axis for K and its linking with K is ± 1. It follows that when $\eta_x = 2$, the Alexander polynomial of a Lissajous knot is congruent to 1 modulo 2.

These two cases together imply that no torus knot is Lissajous. The simplest knot with Arf invariant $= 0$ which is not Lissajous is 8_{10} because it is not strongly positively amphicheiral nor period 2. But 7_5, 8_3 and 8_6 are prime and may or may not be Lissajous.

Conjecture: *Turks head knots, (e.g. the closure of the 3-string braid* $(s_1\bar{s}_2)^{2k+1}$*), are not Lissajous.* Observe that they are strongly positively amphicheiral.

One can define a *racketball* knot (or billiard knot) as the trajectory inside a cube of a ball which leaves a wall at rational angles with respect to the natural frame, and travels in a straight line except for reflecting perfectly off the walls; generically it will miss the corners and edges, and will form a knot. These knots are precisely the same as the Lissajous knots.

It is not clear how to generalize Lissajous knots, but one can generalize racketball knots by changing the shape of the room.

Problem 1.63 Let $C(n)$ be the number of prime knots K in S^3 for which the crossing number $c(K)$ equals n. Similarly, let $U(n)$ (respectively $T(n)$) be the number of prime knots K for which the unknotting number $u(K)$ equals n (respectively, the tunnel number $t(K)$ equals n).

Question: *What is the asymptotic behavior of $C(n)$?, of $U(n)$?, of $T(n)$?, or of ratios of these functions such as $C(n)/U(n)$?*

Remarks: The known values of $C(n)$ are:

n	3	4	5	6	7	8	9	10	11	12	13
$C(n)$	1	1	2	3	7	21	49	165	552	2176	9988

Ernst & Sumners, counting 2-bridge knots of n crossings [284, Ernst & Sumners, 1987, Math. Proc. Cambridge Philos. Soc.], have shown that the number of knots (the mirror image counted, if the knot is different) is at least $(2^{n-2}-1)/3$ for $n \geq 4$, and Welsh used this inequality in [1106, Welsh, 1992] to show that:

$$2.68 \leq \liminf_{n\to\infty} C(n)^{1/n}$$

and also obtained the following inequalities for the number of links ($L(n)$) and for the number of alternating links ($A(n)$) of crossing number n.

$$4 \leq \liminf_{n\to\infty} L(n)^{1/n} \leq \limsup_{n\to\infty} L(n)^{1/n} \leq \frac{27}{2}$$

$$4 \leq \liminf_{n\to\infty} A(n)^{1/n} \leq \limsup_{n\to\infty} A(n)^{1/n} \leq \frac{27}{4}$$

It is not known if any of the limits above exist. It would follow from the following conjectures by Welsh: Both $C(n)$ and $L(n)$ are supermultiplicative functions, that is,

$$C(m+n) \geq C(m)C(n)$$

$$L(m+n) \geq L(m)L(n)$$

Problem 1.64 (de Souza) To move from one projection of a knot K in S^3 to another using Reidemeister moves, one must in general pass through projections with more crossings. If K is an n-crossing knot, let ψ_K be the minimum integer such that any two projections of K with n crossings can be connected by a Reidemeister path through knots of crossing number $\leq \psi_K$. Then let $\psi(n)$ be the maximum of ψ_K taken over all knots with crossing number n.

Question: *Is $\psi(n)$ bounded by a computable function? Is $\psi(n)$ bounded by a polynomial function in n?*

Remarks: A positive solution would imply the validation of some heuristic methods for computing knot invariants.

Observe that the recent proof of the Tait Flyping Conjecture by Menasco & Thistlethwaite [750,1993,Ann. of Math.] implies that one can pass between two minimal alternating projections of an alternating knot by using flypes on tangles, without ever increasing the number of crossings.

Problem 1.65 *Is the crossing number $c(K)$ of a knot K additive with respect to connected sum, that is, is the equality $c(K_1 \# K_2) = c(K_1) + c(K_2)$ true?*

Remarks: Murasugi has shown the conjecture to be valid for alternating knots as a corollary of his proof the Tait conjecture that reduced alternating projections are minimal (also proved by Kauffman [563,1987, Topology], [564,1988,Amer. Math. Monthly] and Thistlethwaite [1043, 1987,Topology] (all independently)). In fact, the proof is valid for a larger class of knots called *adequate* knots.

Problem 1.66 (de Souza) In his first memoir *On Knots* Tait constructed (he thought) amphicheiral knots of every possible even crossing number. He proposed a few forms (and their connected sums) to fill in series of even numbers.

His arguments on the computation of the crossing numbers of the resulting knots were inconclusive for several reasons, one being his assumption that reduced alternating knot projections are always minimal, a theorem that was proved more than a century later by Menasco & Thistlethwaite, [750,1993,Ann. of Math.].

He also tried to prove that every alternating amphicheiral knot has even crossing number, a question that was settled by Murasugi [820, 1987,Topology], as a byproduct of his proof of the Tait conjecture that reduced alternating projections are minimal (also proved by Kauffman [563,1987,Topology], [564,1988,Amer. Math. Monthly] and Thistlethwaite [1043,1987,Topology] (all independently)).

Apparently Tait thought that all even crossing, amphicheiral, prime knots were alternating, but in 1983 van Buskirk [166,Buskirk,1983, Rocky Mountain J. Math.] found a 14-crossing prime, amphicheiral non-alternating knot.

Tait also conjectured that if a knot K is amphicheiral, then its minimal crossing projection has an even number of crossings, [1030,Tait, 6 77,Trans. Royal Soc. Edinburgh; Section 13]. But this conjecture has been disproved recently by Flapan, Liang & Mislow [319,1995] who showed that the 2 component satellite link 9^2_{61} is amphicheiral as an unoriented link (see Figure 1.66.1) and, furthermore, constructed a positively amphicheiral link with minimal crossing number 11 (see Figure 1.66.2), and also by Thistlethwaite, pending independent verification of the knot tables (currently being done by Hoste), with a 15-crossing knot (see Figure 1.66.3 and Figure 1.66.4 to see the amphicheirality).

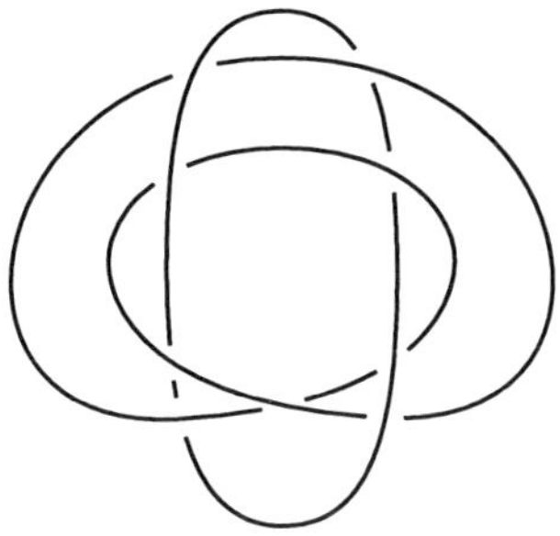

Figure 1.66.1. The link 9^2_{61}

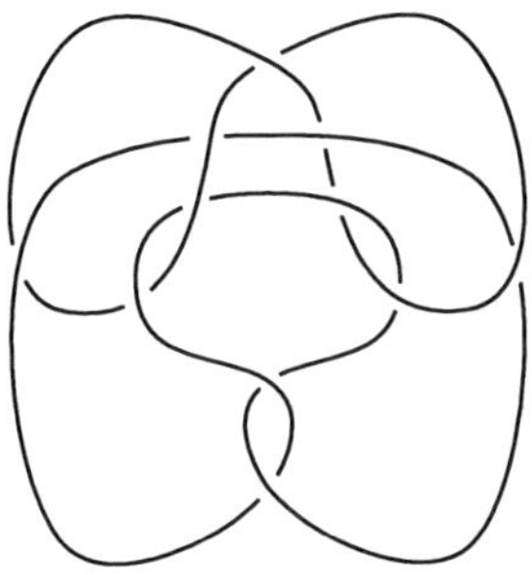

Figure 1.66.2.

Figure 1.66.3. Thistlethwaite's amphicheiral 15-crossing knot

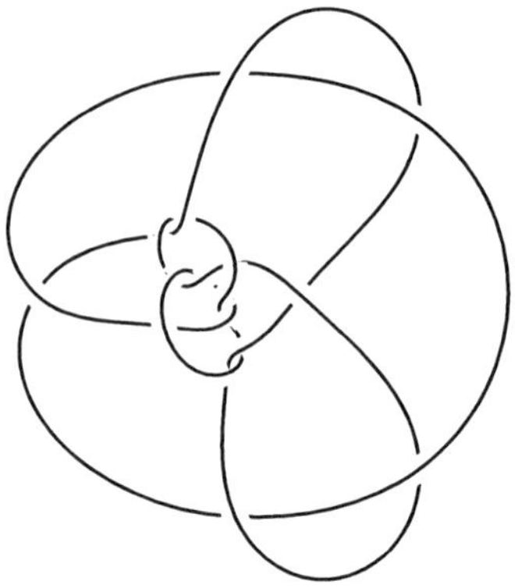

Figure 1.66.4. 16-crossing projection of the same knot, showing amphicheirality

The following questions are still unknown:

(A) Question: *Are there amphicheiral knots of every crossing ≥ 15 ?*

Remarks: *The total number of amphicheiral knots is very small compared to the number of knots.*

n	*3*	*4*	*5*	*6*	*7*	*8*	*9*	*10*	*11*	*12*	*13*
Knots	*1*	*1*	*2*	*3*	*7*	*21*	*49*	*165*	*552*	*2176*	*9988*
Amphicheirals	*0*	*1*	*0*	*1*	*0*	*5*	*0*	*13*	*0*	*58*	*0*

(B) Question: *Are there (prime, alternating) amphicheiral knots with every possible even crossing number?*

Problem 1.67 (de Souza) *Is the crossing number of a satellite knot bigger than that of its companion?*

Remarks: Surely the answer is yes, so the problem indicates the difficulties of proving statements about the crossing number.

Problem 1.68 Let $c(K)$ be the crossing number of a knot K in S^3, and let the asymptotic crossing number of K, $AC(K)$, be defined by

$$AC(K) = inf\{c(K_d)/d^2\}$$

where the infimum is taken over all satellites K_d of homological degree d and all $d = 1, 2, 3, \ldots$.

Conjecture: $AC(K) = c(K)$

Remarks: Obviously $AC(K) \leq c(K)$ and Freedman & He [333,1991, Ann. of Math.] show that $2(genus(K)) - 1 \leq AC(K)$. This question is relevant to energy estimates in magnetohydrodynamics [ibid.]. Note that if dK is d planar parallel copies of K, then $c(dK) \geq d^2 c(K)$ for *adequate* knots K (and links) ([656,Lickorish & Thistlethwaite,1988,Comment. Math. Helv.] and [1044,Thistlethwaite,1988,Invent. Math.]).

Problem 1.69 (de Souza) (A) *Does the connected sum of n knots have unknotting number at least n?*

Remarks: It is not even known that the connected sum of n knots has unknotting number > 2 (but unknotting number one knots are prime [969,Scharlemann,1985b,Invent. Math.]). A yes answer to (A) would follow from a yes answer to the real question:

(B) *Is the unknotting number additive under connected sum, that is, does the equality $u(K_1 \# K_2) = u(K_1) + u(K_2)$ hold?*

(C) (Boileau) *Is the unknotting number of a link invariant under the mutation: take a tangle in a disk with 4 endpoints and rotate it by π.*

Remarks: If so, then the unknotting number of a link is additive.

The unknotting numbers (even though very easy to define) are hard to compute. The unknotting numbers of most knots up to 9 crossings are known. The core of the list has been computed by Y. Nakanishi in [829,1981,Math. Sem. Notes Kobe Univ.] and a few others in [560, Kanenobu & Murakami,1986,Proc. of the AMS], [592,Kobayashi,1989, Kobe J. Math.], [653,Lickorish,1985,Contemp. Math.] and unpublished work by J. R. Rickard, as marked in the footnotes.

Unknotting Numbers			
$u(3_1) = 1$	$u(8_8) = 2^*$	$u(9_8) = 2^*$	$u(9_{29}) = 1$
$u(4_1) = 1$	$u(8_9) = 1$	$u(9_9) = 3$	$u(9_{30}) = 1$
$u(5_1) = 2$	$u(8_{10}) = 1$ or 2	$u(9_{10}) = 2$ or 3	$u(9_{31}) = 2^*$
$u(5_2) = 1$	$u(8_{11}) = 1$	$u(9_{11}) = 2$	$u(9_{32}) = 1$ or 2
$u(6_1) = 1$	$u(8_{12}) = 2^*$	$u(9_{12}) = 1$	$u(9_{33}) = 1$
$u(6_2) = 1$	$u(8_{13}) = 1$	$u(9_{13}) = 2$ or 3	$u(9_{34}) = 1$
$u(6_3) = 1$	$u(8_{14}) = 1$	$u(9_{14}) = 1$	$u(9_{35}) = 2$ or 3
$u(7_1) = 3$	$u(8_{15}) = 2$	$u(9_{15}) = 2^*$	$u(9_{36}) = 2$
$u(7_2) = 1$	$u(8_{16}) = 2^\S$	$u(9_{16}) = 3$	$u(9_{37}) = 2$
$u(7_3) = 2$	$u(8_{17}) = 1$	$u(9_{17}) = 2^*$	$u(9_{38}) = 2$ or 3
$u(7_4) = 2^\dagger$	$u(8_{18}) = 2$	$u(9_{18}) = 2$	$u(9_{39}) = 1$
$u(7_5) = 2$	$u(8_{19}) = 3$	$u(9_{19}) = 1$	$u(9_{40}) = 2$
$u(7_6) = 1$	$u(8_{20}) = 1$	$u(9_{20}) = 2$	$u(9_{41}) = 2$
$u(7_7) = 1$	$u(8_{21}) = 1$	$u(9_{21}) = 1$	$u(9_{42}) = 1$
$u(8_1) = 1$	$u(9_1) = 4$	$u(9_{22}) = 1$	$u(9_{43}) = 2$
$u(8_2) = 2$	$u(9_2) = 1$	$u(9_{23}) = 2$	$u(9_{44}) = 1$
$u(8_3) = 2^*$	$u(9_3) = 3$	$u(9_{24}) = 1$	$u(9_{45}) = 1$
$u(8_4) = 2^*$	$u(9_4) = 2$	$u(9_{25}) = 2^\ddagger$	$u(9_{46}) = 2$
$u(8_5) = 2$	$u(9_5) = 2^*$	$u(9_{26}) = 1$	$u(9_{47}) = 2$
$u(8_6) = 2^*$	$u(9_6) = 3$	$u(9_{27}) = 1$	$u(9_{48}) = 2$
$u(8_7) = 1$	$u(9_7) = 2$	$u(9_{28}) = 1$	$u(9_{49}) = 2$ or 3

†Computed in [653,Lickorish,1985,Contemp. Math.].
*Computed in [560,Kanenobu & Murakami,1986,Proc. of the AMS].
‡Computed in [592,Kobayashi,1989,Kobe J. Math.].
§Computed by J. R. Rickard, but never published.

According to Lickorish, (the only known eye witness) the method used by Rickard (an extension of previous work by Lickorish [653,1985, Contemp. Math.]) consisted of analysing the linking form for the first homology of the double branched cover of a knot when there is more than one homology generator. This method worked for: 7_4, 8_8, 8_{16}, 9_{15}, 9_{17} and 9_{31}. For most of his knots there is an independent (and published) confirmation of his results [560,Kanenobu & Murakami,1986, Proc. of the AMS], but for the knot 8_{16} Rickard's is the only known (and unfortunately unpublished) solution.

Here are the simplest knots whose unknotting numbers are unknown:

8_{10}

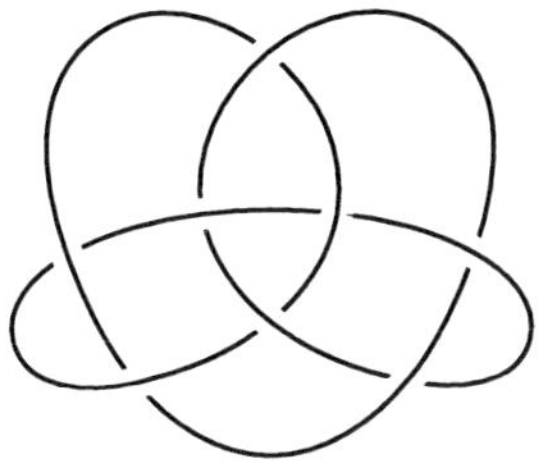

8_{16}

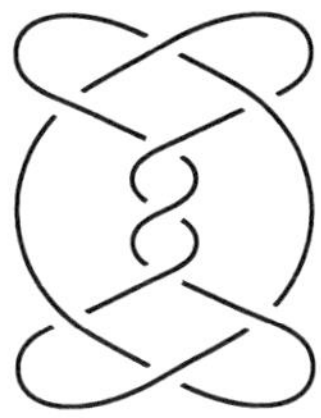

9_{10}

9_{13}

9_{32}

9_{38}

9_{35} according to Alexander–Briggs

9_{35} with its period 3

9_{49} according to Alexander–Briggs

9_{49} with its period 3

Problem 1.70 (A) Conjecture: *The tunnel number of the connected sum of two knots is at least equal to the larger of the tunnel numbers of the two knots.*

Remarks: The tunnel number $t(K)$ of a knot K in S^3 is the minimal number of arcs which must be added to the knot, forming a graph with three edges at a vertex, so that the complement in S^3 is a handlebody (whose boundary will then be a minimal Heegaard splitting of the knot complement). This graph is the simplest graph which (allowing edges to slide over edges) can be moved into a plane, yet contains the knot. This construction amounts to the same thing as boring holes in the complement of K, whence the name tunnel.

T. Kobayashi [594,1994,J. Knot Theory Ramifications] following K. Morimoto [798,1995] proved that for any n, there exist knots K_1 and K_2 for which

$$t(K_1\#K_2) \leq t(K_1) + t(K_2) - n.$$

Moriah & Rubinstein [796,1993] have shown that there exist two knots for which

$$t(K_1\#K_2) = t(K_1) + t(K_2) + 1$$

(the most it could be); also [800,Morimoto, Sakuma, & Yokota, 1994]. H.-Z. Kowng [615,1994] showed that

$$t(K_1) + t(K_2) \leq 3(t(K_1\#K_2) + 1).$$

(B) *Suppose K is a connected sum of n non-trivial knots. Is it true that $t(K) \geq n$?*

(C) *More generally, how does the Heegaard genus behave when two 3-manifolds with boundary are glued together along an annulus?*

Problem 1.71 (Adams) *If a knot or link in S^3 is tunnel number one, then classify the possible tunnels up to isotopy.*

Remarks: Every torus knot has tunnel number one, and the tunnels have been classified [123,Boileau, Rost, & Zieschang,1988,Math. Ann.].

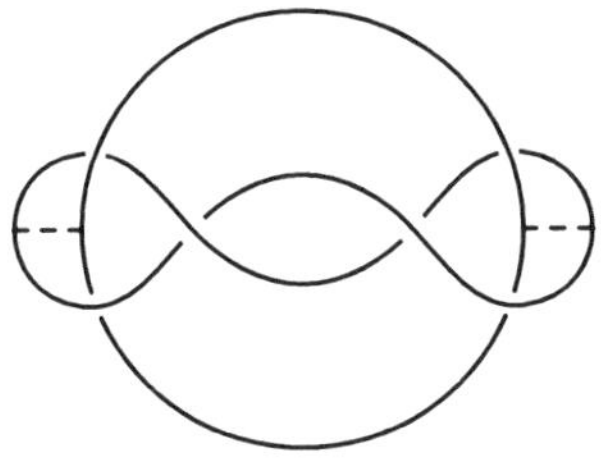

Figure 1.71.1.

Most satellite knots are not tunnel number one, but for those that are, the tunnels have been classified [799,Morimoto & Sakuma,1991,Math. Ann.]. The problem is wide open for the remaining knots, the hyperbolic knots, except that the Heegaard splittings of the figure-8 knot complement are classified [469,Heath,1995], and genus two Heegaard splittings of the 5_2 knot complement are the four well-known ones (Heath).

The link in Figure 1.71.1 is tunnel number one, and has two tunnels, as drawn [7,Adams,1995]. This has been extended to a classification of the tunnels for 2-bridge links (not knots) in [8,Adams & Reid,1994].

Problem 1.72 (Adams) One can extend the notion of tunnel number for knots and links in S^3 to an arbitrary, orientable, 3-manifold M with boundary as follows: let $t(M)$ be the minimum number of properly imbedded, smooth, (thickened) arcs in M such that their complement is a handlebody. Note that this is equivalent to asking for a handlebody decomposition of M with a minimal number of 2-handles (whose cocores form the arcs), and this is the analogue for the case when $\partial(M) \neq \emptyset$ to the question of finding a minimal genus Heegaard splitting of a closed, orientable 3-manifold.

A tunnel belonging to a minimal set should be called an *unknotting* tunnel.

(A) *Now let M be hyperbolic with one cusp. Then M can be thought of as the interior of a compact 3-manifold M' such that $\partial M' = T^2$, and define $t(M)$ to be $t(M')$. Suppose that $t(M) = 1$. Must every unknotting tunnel be isotopic to a geodesic with both ends running out the cusp?*

(B) *Is there an upper bound, independent of M, such that the length of* *any unknotting tunnel is less than the bound?*

Remarks: For a tunnel number one, 2-cusped hyperbolic 3-manifold, any unknotting tunnel must be isotopic to a geodesic with each end running out a cusp, such that the length of the tunnel is bounded by ln(4) [7,Adams,1995]. Of course, the length of a geodesic running out cusps is infinite, so one chooses disjoint cusps, measures the length of the geodesic outside the cusps, and then minimizes over disjoint cusps.

(C) *In general, is a minimal set of unknotting tunnels isotopic to a set of geodesics when M is hyperbolic?*

Problem 1.73 (Bleiler) Let L be a link in a 3-manifold M with link exterior X (= M minus an open disk bundle over L). L is said to have a (g,b)*-presentation* if $\pi_1(X)$ has a presentation with g generators and b meridional generators.

Following Doll [246,1992,Math. Ann.], L is said to have a (g,b)-*decomposition*, or alternatively, to be in *b bridge position with respect to a Heegaard surface F of genus g for M*, if L intersects each component of $M - F$ in b trivial arcs. Obviously, if L has a (g,b)-decomposition, then it has a (g,b)-presentation.

Meridional Generator Conjecture: *If a link L has a (g,b)-presentation, then it has a (g,b)-decomposition.*

Remarks: Cappell & Shaneson's old Problem 1.11 is the special case of this conjecture for $(g,b) = (0,2)$ or $(0,n)$ (still unsolved, but see the Update for progress).

This discussion leads to the definition of the *genus g bridge number* of L in M which is the minimal b for a (g,b)-decomposition (genus 0 bridge number is the classical bridge number). Similarly, one can define the *bridge b genus* of the pair (M,L) to be the minimal genus Heegaard splitting of M for which L has bridge number b. It is easy to show that

tunnel number (L) $\leq$ 1-bridge genus (M,L) $\leq$ tunnel number (L) +1.

The examples of tunnel number one knots with super-additive tunnel number [800,Morimoto, Sakuma, & Yokota,1994] (see Problem 1.70 (A)) are examples of knots where the tunnel number and the 1-bridge genus are different.

Problem 1.74 *Given a prime knot K, is it necessary to know all of its n-fold branched covers in order to distinguish it from all other knots?*

Remarks: There do not exist prime knots K and K' all of whose branched covers are equal (see old Problem 1.27, and [600,Kojima, 1986]); however there may exist a sequence of prime knots $\{K_n\}$ such that the k-fold covers of K and K_n agree for $k \leq n$.

Problem 1.75 (Boileau) Let $K_1, K_2 \subset S^3$ be two hyperbolic knots.

(A) *If K_1 and K_2 have the same 2-fold branched covering, then show that either K_1 and K_2 are mutants, or there is an orientation preserving involution of S^3 which carries K_1 to K_2.*

Remarks: This is true if K_1 and K_2 are π-hyperbolic (meaning that each 2-fold branched covering is hyperbolic and the covering involution is an isometry) [117,Boileau & Flapan,1995,Topology Appl.].

(B) *Is there an example of distinct hyperbolic knots with the same 2-fold and 3-fold cyclic branched coverings?*

Problem 1.76 (Menasco) Conjecture: *A hyperbolic knot does not have a closed, totally geodesic surface imbedded in its complement.*

Remarks: The Conjecture is true for alternating knots and 3-braid knots; but there are hyperbolic links for which the conjecture fails [748, Menasco & Reid,1992].

Problem 1.77 (Gordon) Let M be a compact, orientable 3-manifold with $\partial M \cong T^2$, whose interior admits a complete hyperbolic metric of finite volume. A *slope* on ∂M is an isotopy class of simple, closed essential curves in ∂M. For such a slope r, let $M(r)$ denote the r-Dehn filling of M. The *distance* between two slopes r_1 and r_2, denoted by $\Delta(r_1, r_2)$, is defined to the be minimal geometric intersection number between r_1 and r_2. Let $E(M) = \{\text{slopes r on } \partial M \mid M(r) \text{ is not hyperbolic}\}$. Let $e(M) = |E(M)|$ (which is finite, see Theorem 5.8.2 of [1050,Thurston, 1977]).

When M is the exterior of a hyperbolic knot K in S^3, it is called M_K; (p,q)-Dehn surgery on K is given by $M_K(r)$ where $r = p/q$. In this case, set $E(K) = E(M_K)$ and $e(K) = |E(M_K)|$. Also, let $\mathcal{F}$ be the set of slopes for which $M_K(r)$ has finite fundamental group.

(A) Conjectures: *(1) $e(K) \leq 6$ if K is not the figure-8 knot or the $(-2,3,7)$ pretzel knot;*

(2) $\Delta(r_1, r_2) \leq 4$ for any two slopes $r_1, r_2 \in E(K)$;

(3) If $p/q \in E(K)$, then $|q| \leq 2$, and if $|q| = 2$, then $M_K(p/q)$ contains an essential torus and is not Seifert fibered; furthermore, there is at most one slope $p/q \in E(K)$ with $|q| = 2$;

(4) If $|q| = 2$, then $M_K(p/q)$ contains an incompressible torus iff K is one of the Eudave-Muñoz knots $k(l,m,n,p)$ in [286, Eudave-Muñoz, 1994].

(5) There are at most 3 slopes which can produce manifolds with an essential torus;

(6) $|\mathcal{F}| \leq 4$, the slopes in $\mathcal{F} - \infty$ form a set of consecutive integers, $\mathcal{F}$ contains at most one even integer, and the distance between any two slopes in $\mathcal{F}$ is at most two.

Remarks: $M_K(\infty)$ is, of course, always S^3.

(1) There is no known bound for $e(K)$ which is independent of K, but it is known that if M is hyperbolic, then the number of slopes r such that $M(r)$ admits no metric of (non-constant) negative curvature is at most 24 [104, Bleiler & Hodgson, 1995]. A bound on $e(M)$ would follow from a pinching theorem for negatively curved metrics on closed 3-manifolds.

(2) $M_K(p/q)$ is irreducible if $|q| \geq 2$ [404, Gordon & Luecke, 1987, Math. Proc. Cambridge Philos. Soc.] and contains no essential torus if $|q| \geq 3$ [406, Gordon & Luecke, 1995].

(3) $M_K(p/q)$ has infinite fundamental group if $|q| \geq 3$ [146, Boyer & Zhang, 1995].

(4) If $M_K(p/q)$ is Seifert fibered and $|q| > 1$, then it has exactly 3 exceptional fibers with orbit space the 2-sphere [145, Boyer & Zhang, 1994b].

(5) If K is the figure-8 knot, then

- $M_K(0) = T^2$-bundle over S^1 with monodromy $\left(\begin{smallmatrix}1&1\\1&2\end{smallmatrix}\right)$,
- $M_K(\pm1)$ is the Seifert fibered space

$$(\ O, o, 0 \mid -1; (2,1), (3,1), (7,1)\ ),$$

  in other words, the Brieskorn homology sphere $\Sigma(2,3,7)$, (for the Seifert fibered space notation, see [986,Seifert,1933,Acta Math.]). (Since the figure-8 knot is amphicheiral, $+n$ and $-n$ surgeries are orientation reversing homeomorphic).
- $M_K(\pm2)$ is the Seifert fibered space

$$(\ O, o, 0 \mid -1; (2,1), (4,1), (5,1)\ ).$$

- $M_K(\pm3)$ is the Seifert fibered space

$$(\ O, o, 0 \mid -1; (3,1), (3,1), (4,1)\ ).$$

- $M_K(\pm4)$ is the union of the trefoil knot complement X and the non-trivial I-bundle KI over the Klein bottle (they are glued together along their T^2 boundaries), and therefore contains an incompressible torus. (The gluing diffeomorphism is subtle, so here are the details from Bleiler: coordinatize the torus ∂X for the left handed trefoil knot the usual way for knots in S^3 (with meridian $1/0$ and longitude $0/1$ chosen so that the Seifert fibre of the left hand trefoil is the $6/-1$ curve), and the torus ∂KI by choosing the *meridian* $1/0$ to be the fibre of the circle fibration of KI over the Möbius band and the *longitude* $0/1$ to be the regular fibre of the Seifert fibration of KI over the disc with two exceptional fibres each of index 2. In these coordinates the matrix of the attaching map $\partial KI \to \partial X$ is given by: $\left(\begin{smallmatrix}-4&-5\\1&1\end{smallmatrix}\right)$.
  In particular, the meridian of ∂X goes to the $1/-1$ curve on ∂KI, which is the *meridian* of ∂KI when we consider KI as the exterior of the $(1,2)$ cable of the S^1 factor of $S^2 \times S^1$.) Some of the above calculations can be found in [498,Hodgson, 1986].

The rest (except ∞) are hyperbolic, so e(figure-8 knot) $= 10$.

6. If K is the $(-2,3,7)$ pretzel knot K, then

- $M_K(16)$ contains an incompressible torus,

- $M_K(17)$ is the quotient of S^3 by a finite rotation group isomorphic to $I^* \times \mathbb{Z}/17\mathbb{Z}$, where I^* is the binary icosahedral group (see the groups in Problem 3.37) [104, Bleiler & Hodgson, 1995].
- $M_K(18)$ is the lens space $L(18, 5)$,
- $M_K(37/2)$ contains an incompressible torus,
- $M_K(19)$ is $L(19, 7)$,
- $M_K(20)$ contains an incompressible torus.

$E(K)$ consists of these slopes together with the slope ∞, so $e(K) = 7$.

(B) **Conjectures**: *For general M as above (not necessarily a knot exterior),*

- *$e(M) \leq 7$ if M is not the figure-8 knot complement, or the figure-8 sister or a third manifold (described below);*
- *$\Delta(r_1, r_2) \leq 8$ for any two slopes $r_1, r_2 \in E(M)$;*
- *there are no more than 5 slopes whose Dehn fillings produce finite or infinite cyclic fundamental groups, and the distance between any two of these slopes is at most 3.*

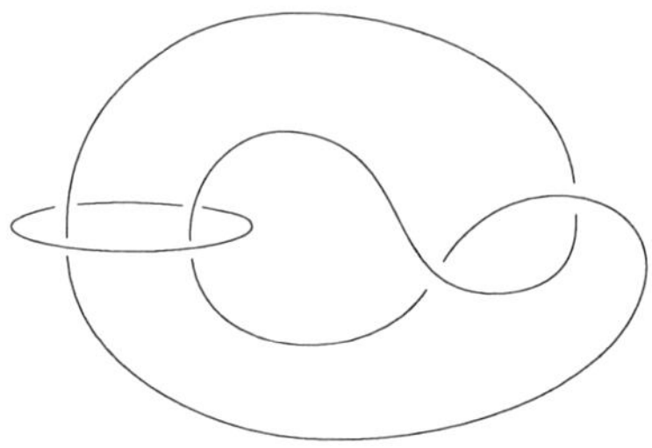

Figure 1.77.1.

Remarks: $e(M) = 10$ for the figure-8 knot complement, as discussed above, and $e(M) = 8$ for two other cases. Each of these 3-manifolds can be constructed from the left-handed Whitehead link (drawn in Figure 1.77.1) by deleting one component and surgering the other with framings +1 (for the figure-8 knot complement), -5 (for the figure-8 sister), and +2 for the third manifold.

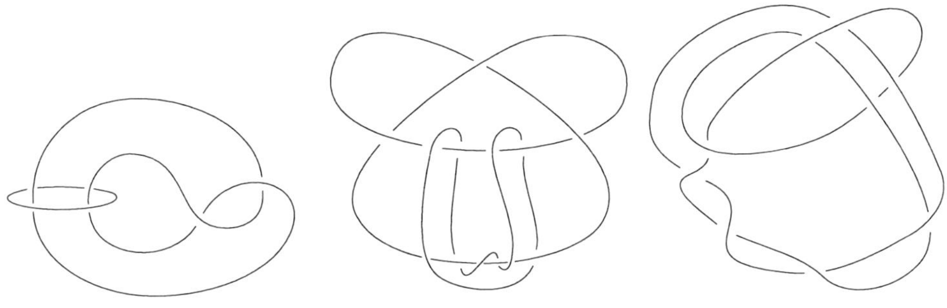

left handed Whitehead link   Whitehead sister 9^2_{50}

Figure 1.77.2.

Each of these manifolds is also a once punctured torus bundle, with respective monodromies:

$$\begin{pmatrix} 1 & 1 \\ 1 & 2 \end{pmatrix}, \begin{pmatrix} -1 & -1 \\ -1 & -2 \end{pmatrix} \text{ and } \begin{pmatrix} 1 & 1 \\ 2 & 3 \end{pmatrix}.$$

(Note that the figure-8 sister can be obtained from the figure-8 knot complement by composing the monodromy of the latter with $-I$; thus they are mutants and have the same volume.)

In Table 1.77.1 below, eleven manifolds are listed with $e(M) \geq 7$ (it is conjectured that these are the only ones). For each manifold, the table gives its SnapPea notation, $mijk$, [1101,Weeks,1995], its approximate hyperbolic volume, a Dehn surgery description, and in the case of those which are once punctured torus bundles the monodromy is given as a composition of the matrices R and L which are $\left(\begin{smallmatrix} 1 & 1 \\ 0 & 1 \end{smallmatrix}\right)$ and $\left(\begin{smallmatrix} 1 & 0 \\ 1 & 1 \end{smallmatrix}\right)$. The left handed Whitehead link, the Whitehead sister, and 9^2_{50} (in Rolfsen's notation) are drawn in Figure 1.77.2. Note that $m016$ is the $(-2,3,7)$ pretzel knot exterior.

Remarks: Evidence for, and discussion of, these conjectures can be found in [104,Bleiler & Hodgson,1995], [144,Boyer & Zhang,1994a,Bull. Amer. Math. Soc.], [146,Boyer & Zhang,1995], [145,Boyer & Zhang, 1994b], [286,Eudave-Muñoz,1994], and [500,Hodgson & Weeks,1995].

SnapPea	$e(M)$	volume	link	framing	monodromy
m003	8	2.03	Wh. link	-5	$-$LR
m004	10	2.03	Wh. link	$+1$	LR
m006	7	2.57	Wh. link	$-5/2$	
m007	7	2.57	Wh. link	$-3/2$	
m009	8	2.67	Wh. link	$+2$	LLR
m016	7	2.83	Wh. sister	$-17/2$	
m017	7	2.83	Wh. link	$-7/2$	
m023	7	2.99	Wh. link	$+3$	LLLR
m035	7	3.18	Wh. link	$-4/3$	
m038	7	3.18	9^2_{50}	$+2$	
m039	7	3.18	Wh. link	$+4$	LLLLR

Table 1.77.1.

Problem 1.78 (Gordon) A *Berge knot* is the name given to any knot K which lies on the boundary F_2 of a genus two handlebody, standardly imbedded in S^3, with the property that K represents an element of a basis of the fundamental group $\mathbb{Z} * \mathbb{Z}$ of each complement of $S^3 - F_2$. Note that any torus knot is a Berge knot because it is the pairwise connected sum $(T^2, (p,q) - curve)\#(T^2, (1,1) - curve)$ along (B^2, B^1). Also note that surgery on S^3 along a Berge knot, with framing given by a parallel copy in F_2, gives a lens space [79,Berge,1995].

Conjecture: *If Dehn surgery on a knot K gives a lens space, then K is a Berge knot.*

Remarks: If surgery on a non-torus knot produces a lens space, then it must be integral surgery [229,Culler, Gordon, Luecke, & Shalen,1987, Ann. of Math.].

Problem 1.79 (Cabling Conjecture) *The only way to get a reducible 3-manifold by surgery on a knot is to surger a cable knot with surgery coefficient equal to the slope of the cabling annulus.*

Remarks: The slope of the cabling annulus is the linking number of the cable with its pushoff into the annulus (equals the torus minus the torus knot), e.g. 6 surgery on the trefoil gives $L(2,1)\#L(3,1)$.

It is known that non-integral surgeries cannot give reducible manifolds by [404,Gordon & Luecke,1987,Math. Proc. Cambridge Philos. Soc.]. It is also known that for a cable knot, only the cabling slope will give a reducible manifold; in fact the conjecture is known for satellite knots [971,Scharlemann,1990,Topology]. The conjecture is also known for strongly invertible knots [285,Eudave-Muñoz,1992,Trans. Amer. Math. Soc.], for alternating knots [749,Menasco & Thistlethwaite,1992, J. Reine Angew. Math.], and for symmetric knots with symmetries of order ≥ 5 [683,Luft & Zhang,1994,Math. Ann.].

Problem 1.80 (Boileau) Let $(M_1, K_1), (M_2, K_2)$ be two pairs where each M_i is an orientable, irreducible, 3-manifold, and each K_i is a null-homotopic knot in M_i.

(A) Conjecture: *If $M_1 - K_1$ is homeomorphic to $M_2 - K_2$, then the pair (M_1, K_1) is homeomorphic to the pair (M_2, K_2).*

(B) Conjecture: *Any non-trivial surgery on K_1 never gives a manifold with the same simple homotopy type as M_1.*

Remarks: Conjecture (B) implies Conjecture (A). Conjecture (B) is true if $\beta_1(M_1) \geq 1$, or if $\pi_1(M_1)$ is infinite and $M_1 - K_1$ contains an incompressible torus which does not cobound a cable space of order 2 with the boundary [116,Boileau, Domergue, & Mathieu,1995].

(C) Conjecture: *A non-trivial surgery on K_1 produces a bundle over S^1 iff K_1 is a fibered knot in M_1 and it is the longitudinal surgery.*

Remarks: If the surgery produces a bundle, then either it is longitudinal surgery or M_1 is a bundle over S^1 [125,Boileau & Wang,1995].

Problem 1.81 (Bleiler) Let M be an oriented 3-manifold with $\partial M = T^2$, and let r be a slope (an isotopy class of unoriented, simple, closed curves) in T^2. Let $M(r) = M \cup B^2 \times S^1$ with ∂B^2 glued to the slope r (this is called an *r-Dehn filling*). One also refers to $M(r_1)$ or $M(r_2)$ as surgeries on $N = M(r_1)$. Call two surgeries *purely cosmetic* if there is an orientation preserving diffeomorphism between $M(r_1)$ and $M(r_2)$, and *chirally cosmetic* if the homeomorphism is orientation reversing.

Call two slopes, as well as the corresponding fillings, *equivalent* if there exists a homeomorphism of M taking one slope to the other.

(A) Cosmetic surgery conjecture: *Two surgeries on inequivalent slopes are never purely cosmetic. Equivalently, if $M(r_1) = M(r_2)$ for inequivalent slopes, then the homeomorphism is orientation reversing.*

Remarks: Gordon & Luecke [405,1989,J. Amer. Math. Soc.] proved that there are no cosmetic (pure or chiral) surgeries on S^3 or $S^2 \times S^1$. However, Mathieu [706,1990] found an infinite set of pairs of inequivalent chirally cosmetic surgeries on the right hand trefoil exterior. Rong [936,1995b] classified the inequivalent cosmetic surgeries on Seifert fibered manifolds M, and noted that all are chiral. Also, there exist hyperbolic manifolds M which have a pair of inequivalent slopes yielding oppositely oriented lens spaces ([105,Bleiler, Hodgson, & Weeks,1995]. These arise from certain 1-bridge braids in $S^1 \times B^2$ which have inequivalent slopes which fill to $S^1 \times B^2$, [78,Berge,1991,Topology Appl.], [357,Gabai, 1989,Topology]).

(B) Conjecture: *There are no cosmetic surgeries, pure or chiral, on hyperbolic manifolds which yield hyperbolic manifolds.*

(C) Conjecture: *Closed geodesics in a hyperbolic 3-manifold are determined by their complements (even allowing orientation reversing diffeomorphisms).*

Remarks: (B) $\Rightarrow$ (C) but they are not equivalent as it may happen that the core of one of the surgeries is not isotopic to a closed geodesic.

One can ask the above questions with homeomorphism replaced by homotopy equivalence or simple homotopy equivalence (see Problem 1.80 for a version which is different because the knots are null-homotopic). There are hyperbolic knot exteriors in lens spaces with a pair of slopes which yield non-homeomorphic but homotopy equivalent lens spaces [105,Bleiler, Hodgson, & Weeks, 1995].

A third equivalent formulation of Conjecture (A) is the

(D) Oriented knot complement conjecture: *If K_1 and K_2 are knots in a closed, oriented 3-manifold M whose complements are* *homeomorphic via an orientation-preserving homeomorphism, then there exists an orientation-preserving homeomorphism of M taking K_1 to K_2.*

Problem 1.82 *Characterize those framed links in S^3 which produce a connected sum of $S^1 \times S^2$'s.*

Conjecture: *Any link which can be obtained from the 0-framed unlink by handle slides (no stabilization by ± 1 unknots necessary).*

Remarks: A 0-framed knot which does not give $S^1 \times S^2$ is said to have Property R; Gabai [355,1987a,J. Differential Geom.] proved that all non-trivial knots have Property R (see Problem 1.17). This problem asks for a generalization of Property R to multi-component links.

This question is related to closed, smooth 4-manifolds with handlebody decompositions having no 1-handles (or, dually, no 3-handles); the boundary of the 0-handle union the 2-handles must be a connected sum of $S^1 \times S^2$'s, as well as surgery on a framed link.

Problem 1.83 (Rudolph) (A) *What is the Grothendieck group, $\mathcal{G}$, of (isotopy classes of) oriented, Seifert fibered surfaces in S^3 under Murasugi sum? In particular, is it finitely generated?*

Remarks: *Murasugi* sum is Gabai's name for an operation (called *star product* by Murasugi [818,1963,Amer. J. Math.], and *plumbing* by Stallings [1010,1978]) which combines two Seifert surfaces, S_1 and S_2, along a disk in each, see Figure 1.83.1, to produce a Seifert surface $S = S_1 \star S_2$ with Milnor number $\mu(S) = \mu(S_1) + \mu(S_2)$ ($\mu(S)$ is the rank of $H_1(S)$). A Seifert fiber surface is a Seifert surface for ∂S which is a fiber for a bundle map $S^3 - \partial(S) \to S^1$. Stallings [ibid.] showed that S is a fiber surface if S_1 and S_2 are, and Gabai [353,1983] proved the converse.

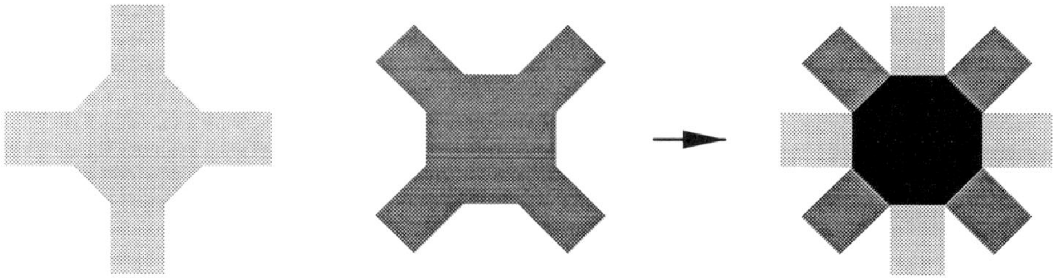

Figure 1.83.1.

(B) *Are all Seifert fibered surfaces equivalent after stabilizing by Hopf plumbing? Probably not!*

Remarks: Hopf plumbing is a special case of Murasugi sum in which one is only allowed to plumb S with a *Hopf band*, a copy of an annulus with one full twist in it. Harer [440,1982a,Topology] showed that all Seifert fibered surfaces are equivalent after stabilizing by Hopf plumbing if twisting (defined by Stallings [ibid.] and generalized by Harer [ibid.]) is performed on the stabilized surfaces.

(C) There is a geometrically defined homomorphism $(\mu, \lambda) : \mathcal{G} \to \mathbb{Z}\oplus\mathbb{Z}$ where μ is the Milnor number, and λ is the *enhancement* from [958,Rudolph,1987,Comment. Math. Helv.] and [840,Neumann & Rudolph,1987,Math. Ann.] (λ is 0 on the fiber surface of a positive Hopf link, and 1 on the negative Hopf link).

Is (μ, λ) *an isomorphism?*

Remarks: Neumann & Rudolph [ibid.], [841,1988], [842,1990, Topology], have studied a generalization of Murasugi sum, called unfolding, defined for fibered knots in all dimensions. In higher (odd) dimensions, in the *simple* case (highly connected fibers), (μ, λ) is an isomorphism from the geometric Grothendieck group $\mathcal{G}$ of fibered knots with respect to unfolding to $\mathbb{Z}\oplus\mathbb{Z}/2$. This is also the algebraic Grothendieck group of Seifert forms with respect to *upper-triangular block sum*, and the Seifert form of a simple fibered knot determines it in $\mathcal{G}$ in all ambient dimensions except 3, where this fails totally.

Problem 1.84 (Birman & Menasco) Markov's Theorem [90, Birman, 1974], [808, Morton, 1986, Math. Proc. Cambridge Philos. Soc.] states that two closed braids $\hat{\beta}_1$ and $\hat{\beta}_2$ are equivalent as links iff $\hat{\beta}_1$ and $\hat{\beta}_2$ are related by a finite sequence of the following moves:

(i) conjugation,

(ii) stabilizing

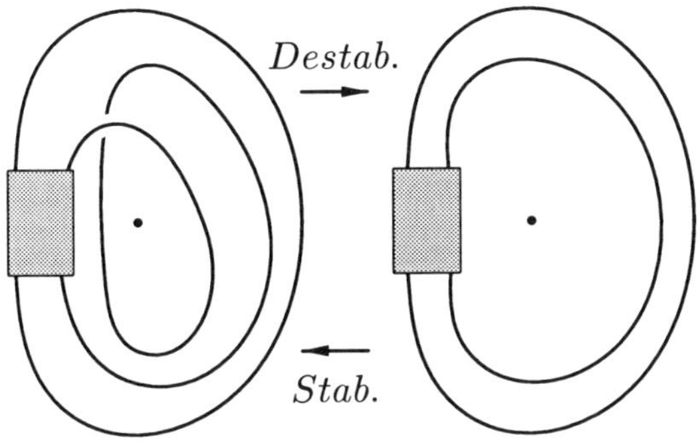

Figure 1.84.1. Stabilization and destabilization

The difficulty in using these simple moves is that arbitrarily many stabilizations (increasing the braid index arbitrarily) may be necessary, and one does not know exactly where the stabilization must be placed (unlike stabilization for equivalence of Heegaard splittings which may take place at any point of a splitting surface).

Birman & Menasco show that the following three moves, which do not increase braid index, suffice to determine whether $\hat{\beta}$ is the unlink [96, Birman & Menasco, 1992, Trans. Amer. Math. Soc.], or is a split link or is a composite link [95, Birman & Menasco, 1990, Invent. Math.].

(1) conjugation

(2) destabilization (see Figure 1.84.1)

(3) exchange move (see Figure 1.84.2)

J. Los [678, 1994, Topology] showed that these three moves sufficed to determine whether $\hat{\beta}$ was a torus link. Birman & Menasco conjecture that moves (1) - (4) determine whether $\hat{\beta}_1$ and $\hat{\beta}_2$ are equivalent links, where (4) is the move:

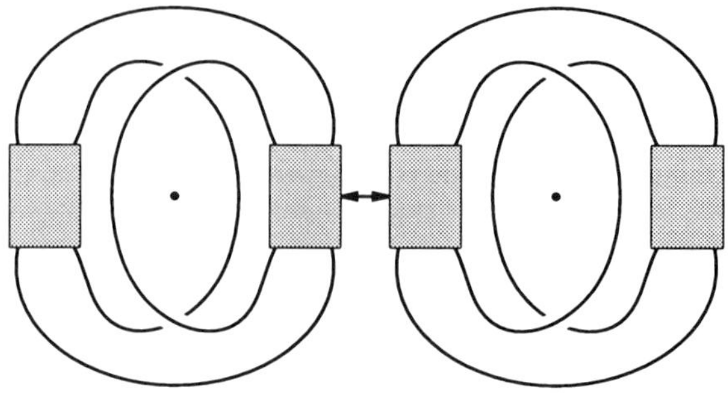

Figure 1.84.2. Exchange Move

(4) braid preserving generalized flype (the simplest type is given in Figure 1.84.3)

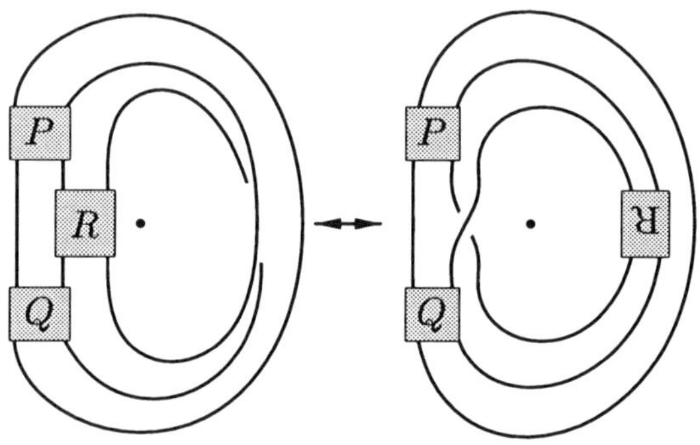

Figure 1.84.3. Simplest generalized flype

Note that (4) includes (3) as a special case. Birman & Menasco have done extensive work [95, Birman & Menasco, 1990, Invent. Math.], [96, Birman & Menasco, 1992, Trans. Amer. Math. Soc.], [97, Birman & Menasco, 1993, Pacific J. Math.] towards verification of the conjecture.

It follows that in using Markov's moves, one does not have to stabilize an n-braid $\hat{\beta}$ to more than a 2n-braid because move (4) can be realized by a sequence of Markov moves with this property. However, one may have to stabilize to, and destabilize from, a $2n$-braid repeatedly.

(A) *The general problem is to find effective algorithms for recognizing whether there is a sequence of the above moves which carries* $\hat{\beta}_1$ *to* $\hat{\beta}_2$.

There is an effective algorithm for recognizing whether two closed braids $\hat{\beta}_1$ and $\hat{\beta}_2$ are conjugates [367,Garside,1969,Quart. J. Math. Oxford Ser. (2)], [1056,Thurston,1988,Bull. Amer. Math. Soc.], [276,El-Rifai & Morton,1994,Quart. J. Math. Oxford Ser. (2)].

(B) *Is there an effective algorithm for determining whether a given braid $\hat{\beta}$, perhaps after conjugation, admits any of the moves (2), (3), or (4)?*

Remarks: One aims to reduce $\hat{\beta}$ to minimal braid index n, so that it may be compared via (1), (3) and (4), to another minimal n-braid.

Note that (B) has been solved for 3-braids [97,Birman & Menasco, 1993,Pacific J. Math.], but it is already unknown for $n > 3$, which is a qualitatively different problem due to the different natures of B_3 and B_n for $n > 3$ (B_3 is a free product with amalgamation of two cyclic groups [ibid.] but nothing like that is true for higher braid groups; also B_3 is linear, with a faithful representation to 2×2 matrices (over a polynomial ring) [90,Birman,1974; Chap. 3], whereas the higher braid groups are not known to be linear).

In attempting to solve (A) and (B), Menasco has been developing the following approach. There is a one-to-one correspondence between conjugacy classes of braids β and isotopy classes of homeomorphisms f of B^2 which keep a set of n points invariant.

Bestvina & Handel in [83,1995,Topology] and J. Los [677,1993,Proc. London Math. Soc.] gave effective algorithms for deciding whether the isotopy class of f contained a

(a) periodic diffeomorphism, or

(b) a reducible diffeomorphism (there exists an invariant closed 1-manifold and the diffeomorphism restricted to each complementary region is either periodic or pseudo–Anosov), or

(c) a pseudo–Anosov diffeomorphism.

The algorithm in (b) produces the 1-manifold, and in (c) produces the invariant train track.

It follows that braids fall into the three categories of

(a) periodic,

(b) reducible, or

(c) pseudo–Anosov.

Note that if β is periodic of order p, then β^q is equal to p full twists in the n strands for some p and q with $q \geq 1$.

(C) Conjecture: *If β is periodic and $|p/q| > 1$, then $\hat{\beta}$, up to conjugacy, does not admit moves (2), (3), or (4).*

If β is reducible, then $\hat{\beta}$ reduces to the study of simpler braids.

If β is pseudo–Anosov, there is an invariant train track, and its thickening has an outside boundary component which is S^1; this S^1 becomes a torus T^2 under suspension. The outside cusps of the train track produce a p/q-torus link in the T^2 (see Figure 1.84.4).

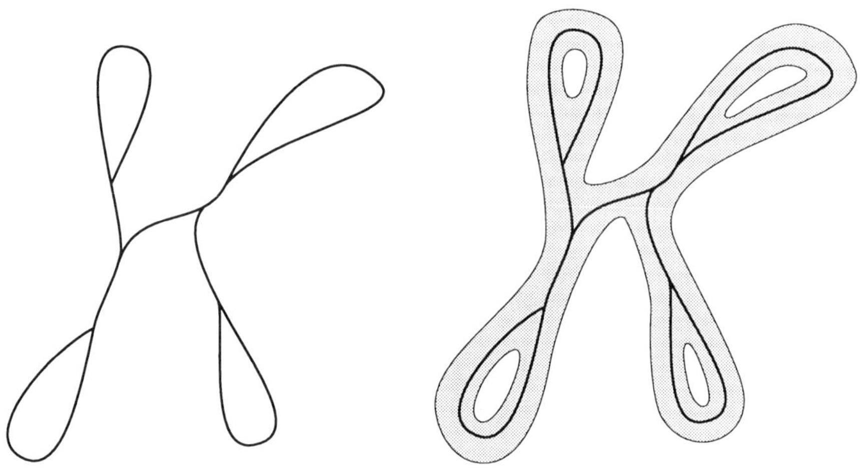

A possible train track for $D^2 - 4\text{pt}'\text{s}$ Fibered regular neighborhood of train track

Figure 1.84.4.

(D) Conjecture: *if $\hat{\beta}$ up to conjugacy admits a move (2), then $|p/q| \leq 1$.*

(E) Conjecture: *if $\hat{\beta}$ up to conjugacy admits a move (3), then* $|p/q| \leq 2$.

(F) Conjecture: *if $\hat{\beta}$ up to conjugacy admits a move (4), then* $|p/q| \leq 4$.

Remarks: For 3-braids, there are examples for which the above inequalities are sharp.

Problem 1.85 (Whitten) *Is the commutator subgroup of a knot group, $\pi_1(S^3 - K)$, Hopfian?*

Remarks: The answer is yes if the commutator subgroup is finitely generated, since then it is free of finite even rank equal to the genus of the knot (see Theorem 4.5.1, page 29 in [844,Neuwirth,1965]). Such a subgroup is always residually finite [472,Hempel,1976]. For the definitions of Hopfian and residually finite, see Problem 3.33.

Problem 1.86 (Cooper) *Let K be a smooth knot in S^3 and let $X = S^3 - K$. Is there always an irreducible representation of $\pi_1(X)$ into $SL(2, \mathbb{C})$?*

Remarks: Note that we may replace $SL(2, \mathbb{C})$ by $PSL(2, \mathbb{C})$ in this problem because the obstruction to lifting a representation lies in $H^2(X; \mathbb{Z}/2\mathbb{Z}) = 0$. If K is a hyperbolic knot, then the hyperbolic structure gives an irreducible representation (irrep) of $\pi_1(X)$ into $PSL(2, \mathbb{C})$. Thurston [1050,1977] shows that in fact there is an affine curve of representations and all but finitely many of them are irreducible.

Suppose that K is not hyperbolic and is not the unknot; then K is either a torus knot (and it is easy in this case to find irreps), or is a satellite knot. In the latter case, X contains an essential non-boundary parallel torus. If the algebraic winding number of the knot in the torus is non-zero, and if the complement of the torus admits an irrep, then this irrep can be extended over X. For more information, see [227, Cooper, Culler, Gillett, Long, & Shalen,1994,Invent. Math.].

Problem 1.87 (Lickorish) It is known, using the Seifert form, that the Alexander polynomial of a boundary link (a link that bounds a disconnected orientable surface in S^3) is zero.

Prove this using only the Conway skein formula.

Remarks: The right sort of proof (recreating the Seifert surface in skein theory is not cricket) ought to give new information on the HOMFLY polynomial of such links. Even though the skein formula completely characterizes the HOMFLY type polynomials, there may be some facts that cannot be proved in a straightforward way; this is a test case.

Problem 1.88 (Kuperberg) A prime knot is either a hyperbolic knot, or is a satellite or torus knot.

(A) *Is the Jones polynomial of a satellite knot always non-trivial?*

Remarks: For the (p, q)-torus knot the Jones polynomial is

$$(t^{(p-1)(q-1)/2})(1 - t^{(p+1)} - t^{(q+1)} - t^{(p+q)})/(1 - t^2)$$

where the polynomial is normalized as 1 for the unknot [551, Jones, 1987, Ann. of Math.] .

(B) *Does the Jones polynomial (using cables)* determine *the volume of a hyperbolic knot?* Determine *should at least mean that knots with different volume have different polynomials.*

(C) (Jones) *Does the Jones polynomial distinguish the unknot? Or, as the questions is often put, is there a* Jones knot, *a knot with the same Jones polynomial as the unknot (see Problem 3.108)?*

Problem 1.89 (Stanford) Given a positive integer n, consider the finite set $S(n)$ of oriented knots with diagrams of less than n crossings. Let $f(n)$ be the smallest integer with the following property: Whenever $K, K' \in S(n)$ are two knots such that $v(K) \neq v(K')$ for some Vassiliev invariant v, then there exists a Vassiliev invariant w of order less than $f(n)$ such that $w(K) \neq w(K')$.

(A) *What is the asymptotic behavior of $f(n)$?*

Remarks: Little is known in general about f, except that it is bounded below by $log_2(n) - 1$.

(B) *Do Vassiliev invariants distinguish knots from their reverses?*

Remarks: The first oriented knot which is different from its reverse (knot with opposite orientation) is 8_{17}. It is known that no Vassiliev invariant of order ≤ 9 can distinguish a knot from its reverse, and so far there is no known higher order Vassiliev invariant which will distinguish a knot from its reverse. So, for example, a result like $f(n) \leq n$ would show that Vassiliev invariants do not distinguish all oriented knots from their reverses (see Problems 1.61 and 1.92).

Vassiliev invariants were introduced in [1072, Vassiliev, 1990], and good further references are [94, Birman & Lin, 1993, Invent. Math.] and [62, Bar-Natan, 1995a, Topology].

Problem 1.90 (Przytycka & Przytycki) The problem of computing the Jones, skein and Kauffman polynomials and most of their substitutions is NP-hard which (up to the famous conjecture that NP is not P) means that they cannot be computed in polynomial time. A few values of these polynomials can be calculated in polynomial time (see below); if the polynomial is expanded as a series about one of these points, can the coefficients in the series be calculated in polynomial time?

In what follows we use the skein relations:

for the Jones polynomial:

$$\frac{1}{t}V_{L_+}(t) - tV_{L_-}(t) = (\sqrt{t} - \frac{1}{\sqrt{t}})V_{L_0}(t);$$

for the skein (Homflypt) polynomial, [341, Freyd, Yetter, Hoste, Lickorish, Millet, & Ocneanu, 1985, Bull. Amer. Math. Soc.], [904, Przytycki & Traczyk, 1987, Kobe J. Math.]:

$$aP_{L_+} + a^{-1}P_{L_-} = zP_{L_0};$$

for the Kauffman polynomial of unoriented framed links:

$$\Lambda_{L_+}(a,z) + \Lambda_{L_-}(a,z) = z(\Lambda_{L_0}(a,z) + \Lambda_{L_\infty}(a,z)),$$

and Λ $(a,z) = a\Lambda$ (a,z).

We assume that the number of components of a link is known or bounded uniformly from above.

(1) Conjecture:

(a) *For any fixed k, the Jones polynomial $V_L(t) \bmod (t+1)^k$ can be computed in polynomial time.*

(b) *Let $V_L(t) = \Sigma_{i=0} a_i(t+1)^i$; then a_i can be computed in n^{3+2i}-time, where n denotes the number of crossings of a diagram.*

Remarks: a_0 is the determinant of a link, so it can be computed in n^3 time (in fact even quicker as it can be computed as a determinant). The above expansion of the Jones polynomial is related (via Listing–Tait translation of knots to graphs) to the rank polynomial of the graph [892,Przytycka & Przytycki,1993].

One can compute the Jones polynomial at $t_0 = \pm 1, \pm i, \pm e^{2\pi i/3}$ and $\pm e^{4\pi i/3}$ in polynomial time and, as shown in [1075,Vertigan,1995], computing all other substitutions is NP-hard; compare [1107,Welsh,1993].

(2) Conjecture: *The kth derivative of $V_L(t)$ at t_0 can be computed in polynomial time (for fixed k).*

Remarks: This is true for $t_0 = 1$ [892,Przytycka & Przytycki, 1993] and the kth derivative can be interpreted as the kth Vassiliev invariant related to the Jones polynomial.

(3) Conjecture:

(a) Consider the partial derivative $\partial^{i+j} P_L(a,z)/\partial^i a \partial^j z$. For fixed i and j one can compute this derivative in polynomial time for: $z = \pm(a_0 + a_0^{-1})$, $(a,z) = (\pm 1, \pm\sqrt{2})$, or $(\pm 1, \pm 1)$, or $(\pm e^{\pm \pi i/6}, \pm 1)$, where all the signs are independent.

(b) $\partial^i P_L(a,z)/\partial^i a$ can be computed in polynomial time at $a = \pm i$.

Remarks: Vertigan [ibid.] has proved that computing substitutions of the skein polynomial is NP-hard except for the substitutions listed above (when they have polynomial complexity). Part (b) is the most interesting as $P_L(i,z)$ is the Alexander polynomial of L and if the conjecture holds one would have sensibly defined the Alexander–Vassiliev expansion of the skein polynomial. $z^{\#|L|-1}P_L(a,z)$ is a polynomial in the variable z (coefficients are Laurent polynomials in the variable a). Vertigan [891,Przytycka & Przytycki,1992] proved the remarkable result that derivatives of this polynomial with respect to z can be computed in polynomial time (in other words any fixed coefficient of expansion of $P_L(a,z)$ at $z=0$ can be computed in polynomial time).

(4) Conjecture:

(a) *Consider the partial derivative of the Kauffman polynomial of a framed unoriented link* $\partial^{i+j}\Lambda_L(a,x)/\partial^i a \partial^j x$. *For fixed* i *and* j *one can compute this derivative in polynomial time for:* $(a,x) = (-q^{\pm 3}, q+q^{-1})$ *where* $q^{16}=1$ *or* $q^{24}=1$ *but* $q \neq \pm i$, *or* $(a,x)=(q^{\pm 3}, q+q^{-1})$ *where* $q^8=1$ *or* $q^{12}=1$ *but* $q\neq \pm i$, *or* $(a,x)=(-q^{\pm 1}, q+q^{-1})$ *where* $q^{16}=1$ *but* $q\neq\pm i$, *or* $(a,x)=(-q^{\pm 1}, q+q^{-1})$ *where* $q^5=1$, *or* $x=\pm(a+a^{-1})$ *where the number of components of* L *is fixed.*

(b) $\partial^i\Lambda_L(a,x)/\partial^i a$ *can be computed in polynomial time at* $a=\pm i$.

Problem 1.91 (Przytycki) Denote the reverse of K by $-K$ (the knot obtained from an oriented knot K by changing its orientation, see Problem 1.61). Let $s(K)$ denote the satellite of K with pattern s (see Problem 1.13). It has been shown in [654,Lickorish,1988] and [895,Przytycki, 1989,Canad. J. Math] that the skein (Homflypt) and Kauffman polynomials of links which are satellites (with given pattern) of $K_1\#K_2$ and $K_1\#-K_2$ are the same. In particular if two knots K' and K'' are satellites (with given pattern) of $K_1\#K_2$ and $K_1\#-K_2$ then the skein and Kauffman polynomials of links which are satellites of them have the same skein and Kauffman polynomials.

(1) Conjecture:

(a) *If K is a simple (the complement has no, non-parallel to the boundary, incompressible tori or annuli), prime knot and for a knot K' the skein and Kauffman polynomials of any satellite (with the same pattern) of K and K' are the same then $K = \pm K'$.*

(b) *Consider any pair of oriented knots K and K'. The skein and Kauffman polynomials of any satellite (with the same pattern) of K and K' are the same if and only if*

$$K = s(\epsilon_1 K_1 \# \epsilon_2 K_2 \# \cdots \# \epsilon_n K_n) \quad \textit{and}$$

$$K' = s(\epsilon'_1 K_1 \# \epsilon'_2 K_2 \# \cdots \# \epsilon'_n K_n),$$

where ϵ_i, ϵ'_j are $+1$ or -1.

Remarks: Conjecture 1(a) was first formulated in [893,Przytycki, 1986] and published in [809,Morton,1988] (Problem 16); however the assumption that K is simple was wrongly omitted. Conjecture **(1)** is closely related to the Bar-Natan conjecture that Vassiliev invariants are as powerful as skein and Kauffman polynomials together with cablings (see Conjecture 2.13 of [903,Przytycki, 1994c]).

It has been proven in [810,Morton & Traczyk,1988] that if two knots differ by mutation then they cannot be distinguished by the Jones polynomial of their satellites.

(2) Question: (Przytycki) *Let K be a prime, simple, unoriented knot. Is there any knot, other than mutations of K, which cannot be distinguished from K by the Jones polynomial of K and its satellites?*

(3) Question: (Kanenobu) *Are there infinitely many different knots with the same Kauffman polynomial?*

Remarks: Kanenobu [559,Kanenobu,1986,Proc. Amer. Math. Soc.] found infinitely many different knots with the same skein (Homflypt) polynomial.

(4) **Question:**

(a) **(Rong 1991)** *Is the homology group $H_1(M_L^{(2)})$ determined by the Kauffmann polynomial of L (where $M_L^{(2)}$ is the 2-fold branched cover of S^3 along L)?*

(b) *Is it determined by the Jones polynomials of cables of L?*

Remarks: The absolute value of $V_L(-1)$ (the determinant of L) is the order of $H_1(M_L^{(2)})$ (order 0 means that the group is infinite).

$$log_3(|V_L^2(e^{\pi/3})|) = rank(H_1(M_L^{(2)}); \mathbb{Z}/3\mathbb{Z}),$$

[655,Lickorish & Millett,1986,Comment. Math. Helv.].

$$log_5(|F^2(1, 2cos(2\pi/5))|) = rank(H_1(M_L^{(2)}); \mathbb{Z}/5\mathbb{Z}),$$

[552,Jones,1989,Comm. Math. Phys.]

Rong [933,1991,Indiana Univ. Math. J.] has observed that $H_1(M_L^{(2)}; \mathbb{Z}/5\mathbb{Z})$ is not determined by the Jones and skein polynomials (more generally skein equivalence class). He uses the Kanenobu examples [559,Kanenobu,1986,Proc. Amer. Math. Soc.], the simplest being $4_1\#4_1$ and 8_9 ($M_L^{(2)}$ is equal to $L(5,2)\#L(5,2)$ and $L(25,7)$ respectively).

Problem 1.92 (Przytycki) Skein modules are quotients of free modules over isotopy classes of links (possibly framed or oriented) in a 3-manifold by properly chosen local (skein) relations. In the choice of relations we are guided by polynomial invariants of links in S^3. Very little is known about skein modules, but they should become the main objects of *algebraic topology based on knots.* The first main goal should be to find an analogy to the Mayer–Vietoris or Seifert–van Kampen theorems (methods of TQFT should be of use); also criteria for knot periodicity give some hope for an analogy to the Smith theory of homology of (branched) coverings. There are now more problems than answers so here are a few reasonable conjectures (also see [508,Hoste & Przytycki,1992], [897,Przytycki,1991,Bull. Polish Acad. Sci. Math.], [1069,Turaev,1990,J. Soviet Math.]).

Part I: *The q-deformation of the fundamental group.*

Let M be oriented 3-manifold and $\mathcal{L}^{fr}$ denote the (ambient) isotopy classes of oriented framed links in M. Let $R = \mathbb{Z}[q^{\pm 1}]$ and S^{fr} denote the submodule of the free module $R\mathcal{L}^{fr}$ (i.e. free module span by $\mathcal{L}^{fr}$), generated by skein expressions $L_+ - q^2 L_-$ and $L^{(1)} - qL$ where $L^{(1)}$ denotes a framed link obtained from L by adding one positive twist to its framing (compare Figure 1.92). We get our skein module as the quotient:

$$\mathcal{S}^{fr}(M) = R\mathcal{L}^{fr}/S^{fr}.$$

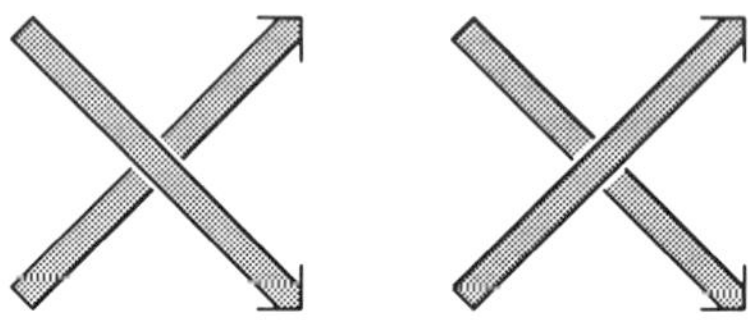

L_+ L_-

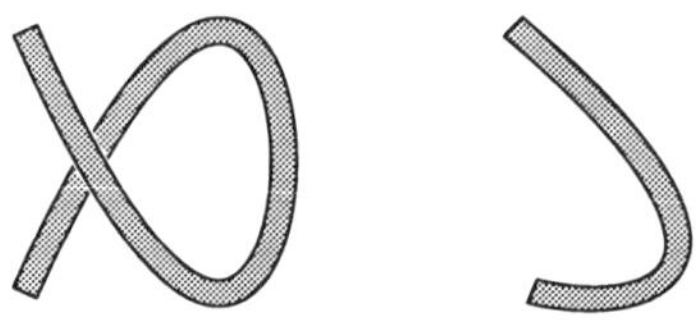

L_+ L_-

(A) *Compute $\mathcal{S}^{fr}(M)$ for any oriented 3-manifold M.*

Remarks: If $q = 1$ (i.e. $R = \mathbb{Z}$) and we allow the empty knot then our skein module become the symmetric tensor algebra over free module of conjugacy classes of the fundamental group of M, $S\mathbb{Z}\hat{\pi}_1(M)$. For general R, and M a rational homology sphere or a compact submanifold, $\mathcal{S}^{fr}(M) = SR\hat{\pi}_1(M)$. The same holds true if M has no nonseparating 2-torus or 2-sphere, since otherwise the skein module has torsion. For example, for a knot K which cuts some 2-sphere k times (algebraically), one has $(q^{2k} - 1)K = 0$.

The case of the skein module being a q-deformation of the first homology group (with the skein relation $L_+ - qL_0$) is fully computed in [902,Przytycki,1994b,Abstracts Amer. Math. Soc.].

Part II: *The skein module based on the Homflypt skein relation.*

Let M be an oriented 3-manifold, $\mathcal{L}$ the set of all oriented links in M up to ambient isotopy of M, $R = \mathbb{Z}[v^{\pm 1}, z^{\pm 1}]$, $R\mathcal{L}$ the free R-module generated by $\mathcal{L}$, and $\mathcal{M}_3$ the submodule of $R\mathcal{L}$ generated by the skein expressions $v^{-1}L_+ - vL_- - zL_0$. For convenience we allow the empty knot, $\emptyset$, and add the relation $v^{-1}\emptyset - v\emptyset - zT_1$, where T_1 denotes the trivial knot.

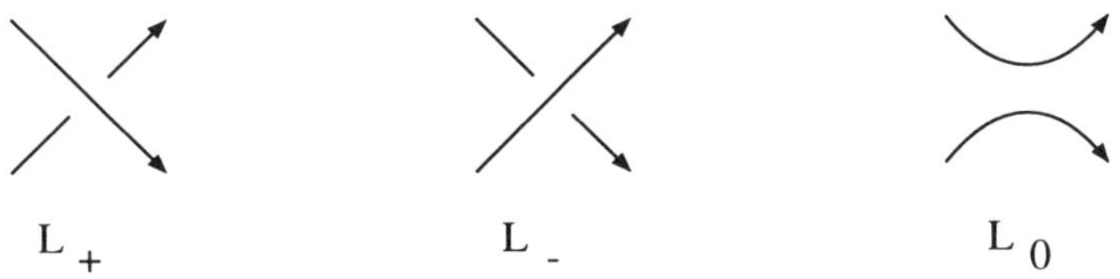

Figure 1.92.1.

Then the third skein module of M is defined to be:

$$\mathcal{S}_3(M) = \mathcal{S}_3(M; \mathbb{Z}[v^{\pm 1}, z^{\pm 1}], v^{-1}L_+ - vL_- - zL_0) = R\mathcal{L}/\mathcal{M}_3.$$

(B) Conjecture: *If M is a rational homology 3-sphere or a compact, connected 3-dimensional submanifold of a rational homology sphere, then $\mathcal{S}_3(M)$ is isomorphic as an R-module to the symmetric tensor algebra $S(R\hat{\pi}^0)$, where $\hat{\pi}^0 = \hat{\pi} - \{1\}$, and $\hat{\pi}$ is the set of conjugacy classes of the fundamental group $\pi = \pi_1(M)$.*

Remarks: The Conjecture holds for S^3, for then it is equivalent to the existence of the skein polynomial. More generally it holds for a product of a surface and an interval [899,Przytycki, 1992b]. In this case the skein module has a structure of a Hopf algebra, [508,Hoste & Przytycki,1992], [898,Przytycki,1992a], [1069,

Turaev, 1990, J. Soviet Math.], [1070, Turaev, 1991, Ann. Sci. École Norm. Sup. (4)].

(C) Conjecture: *Let F be an incompressible surface in an oriented 3-manifold M. Then the map of skein modules, $\mathcal{S}_3(M - F) \to \mathcal{S}_3(M)$, generated by the inclusion, $M - F \hookrightarrow M$, is a monomorphism.*

(D) Questions:

(a) *If M is compact and irreducible and does not allow a non-separating torus, is $\mathcal{S}_3(M)$ free?*

(b) *If M is compact and irreducible, is $\mathcal{S}_3(M)$ always free?*

(c) *If M is irreducible, is $\mathcal{S}_3(M)$ always torsion free?*

Remarks: Non-separating 2-spheres in M produce torsion in $\mathcal{S}_3(M)$. The simplification of this skein module (the one with relation $q^{-1}L_+ = qL_-$ for framed oriented links), has torsion related to a nonseparating torus.

The third skein module of the Whitehead manifold is not free (but possibly torsion free as in the case of the skein module based on the Kauffman bracket, [510, Hoste & Przytycki, 1995]). The third skein module of a possible counterexample to the Poincaré Conjecture is unlikely to be free [158, Bullock, 1994, Topology Appl.].

Part III: *The skein module based on the Kauffman bracket skein relation.*

Let M be an oriented 3-manifold, $R = \mathbb{Z}[A^{\pm 1}]$ and $S_{2,\infty}(M)$ denote the submodule of $R\mathcal{L}_2^{fr}$ generated by skein expressions $L = AL_A + A^{-1}L_B$ and $L^{(1)} = -A^3 L$; see Figure 1.92.2 for the triple L, L_A, L_B.

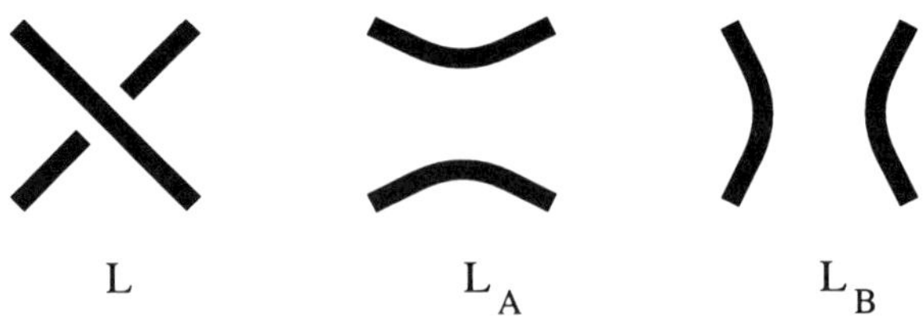

Figure 1.92.2.

The Kauffman bracket skein module is defined as: $\mathcal{S}_{2,\infty}(M) = R\mathcal{L}_2^{fr}/S_{2,\infty}(M)$.

(E) Conjecture: *If M is irreducible and has no incompressible non-parallel to the boundary, closed surfaces then $\mathcal{S}_{2,\infty}(M)$ is torsion free. If M is also compact, then $\mathcal{S}_{2,\infty}(M)$ is a free module.*

Remarks: Conjecture (E) is proven for $F \times [0,1]$ [897,Przytycki,1991,Bull. Polish Acad. Sci. Math.], for lens spaces [509, Hoste & Przytycki,1993,J. Knot Theory Ramifications], the classical Whitehead manifold [510,Hoste & Przytycki,1995] and for the complement of $(2,k)$ torus knots [160,Bullock,1995b]. There are examples of irreducible 3-manifolds containing an incompressible torus such that $\mathcal{S}_{2,\infty}$ has a torsion element (e.g. 3-torus or the double of the complement of the figure eight knot).

(F) Conjecture: *If $M = M_1 \# M_2$, where M_i is not equal to S^3 possibly with holes, then $\mathcal{S}_{2,\infty}(M)$ has a torsion element.*

Remarks: Conjecture (F) holds when $H_1(M_1;\mathbb{Z})$ and $H_1(M_2;\mathbb{Z})$ are not 2-torsion.

(G) Questions:

(i) Is the skein module $\mathcal{S}_{2,\infty}(M)$ free for a compact, irreducible 3-manifold without a incompressible non-parallel to the boundary torus (but with, possibly, higher genus, closed, nonseparating surfaces)?

(ii) Is the skein module $\mathcal{S}_{2,\infty}(M)$ free for a closed irreducible non-Haken 3-manifold?

(H) Conjecture: (Traczyk) *If M is a simple connected 3-manifold other than S^3 (possibly with holes), then $\mathcal{S}_{2,\infty}(M)$ is infinitely generated.*

Remarks: Conjecture (H) holds for some Whitehead type manifolds (including the classical Whitehead manifold) [510,Hoste & Przytycki,1995].

For $M = F \times I$, the Kauffman bracket skein module is an algebra, where $L_1 * L_2$ is obtained by placing L_1 above L_2 in $M = F \times I$, and the empty knot, $\emptyset$, is the unit of the algebra.

(J) Question: (Bullock, Przytycki) *Find the structure of the algebra.*

Remarks: Bullock [159,1995a] proved that the algebra is always finitely generated. It is abelian for $F = B^2$, annulus or a disk with 2 holes; in the last case the skein module is algebra isomorphic to $\mathbb{Z}[A^{\pm 1}][x,y,z]$. The first interesting case occurs when F is a torus. Then $\mathcal{S}_{2,\infty}(T^2 \times I)$ is a *quantization* of the free commutative algebra on three generators, x, y, z modulo the relation $xyz + x^2 + y^2 + z^2 = 4$.

Part IV: *The fourth, unoriented, skein module.*

Let M be an oriented 3-manifold and $\mathcal{L}^{fr}$ denote the set of ambient isotopy classes of unoriented framed links in M. Let $R = \mathbb{Z}[q^{\pm 1}]$ and S_4^{fr} denote the submodule of the free module $R\mathcal{L}^{fr}$, generated by $L^{(1)} - qL$ and skein expressions

$$v_3(q)L_{+++} + v_2(q)L_{++} + v_1(q)L_+ + v_0(q)L_0,$$

where $v_3(1) = -v_0(1) = 1$ and $v_2(1) = v_1(1) = 0$. We get our skein module as the quotient:

$$\mathcal{S}_4^{fr}(M) = R\mathcal{L}^{fr}/S_4^{fr}.$$

It is very difficult to analyze this skein module in general, so we will limit our problems to the case of $M = S^3$.

(K) Questions:

(i) Is $\mathcal{S}_4^{fr}(S^3)$ generated by trivial links?
(ii) Is it a free module?
(iii) Is the submodule generated by links of fixed braid index, finitely generated?

Remarks: For $q = 1$, a positive answer for (i) is equivalent to the Montesinos–Nakanishi conjecture (see Problem 1.59). More generally, if the Montesinos–Nakanishi conjecture holds then $\mathcal{S}_4^{fr}(S^3)$ modulo the ideal $(q-1)^{k+1}$ is generated by trivial links (this reduction of the skein module can be treated as its kth degree Vassiliev part). (iii) holds for braid index ≤ 5, by [228,Coxeter,1957].

Part V: *The Vassiliev–Gusarov skein modules.*

Let $\mathcal{K}^{sg}$ denote the set of singular oriented knots in M where we only allow imbeddings and immersions of S^1 with double points, up to ambient isotopy; additionally, for any double point we choose an orientation for a small ball around it (if M is oriented the chosen orientation of the ball agrees with that of M). Let R be a commutative ring with 1 (e.g. $\mathbb{Z}$ or $\mathbb{Q}$). In $R\mathcal{K}^{sg}$ we consider resolving singularity relations $\sim$: $K_{cr} = K_+ - K_-$; see Figure 1.92.3.

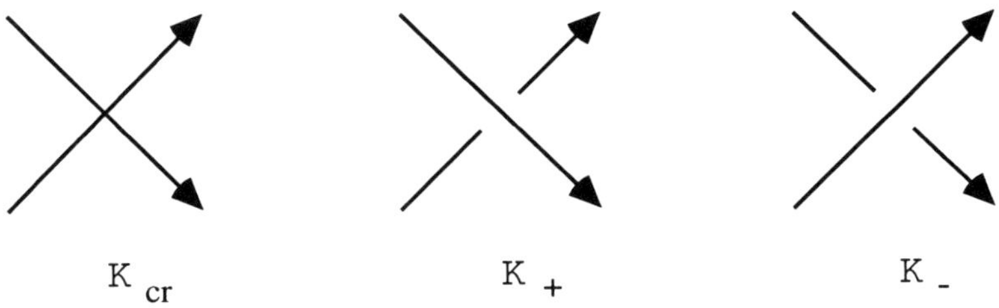

Figure 1.92.3.

$R\mathcal{K}^{sg}/\sim$ is obviously R-isomorphic to $R\mathcal{K}$. Let C_m be the submodule of $R\mathcal{K}^{sg}/\sim = R\mathcal{K}$ generated by immersed knots with m double points. The mth Vassiliev–Gusarov skein module $W_m(M,R)$ is defined by $W_m(M,R) = R\mathcal{K}/C_{m+1}$. We have the filtration:

$$\cdots \subset C_m \subset \cdots \subset C_1 \subset C_0 = R\mathcal{K}$$

and therefore we have also the sequence of epimorphisms $\{1\} \leftarrow W_0 \leftarrow W_1 \leftarrow W_2 \leftarrow \cdots \leftarrow W_m \leftarrow \cdots$ We define the V–G skein module $W_\infty(M,R)$ as the inverse limit $W_\infty(M,R) = lim_{\leftarrow} W_m(M,R)$. Equivalently the V–G skein module is the completion $\widehat{R\mathcal{K}}$ of $R\mathcal{K}$ with respect to the topology yielded by the sequence of descending submodules C_i. The kernel of the natural R-homomorphism $\theta : R\mathcal{K} \to \widehat{R\mathcal{K}}$ is equal to $\bigcap_{i=0}^{\infty} C_i$.

A V–G invariant of degree m of knots is defined as an element of the dual space $V^m(M,R) = W_m^*(M,R) = Hom(W_m(M,R),R)$ (sometimes it is defined as an element of $Hom_{\mathbb{Z}}(W_m(M,\mathbb{Z}),A)$, where A is an abelian group).

Computing the V–G skein module of any 3-manifold is a very difficult problem, so we restrict to the case of $M = S^3$.

There are two competing conjectures:

(L) Conjecture: (Vassiliev) *The V–G invariants classify oriented knots.*

(M) Conjecture:

(a) **(Lin).** *No V–G invariant distinguishes K from $-K$, where $-K$ denotes the knot obtained from K by reversing its orientation.*

(b) **(Bar-Natan)** [62,1995a,Topology]. *The V–G invariants are precisely as powerful as the skein (Homflypt) and Kauffman polynomials of knots and all of their cablings (in particular (a) holds).*

(c) **(Przytycki)** *The V–G invariants of degree 10 or less can be deduced from the skein (Homflypt) and Kauffman polynomials of knots and their 2-cables.*

(d) **(Przytycki)** *The R-algebra $\bigcap C_i$ is the smallest subalgebra of $R\mathcal{K}$ containing the expressions $K-(-K)$ and closed under the operation of taking satellites (including connected sum).*

Remarks: If Conjecture **(1)** in Problem 1.91 holds, then (b) follows from (c).

(N) Conjecture: (Gusarov) *Vassiliev–Gusarov skein modules of S^3 are torsion free.*

Remarks: (N) has been checked by Gusarov [424,1994,Adv. Soviet Math.] up to degree 6, and Bar-Natan has checked that they have no 2-torsion up to degree 9.

Consider the skein module $A_k(M)$ of an oriented 3-manifold M built in the same way as the Vassiliev–Gusarov skein modules, but using oriented links, $R = \mathbb{Z}[z^{\pm 1}]$, and with resolving singularity equation:

$$L_x = L_+ - L_- - zL_0.$$

(P) Conjecture: (Przytycki) *For fixed k, and any generating set of $A_k(S^3)$, any link can be expressed in the generators in polynomial time with respect to the number of crossings.*

Remarks: $A_0(S^3)$ is freely generated by the unknot, and a coefficient of any link is its Alexander polynomial; thus Conjecture (P) holds for k=0. The structure of $A_1(S^3)$ follows from results in [935,Rong,1995a]. Generally $A_k(S^3)$ is finitely generated. Conjecture (P) holds for the standard Vassiliev–Gusarov skein modules (of S^3), and the polynomial complexity of Vassiliev invariants is, possibly, their most important feature.

Problem 1.93 Conjecture: *The untwisted double of a knot is ribbon iff the knot is ribbon.*

Remarks: This problem may be easier than the corresponding problem for *slice* (Problem 1.38) because it could be a 3-dimensional issue rather than 4-dimensional. Recall that a knot is ribbon if it bounds an immersed disk in S^3 such that the singular set consists entirely of *ribbons*, as in Figure 1.93.1. A slice knot is one which bounds a smoothly imbedded disk in B^4.

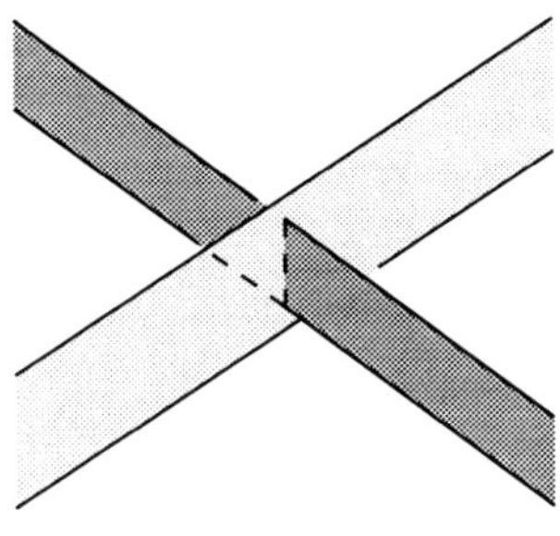

Figure 1.93.1.

Problem 1.94 *Does every element of order 2 in the group, C_1^3, of concordance classes of knots in S^3 contain a representative which is isotopic to its inverse?*

Remarks: The point to the question is that C_1^3 may be very simple in that every knot is of infinite order or is concordant to a knot K which is isotopic to $-K$, its inverse (for the definition, see Problem 1.61; also Problem 1.32).

Problem 1.95 (Akbulut) *Does there exist a homology 3-sphere Σ, other than S^3, with the following property: any knot K, representing $0 \in \pi_1(\Sigma)$, which is slice in some contractible 4-manifold W which Σ bounds, is already slice in $\Sigma \times [0,1]$?*

Remarks: It is likely that S^3 satisfies the property (W is homeomorphic to B^4, but it could have a smoothing in which a knot K is slice even though K is not slice in $S^3 \times [0,1]$).

Problem 1.96 (Cochran) (A) *Give a geometric description of those links in S^3 for which all of Milnor's $\bar{\mu}$-invariants are zero.*

Remarks: The original paper is [765,Milnor,1957b]. The types of links known to have vanishing μ-invariants are boundary links, slice links, homology boundary links [483,Hillman,1981], fusions of boundary links [214,Cochran,1990], and E-links [213,Cochran, 1987,Invent. Math.]. All of the examples known are fusions of boundary links, so:

(B) *Is every classical link with $\bar{\mu} = 0$ isotopic to (or concordant to) a fusion of a boundary link?*

Remarks: Since Milnor's invariants, properly interpreted, have been shown to be Vassiliev invariants of finite type [63,Bar-Natan,1995b] [660,Lin,1992], [661,Lin,1995], these questions are related to the problem of finding links all of whose Vassilliev invariants are the same as a trivial link.

Problem 1.97 (Koschorke) *Given a link $f = f_1 \amalg \ldots \amalg f_r : S^1 \amalg \ldots \amalg S^1 \to \mathbb{R}^3$, is its link homotopy class $[f]$ completely determined by the (standard) homotopy class*

$$\kappa[f] := [f_1 \times \ldots \times f_r] \in [(S^1)^r, \widetilde{C}_r(\mathbb{R}^3)]$$

(where $\widetilde{C}_r(\mathbb{R}^3)$ denotes the space of ordered configurations of r pairwise distinct points in $\mathbb{R}^3$).

Remarks: $\kappa[f]$ is trivial iff all μ-invariants of f are defined and vanish [604,Koschorke,1985; page 128]. Hence the kernel of κ (in the set theoretic sense) is trivial (see [763,Milnor,1954,Ann. of Math.; page 190]), but one does not know whether κ is injective in general.

Problem 1.98 (Eliashberg) A knot K in S^3 is said to be transversal to the standard contact structure on S^3 if it is transverse to the plane field (which can be taken to be the planes orthogonal to the Hopf fibration on S^3). Let $\mathcal{T}$ be the transversal isotopy classes of transversal knots, and let $\mathcal{K}$ be (smooth) isotopy classes of knots.

If $K \in \mathcal{T}$, define its self linking, $\lambda(K)$, to be the linking number $\lambda(K, K')$ where K' is a push off of K along the vector field lying in the plane field above (e.g. $\sqrt{-1}$ times the Hopf vector field on $S^3 \subset \mathbb{C}^2$). Thus if K is a Hopf fiber, $\lambda(K) = -1$.

Let $F : \mathcal{T} \to \mathcal{K} \times \mathbb{Z}$ be given by $F(K) = (K_{TOP}, \lambda(K))$, where K_{TOP} is merely the isotopy class of K.

(A) *Is F injective?*

Remarks: Yes if K_{TOP} is the unknot [279,Eliashberg,1993] .

(B) *What is the image of F?*

Remarks: Any knot K is isotopic to a transversal knot, and putting a small *kink* in K decreases $\lambda(K)$ by 1, so the real question is, given K, what is

$$\bar{\lambda}(K) = max\{\lambda(K') \mid K' \in \mathcal{T}, \mathcal{K}'_{TOP} = \mathcal{K}\}?$$

The Bennequin [76,1983,Astérisque], and Kronheimer & Mrowka [623,1993,Topology], [627,1995b,Topology] inequality is

$$\bar{\lambda}(K) \leq 2g(K) - 1$$

where $g(K)$ is the minimal genus of a smooth surface in B^4 whose boundary is K. The inequality is exact for algebraic knots (this follows from the adjunction formula). However, for the left-handed trefoil knot, $g = 1$ whereas for all known transversal representations, $\lambda(K) \leq -4$.

(C) *Can any link of unknots be represented by a transversal link with all self-linking equal to* -1*?*

Problem 1.99 (Eliashberg) *Are knot type, Bennequin number, and rotation number a set of complete invariants for Legendrian knots in* S^3*?*

Remarks: A knot K is said to be *Legendrian* if its tangent vectors all lie in the contact plane field. Push K off itself by pushing it slightly along the vector field normal to the contact structure; the *Bennequin number* of K is the linking number between K and its pushoff.

Choose two linear independent vector fields on S^3 which are tangent to the contact structure. Then the tangent vector field to the knot can be though of in terms of this trivialization as a map $S^1 \to S^1$. The degree of this map is independent of the trivialization and is called the *rotation number*, (sometimes the *Maslov number*), of the knot K. Notice that the rotation number is an invariant of the oriented knot while the Bennequin number of a knot K is independent of its orientation.

The answer is yes if K is the unknot [279,Eliashberg,1993]. A similar question for some other contact 3-manifolds, for instance for the connected sum of 2 copies of $S^2 \times S^1$, has a negative answer [327,Fraser, 1994].

Problem 1.100 (Boileau & Rudolph) Let $V \subset C^2$ be a smooth algebraic curve passing through the origin. Consider $L_t = V \cap S_t$ where $S_t \subset C^2$ is the sphere of radius t. Except for a finite set of values of t where the intersection is not transverse, L_t is a smooth link imbedded in S_t, and such links are called *C-transverse links.* Define the big genus of L_t by $bg(L_t) = min\{genus(F)\}$ where F is formed by gluing (along L_t) a planar surface to a Seifert surface in S_t for L_t (see [126,Boileau & Weber,1983,L'Enseign. Math.]).

(A) *Is the function* $t \to bg(L_t)$ *non-decreasing?*

Remarks: This is true for the smooth, 4-ball big genus by the proof of the local Thom conjecture ([623,Kronheimer & Mrowka, 1993,Topology] and [627,Kronheimer & Mrowka,1995b,Topology], or [625,Kronheimer & Mrowka,1994b,Math. Res. Lett.]).

(B) *If L_t is a trivial link for a given t, does it follow that $L_{t'}$ is a trivial link for $t' \leq t$.*

Remarks: Of course (B) is a special case of (A). If L_t is a trivial knot, and $L_{t'}$ is a knot with $t' \leq t$, then $L_{t'}$ is a trivial knot [403, Gordon,1981b,Math. Ann.]. Moreover, by work of Eliashberg [280,1995], the complex disk $V \cap B_t$ is trivially imbedded in B_t.

(C) **Conjecture (Rudolph)**: *Every C-transverse link is a quasipositive link (i.e. can be realized as a quasipositive closed braid).*

Remarks: A quasipositive closed braid is the closure of a braid which is the product of conjugates of positive half twists (i.e. $\omega\sigma_i\omega^{-1}$ for some word ω and for a positive half twist σ_i between strands i and $i+1$).

The proof of the local Thom conjecture [ibid.] shows that many knots are not C-transverse, for example, the figure-8 knot, [960, Rudolph,1993,Bull. Amer. Math. Soc.], some doubles, some pretzel knots, and some iterated torus knots [124,Boileau & Rudolph, 1995].

(D) *Given V, there are only a finite number of topological types for the pairs (S_t, L_t). Do they determine the topological type of the pair (C^2, V)?*

Remarks: Neumann [835,1989,Invent. Math.] showed that the link at infinity (the pair (S_t, L_t) for $t \gg 1$) determines the pair $(\mathbb{C}^2, V)$ when V is regular at infinity (i.e. if $V = \{(z_1, z_2) \in \mathbb{C}^2 | f(z_1, z_2) = 0\}, f =$ polynomial, then V is regular at infinity if there exists $\epsilon \geq 0$, $N \geq 0$ such that $f|(f^{-1}(D^1(\epsilon)) - D^2(N))$ is a C^∞ fibration over $D^1(\epsilon)$, where $D^k(r)$ is the ball of radius r in $\mathbb{R}^k$). However if V is not regular at infinity, this is no longer true [45,Artal-Bartolo,1993,L'Enseign. Math.] .

(E) *If V is not regular at infinity, is the diffeomorphism type of V determined by the pairs (S_t, L_t), or even just by the link at infinity?*

Remarks: The link at infinity determines the Euler characteristic of V, but this does not suffice because V may not be connected (the number of components is not known) [Neumann, ibid.].

Problem 1.101 (Rudolph) Consider germs of C^∞ maps $(\mathbb{R}^2, 0) \xrightarrow{f} (\mathbb{R}^4, 0)$ with an isolated singularity at the origin (i.e. the differential of

f is of maximal rank except at the origin).

Question: *What knot types can occur as the link of the singularity (i.e $\epsilon S^3 \cap f(\mathbb{R}^2)$ for a small sphere of radius ϵ) when f is the germ of a symplectic map? Of a Lagrangian map?*

Remarks: A map is Lagrangian if the standard symplectic 2-form on $\mathbb{R}^4$ vanishes on the image of the differential of f.

Problem 1.102 (Rudolph) Consider proper imbeddings $\mathbb{C} \xrightarrow{f} \mathbb{C}^2$. If f is polynomial, then $f(\mathbb{C})$ is C^∞ unknotted. In fact, f is conjugate by a polynomial automorphism to a linear map of $\mathbb{C}$ into $\mathbb{C}^2$, and the group of polynomial automorphisms is connected, so f is isotopic through polynomial maps to the unknot ([1,Abhyankar & Moh,1975,J. Reine Angew. Math.] for an algebraic proof, [955,Rudolph,1982,J. Reine Angew. Math.] for a knot-theoretical proof, and [840,Neumann & Rudolph,1987,Math. Ann.] for the easiest proof).

(A) *If f is an entire function, is the imbedding C^∞ unknotted? Through entire imbeddings?*

(B) *If f is an entire imbedding of the open unit disk in $\mathbb{C}$ to $\mathbb{C}^2$, is the imbedding C^∞ unknotted?*

(C) *If $\mathbb{R}^2 \xrightarrow{f} \mathbb{R}^4$ is a proper imbedding whose image S is a complete minimal surface (in the differential geometric sense) in flat $\mathbb{R}^4$, is S topologically unknotted? C^∞ unknotted?*

Remarks: Complex analytic curves are minimal surfaces , so (C) is a proper generalization of (A) and (B). The total curvature of an algebraic curve is finite, so (C) could be specialized to include the hypothesis of finite total curvature.

There are C^∞ ribbon imbeddings of $\mathbb{R}^2$ in $\mathbb{R}^4$ which are topologically unknotted but smoothly knotted; this follows because the double cover of $\mathbb{R}^4$ branched along the ribbon $\mathbb{R}^2$ is an exotic $\mathbb{R}^4$ (the first such example is due to Freedman, and it and others appear in [395,Gompf, 1993,J. Differential Geom.]).

Can Hass's theorem [454,1983,Math. Proc. Cambridge Philos. Soc.], that every ribbon surface in B^4 is isotopic to a minimal surface, be extended to proper ribbon imbeddings in $\mathbb{R}^4$?

Problem 1.103 *Let $D \subset B^4$ be the ribbon disk of a ribbon knot. Is $B^4 - D$ aspherical?*

Remarks: Recall that an arcwise connected X is aspherical iff X is a $K(\pi_1(X), 1)$ iff $\pi_n(X) = 0$ for all $n \neq 1$. For a thorough discussion of these issues, see Bogley's Chapter X in [501,Hog-Angeloni, Metzler, & Sieradski,1993].

Problem 1.104 (Cochran) *Is every link of two 2-spheres in S^4 link homotopic to the unlink?*

Remarks: A link homotopy allows a component to intersect itself but no other components.

Problem 1.105 (Cochran & Melvin) Let $S^2 \to S^4$ be a 2-knot K^2. Every 2-knot is slice (or null-concordant), i.e. the boundary of a smooth B^3 imbedded in B^5 [571,Kervaire,1965,Bull. Soc. Math. France] (see Problem 1.56 for the question of whether 2-links are slice).

A 2-knot K^2 is said to be a *0-slice* knot if there exists a B^3-slice whose intersections with all spheres of radius r, rS^4, $r < 1$, only consist of unions of 2-spheres. K^2 is said to be *k-slice* if k is the minimum over all B^3-slices of the maximum genus of any component of $(B^3 \cap rS^4)$ for $r < 1$. K^2 is said to be *k-null-bordant* if k is minimized, as above, over all orientable 3-manifolds M^3 in B^5 with $\partial M^3 = K^2$.

(A) Question: *Is every 2-knot 0-slice, or are there 1-slice 2-knots? Or even k-slice 2-knots for each $k \geq 1$?*

(B) *Same question with slice replaced by null-bordant.*

Remarks: If K^2 is k-slice and j-null-bordant, then obviously $j \leq k$.

0-null-bordant 2-knots are particularly interesting because the Gluck construction on such knots gives S^4 [745,Melvin,1977] (see Problem 4.24).

Ribbon 2-knots are obviously 0-slice, where a ribbon 2-knot is one which bounds an immersed B^3 in S^4 with singular set restricted analogously to the case of classical ribbon knots, or equivalently, the ribbon 2-knot can be built in B^5 from the unlink only by fusing [1125, Yajima, 1964, Osaka J. Math.]. Thus a 0-slice 2-knot is built from some 0-handles by adding only 1-handles which reduce the number of components, and possibly 2- and 3-handles; whereas a ribbon 2-knot is the same but without any 2- or 3-handles.

It is known that there are 0-null-bordant 2-knots which are not ribbon [207, Cochran, 1983, J. London Math. Soc.].

Chapter 2

Surfaces

- Problems 2.1–2.8 (1977), 2.9–2.20 (new).
- Mapping class group, 2.1, 2.2, 2.4, 2.9–2.15.
- Hyperbolic geometry, 2.6–2.7.

Introduction

The following problems use these definitions: F_g denotes a closed, orientable surface of genus g, which bounds a handlebody N_g ($= B^3 \cup g$ 1-handles); given a surface S, its *mapping class group*, Γ_S, is the group of isotopy classes of orientation preserving homeomorphisms; $\Gamma^p_{g,q}$ denotes the mapping class group of an oriented surface of genus g with p marked points and q boundary components (diffeomorphisms modulo isotopies which fix each marked point and fix each boundary point); $Sp(2g, \mathbb{Z})$ is the group of $2g \times 2g$ symplectic integer matrices; $\lambda : \Gamma_g \to Sp(2g, \mathbb{Z})$ is the natural homomorphism mapping each element of Γ_g to its induced automorphism on $H_1(\Gamma_g; \mathbb{Z})$; κ_g is the kernel of λ, and is called the Torelli group.

Problem 2.1 (Birman) (A) *Is κ_g finitely generated? (Conjecture: No.)*

Remarks: It is known [888, Powell, 1978, Proc. Amer. Math. Soc.] that κ_g is generated by isotopy classes of

(1) Dehn twists about separating curves, and
(2) if $g \geq 3$, Dehn twists $t_c t_{c'}^{-1}$ about pairs c, c', of homologous, disjoint, non-separating curves.

(B) *Is the subgroup of κ_g which is generated by maps of type (1) of finite index in κ_g for any $g \geq 3$? (For $g = 2$, it is the full group.)*

Remarks: Chillingworth [201, 1972, Math. Ann.; page 152] gives examples of homeomorphisms in κ_g not generated by type (1) maps.

Update: The Torelli group, κ_g, is finitely generated for $g > 2$ [546, Johnson, 1983, Ann. of Math.]. For $g = 2$, the Torelli group is infinitely generated [730, McCullough & Miller, 1986b, Topology Appl.] and is in fact freely generated by an infinite set of Dehn twists on separating curves [752, Mess, 1992, Topology].

The quotient of κ_g by the maps of type (1) is infinite for $g > 2$ and can be identified with a quotient of the third exterior power of $H_1(F_g; \mathbb{Z})$ [547, Johnson, 1985a, Topology]. The group generated by maps of type (1) contains the commutator subgroup of the Torelli group as a finite index subgroup [548, Johnson, 1985b, Topology].

Problem 2.2 (Birman) *Find explicit representations of κ_g or $\mathcal{M}_g$ which do not factor through the homomorphism λ.*

Remarks: Since $\mathcal{M}_g$ is residually finite [418, Grossman, 1974, J. London Math. Soc.], such representations surely exist. In particular, κ_g has nontrivial homomorphisms onto $\mathbb{Z}/2\mathbb{Z}$, but these representations are not straightforward [91, Birman & Craggs, 1978, Trans. Amer. Math. Soc.].

Update: Jones–Witten theory in principle gives many such representations. Kohno [599, 1992, Topology], using [784, Moore & Seiberg, 1989, Comm. Math. Phys.], gives many examples. Wright [1120, 1994, J.

Knot Theory Ramifications] calculated the Reshetikhin–Turaev representation from $\mathcal{M}_g$ to $PGL(n,\mathbb{C})$ for the root of unity $e^{\pi i/2}$; this representation does not factor through λ and, restricted to κ_g, is the direct sum of the Birman–Craggs homomorphisms. Johnson [545, Johnson, 1980, Trans. Amer. Math. Soc.] had shown that the space of Birman–Craggs homomorphisms spans the space of cubic polynomials on the $\mathbb{Z}/2\mathbb{Z}$-affine space of spin structures on F_g.

Problem 2.3 (D. Johnson) *If $h : F_g \to F_g$ is a homeomorphism with $h_* = id$ on $H_1(F_g)$, does h extend over some 3-manifold?*

Remarks: For $g = 1$, h extends iff trace $h_* = 2$ iff h_* is conjugate to $\left(\begin{smallmatrix}1&0\\k&1\end{smallmatrix}\right)$ iff h is conjugate to a k-fold Dehn twist around some curve.

Update: In general, h does not extend over some 3-manifold. After earlier work of Casson, D. Johnson, Johannson, Scharlemann (and perhaps others), the definitive treatment is [128, Bonahon, 1983a, Ann. Sci. École Norm. Sup. (4)].

Problem 2.4 (Birman) Let α be the obvious homomorphism

$$\mathcal{N}_g \xrightarrow{\alpha} \mathrm{Aut}(\pi_1(N_g))$$

where $\mathcal{N}_g$ is the group of isotopy classes of orientation preserving homeomorphisms of N_g.

Is kernel(α) finitely generated? (Conjecture: *No.)*

Remarks: $\mathcal{N}_g$ is known to be finitely generated [1023, Suzuki, 1977, Canad. J. Math]. E. Luft has proved that kernel(α) is generated by Dehn twists on properly imbedded 2-balls in N_g.

Update: The answer is no for a genus two handlebody [616, Kramer, 1983], and for all handlebodies of genus > 1 [725, McCullough, 1985, Topology].

Problem 2.5 (Birman) (A) Suppose that F_g is imbedded in S^3 as a Heegaard surface, with $S^3 = U \cup V$ and $U \cap V = F_g$. Let A (resp. B) be the subgroup of $\mathcal{M}_g$ of maps which extend over U (resp. V). Let C be the centralizer of $A \cap B$ in A.

Can we express every element $\alpha \in A$ as $\alpha = \gamma\delta$ where $\gamma \in C$ and $\delta \in A \cap B$?

(B) Suppose $\mathcal{P}$, $\mathcal{P}' \subset A \cap B$ have finite order p. Suppose also that $\mathcal{P}' = \alpha\mathcal{P}\alpha^{-1} = \beta\mathcal{P}\beta^{-1}$ for some $\alpha \in A$, $\beta \in B$.

Conjecture: *There exists $\delta \in A \cap B$ such that $\mathcal{P}' = \delta\mathcal{P}\delta^{-1}$.*

Remarks: The conjecture is true for $p = 2$. A positive answer for any $p \geq 3$ would imply the Smith conjecture (see Problem 3.28) for period p (Birman). Note that an affirmative answer to part (A) would imply an affirmative answer to the conjecture for every p.

Update:

(A) The answer is no for genus > 2 because then the centralizer is trivial.

(B) Interest has died now that the Smith conjecture has been proved.

Problem 2.6 (Thurston) *Can the group of symmetries of $\pi_1(F) = \pi_1(F_g)$ be represented geometrically as a group of homeomorphisms of F? More precisely, since*

$$\pi_0(Homeo F) \cong Aut(\pi_1 F)/InnAut(\pi_1 F) = OutAut(\pi_1 F),$$

is there a right inverse σ for π_0:

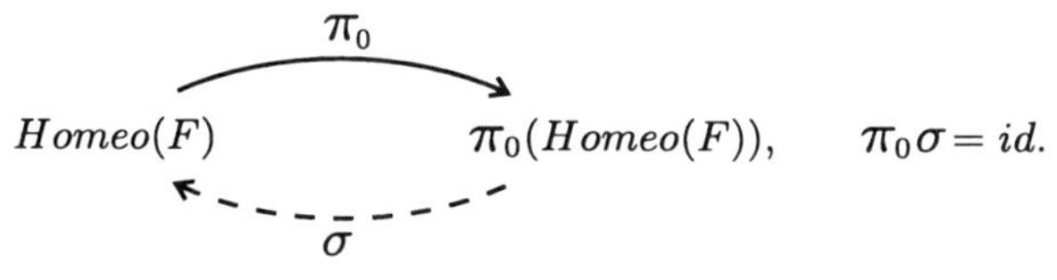

Remarks: If we replace F by its unit tangent bundle TF, this geometric representation is possible, i.e., there is a natural representation $\pi_0(\text{Homeo}(F)) \to \text{Homeo}(TF)$ (consider the Cheeger homeomorphism in [412, Gromov, 1975]).

The problem is also interesting (and well known) for subgroups (particularly finite) of $\pi_0(\text{Homeo}(F))$. It is known for solvable, finite groups and for infinite groups with two ends.

Update: For an oriented, closed surface F_g, the map $\pi_0 : \text{Diff}(F_g) \to \pi_0(\text{Diff}(F_g)) = \mathcal{M}_g$ does not have an inverse σ (at least if $g \geq 86$) [802,Morita,1987,Invent. Math.], but for any finite group in $\mathcal{M}_g$, there is an inverse σ [570,Kerckhoff,1983,Ann. of Math.]. For the case of Homeo(F_g), Morita's proof does not apply, but the existence of an inverse σ seems unlikely.

Problem 2.7 (Thurston) *Characterize the subgroups of π_0(Diff(F)) which act as translations of some complex geodesic in Teichmüller space.*

Update: It is known by work of Masur [705,1986,Duke Math. J.] that these subgroups can't act cocompactly on the complex geodesic. This together with the fact that there is a bound on the orders of finite elements in the mapping class group implies that these groups are virtually free. It is still not known if they are necessarily finitely generated.

Thurston gave examples in which each of the pseudo–Anosov elements in the subgroup has a stretching factor which is a quadratic integer, and such that the group is commensurable with $PSL(2,\mathbb{Z})$ as a group of isometries of the hyperbolic plane (see Exposé 13 in [296, Fathi, Laudenbach, & Poénaru,1979]. Veech found examples which are not commensurable with $PSL(2,\mathbb{Z})$, but are lattices in $PSL(2,\mathbb{R})$ [1073,Veech,1989,Invent. Math.] (for further work concerning these examples, see [1074,Veech,1992,Geom. Funct. Anal.] and [453,Harvey, 1993]). Smillie [1006,1995] showed that the group of translations of a complex geodesic is a lattice in $PSL(2,\mathbb{R})$ if and only if the image of the geodesic in moduli space is a closed set.

Also, for an analysis of parabolic elements which stabilize Teichmüller discs, see [992,Shiga & Tanigawa,1989,Kodai Math. J.], and for the definitive answer to this aspect of the question, see [264,Earle & Sipe, 1995] and [715,Matsumoto & Montesinos-Amilibia,1994,Bull. Amer. Math. Soc.].

Problem 2.8 (Meeks) *For what genus g does every periodic diffeomorphism $h : F_g \to F_g$ have an invariant circle?*

Remarks: Yes for $g \leq 10$, no for $g = 11$, yes for some $g > 11$, and no for an infinite number of g. For $g = 11$, there is *essentially* one diffeomorphism h, of order 30, with no invariant circle [740,Meeks, III, 1979,J. Differential Geom.].

Update: Still open.

NEW PROBLEMS

Problem 2.9 (Mess) (A) *Find a finite presentation for the Torelli group, κ_g, $g \geq 3$.*

(B) *Given g, what is the largest k for which κ_g admits a classifying space with finite k-skeleton.*

Remarks: Perhaps (Vogtman) $k = g - 2$ so the Torelli groups is only finitely presented for $g \geq 4$. This guess is supported by analogy with the behavior of S-arithmetic groups and solvable groups, as investigated by H. Abels, R. Bieri, R. Strebel and others.

As in the Update to Problem 2.1, κ_1 is trivial, κ_2 is infinitely generated [730,McCullough & Miller,1986b,Topology Appl.] and in fact free on infinitely many generators [752,Mess,1992,Topology], and κ_g, $g \geq 3$, is finitely generated [546,Johnson,1983,Ann. of Math.]. κ_3 does not have a classifying space with finite 3-skeleton, as shown by Mess expanding on a result of Johnson & Millson.

Problem 2.10 (Congruence Subgroup Conjecture) *Every subgroup of finite index of Γ_S contains a congruence subgroup, where S is a compact, orientable surface, perhaps with boundary.*

Remarks: Using the analogy between mapping class groups and arithmetic groups (see [530,Ivanov,1987b,Успехи Мат. Наук] for a discussion) as a guide, define a congruence subgroup of Γ_S as follows: first recall that a subgroup H of a group G is called characteristic if it is invariant under all automorphisms of G. If H is a characteristic subgroup of $\pi_1(F_g)$ of finite index, then the kernel K_H of the homomorphism $Out(\pi_1(F_g)) \to Out(\pi_1(F_g)/H)$ is a subgroup of Γ_S of finite index. The subgroups K_H which arise from subgroups H are called *congruence subgroups.* Note that since $\pi_1(F_g)$ is finitely generated, every subgroup H' of $\pi_1(F_g)$ of finite index contains a characteristic subgroup of finite index.

Problem 2.11 (Ivanov) (A) *Is it true that $H^1(G) = 0$ for any subgroup G of finite index in Γ_S ?*

Remarks: It is known that $H^1(\Gamma_S) = 0$. Γ_S is residually finite [418,Grossman,1974,J. London Math. Soc.] (also exercise 1 in [532,Ivanov,1992]), so there are many subgroups of finite index which are not well understood.

(B) *Does Γ_S satisfy Kazhdan's property T?*

Remarks: A positive answer to (B) implies the same for (A). A good reference for property T is [237,de la Harpe & Valette,1989,Astérisque], but the definition is technical and is not included here.

Problem 2.12 (A) (Penner) *Is it possible that all nontrivial (i.e. $\neq 1$) elements of a normal subgroup of Γ_S are pseudo–Anosov diffeomorphisms?*

Remarks: Of course, conjugates and powers of pseudo–Anosovs are also pseudo–Anosov, but the product of two is not in general. It is known how to find big free subgroups consisting entirely of pseudo–Anosov elements, but not how to find one which is also normal.

(B) (Ivanov) If the subgroup H of $\pi_1(S)$ is characteristic, then the kernel of the natural homomorphism

$$\Gamma_S = Out(\pi_1(S)) \to Out(\pi_1(S)/H)$$

is a normal subgroup. Call the class of normal subgroups of Γ_S that arise by this construction C.

Question: *Is it true that any normal subgroup of Γ_S is commensurable with a normal subgroup in C?*

Remarks: In general the subgroups in C have infinite index, e.g. the Torelli group which is the subgroup of Γ_S which acts trivially on $H_1(S)$.

Problem 2.13 (Mess) In the mapping class group Γ_g, $g \geq 3$, of a closed surface of genus g, a Dehn twist τ on a non-separating curve is a product of 3 commutators (by the lantern relationship [888,Powell, 1978,Proc. Amer. Math. Soc.]).

(A) *Is 3 the minimum possible?*

(B) *What is the genus of* τ^n *as a function of* n*, where the genus is the number of commutators needed to express* τ^n*?*

(C) *For genus 2, the abelianization of the mapping class group* Γ_g *is* $\mathbb{Z}/10\mathbb{Z}$*, generated by* τ*. So* τ^{10k} *is a product of commutators; how many as a function of* k*?*

(D) *What is the genus of a Dehn twist about a separating curve on a surface of genus at least 3? This might depend on the genus of the surface bounded by the separating curve (on a genus 2 surface, a Dehn twist on a separating curve has order 5 in the first homology group).*

Problem 2.14 (Mess) *Suppose there is an inclusion of* $\pi_1(F_g)$, $g \geq 2$, *into a mapping class group* Γ_h^1 *of a one pointed surface of genus* h*. Is it ever the case that this inclusion lifts to* $\Gamma_{h,1}$*, the mapping class group of a surface of genus* h *with one boundary component (diffeomorphisms modulo isotopies which fix the boundary)?*

Remarks: One way to construct inclusions of $\pi_1(F_g)$ into the mapping class group is to consider a fixed complex structure on a surface H of genus g', and the family of hyperbolic orbifolds obtained by making a point of H into a cone point of order n. Holding n fixed and varying the point, one obtains a π_1-injective inclusion of H into the moduli space $\mathcal{M}_{g'}^1$ of hyperbolic orbifolds with genus g' and one cone point of order n. A finite cover of H is then included π_1-injectively (in the sense appropriate to the orbifold $\mathcal{M}_{g'}^1$) in the moduli space of surfaces of genus g, where the surface of genus g is a finite cover (in the sense of orbifolds) of the orbifold of genus g' with one cone point of order n. (The resulting imbedding of the unit disc into Teichmüller space is not a Teichmüller geodesic.)

If the answer to the problem is yes, i.e. the inclusion lifts to $\Gamma_{h,1}$, then $\pi_1(F_g) \times \pi_1(F_g)$ is a subgroup of Γ_{2h}. The converse was proved in [754, Mess, 1995b].

Problem 2.15 (Ivanov) (A) Conjecture (Mostow–Margulis superrigidity)*: If* G *is an irreducible arithmetic group of rank* ≥ 2*, then any homomorphism* $G \to \Gamma_S$ *has finite image.*

Remarks: This is true for many arithmetic groups. An interesting open case is that of cocompact lattices in $SU(p,q)$ (G. Prasad).

By a deep result of Margulis [695,Margulis,1991] about finiteness of quotient groups of lattices, this conjecture is essentially equivalent to the following:

(B) Conjecture: *Irreducible lattices in isometry groups of symmetric spaces of R-rank ≥ 2 never occur as subgroups of mapping class groups.*

Remarks: A lattice Γ in a Lie group G is a discrete subgroup such that G/Γ has finite volume with respect to Haar measure. Both non-cocompact and cocompact lattices in $PSL(2,\mathbb{R})$ do occur in mapping class groups of surfaces of sufficiently large genus by the remark to the previous problem. The space $SL(n,\mathbb{R})/SO_n(\mathbb{R})$ of shapes of ellipsoids is a symmetric space of $R - rank \geq 2$ when $n \geq 3$, and is a typical example. A reducible lattice is one that has a finite index subgroup which is a direct product of two infinite subgroups, and an irreducible lattice is one that is not reducible. There are reducible lattices which act on the product of two copies of the hyperbolic plane, and there are also irreducible lattices, e.g. $PSL(2,\mathbb{Z}[\sqrt{2}])$ which do so.

If such lattices occurred as subgroups of the mapping class groups, one would expect a holomorphic map of the symmetric space to the Teichmüller space of the surface, and these are not expected.

[Margulis, ibid.] is the definitive treatment of lattices.

Problem 2.16 (Ivanov) *Let $d_W(\ ,\ )$ be the word metric on Γ_S with respect to some given finite set of generators and let $\tau \in \Gamma_S$ be a Dehn twist. What is the growth rate of $d_W(t^n, 1)$?*

Remarks: It is expected that either the growth rate is linear, or $d_W(t^n, 1) = O(\log n)$. In the arithmetic groups case, logarithmic growth corresponds to virtually unipotent elements of arithmetic groups of rank ≥ 2 [680,Lubotzky, Moses, & Raghunathan,1993,C.R. Acad. Sci. Paris Sér. I Math.].

Problem 2.17 (Mess) *Consider surface bundles over surfaces where both fiber F and base B have genus ≥ 2 and where $\pi_1(B)$ injects in the* *mapping class group of the fiber. Does such a bundle have a multisection (a submanifold which finitely covers the base)?*

Remarks: One can construct such bundles as follows: Take a surface F and consider the bundle of pointed surfaces $F \times F$ with fiber over p equal to (F,p). Turn this into a bundle of orbifolds where the fiber over p is F with a cone point of order n at p. This 4-dimensional orbifold has a finite orbifold cover which is a fiber bundle of surfaces as desired.

Problem 2.18 (Mess) (A) *Let E be a surface bundle over a surface, with base F_g a closed, oriented surface of genus g and fiber $F_{g'}$. What is the minimum g for which $\sigma(E) \neq 0$?*

Remarks: There is an example for $g = 129$ [50,Atiyah,1969], and this can be lowered (Mess) to $g = 42$ using Wajnryb's presentation of the mapping class group [1083,Wajnryb,1983,Israel J. Math.]. $\sigma(E) = 0$ if $g' < 3$ because the second rational cohomology of the mapping class group vanishes for genus < 3.

(B) *Let $g(n)$ be the least genus for which $\sigma(E) = 4n$. The* $\lim_{n\to\infty} \frac{g(n)}{n}$ *exists, is finite, and is non-zero. What is the limit?*

Remarks: Since $g(m+n) \leq g(m)+g(n)$ by the obvious construction, the existence of the limit follows from a standard elementary analysis result. [67,Barge & Ghys,1992,Math. Ann.] may be useful.

Note that the same two questions can be asked when E is restricted to be a complex bundle.

Problem 2.19 (Cooper & Mess) *Does every torsion free 3-manifold group act faithfully by homeomorphisms of $\mathbb{R}^2$?*

Remarks: The binary polyhedral groups do not act because of the theorem of Kerékjártó and Eilenberg ([1081,von Kerékjártó,1934b,Acta Sci. Math. (Szeged)], [1080,von Kerékjártó,1934a,Acta Sci. Math. (Szeged)], and [275,Eilenberg,1934,Fund. Math.]) that a finite order homeomorphism of $\mathbb{R}^2$ is conjugate to a rotation or a reflection, and therefore the only finite groups that act faithfully are cyclic or dihedral.

The free abelian group of rank 2 acts faithfully on the unit disc, acting as the identity on the boundary. It follows that that a semidirect product with normal subgroup free abelian of rank 2 and infinite cyclic quotient acts faithfully on the plane, setwise preserving a collection of disjoint discs, one centered at each integral point.

For a different sort of action, the fundamental group G of a Solv manifold imbeds as a subgroup of the direct product of two copies of the $ax+b$ group of affine motions of the line, and it follows easily that G acts on $\mathbb{R}^2$, leaving the region $\|x\| \geq \|y\|$ pointwise fixed, and preserving each of the line segments $y = c$, $\|x\| \leq c$.

The full automorphism group of a surface of negative Euler characteristic acts faithfully on the circle; and therefore so do its subgroups, which include the fundamental groups of all 3-manifolds which fiber over the circle with fiber of negative Euler characteristic but are not mapping tori of finite order diffeomorphisms. Groups which act effectively on the circle extend radially to groups which act on the plane. So in particular many fundamental groups of compact hyperbolic 3-manifolds act effectively on the plane. Suppose F is a hyperbolic surface and $f : F \to F$ is a finite order diffeomorphism. Then the mapping torus of f is a 3-manifold M_f with fundamental group a semidirect product $G = \langle f \rangle \ltimes \pi_1 F$, and the quotient of G by its center acts effectively on the circle. Although G does not act effectively on the circle in any *natural* way, G does act effectively on the plane, preserving the foliation by concentric circles centered at the origin. So the fundamental groups of all 3-manifolds which are fiber bundles over the circle act effectively on the plane, and the argument extends to Stallings fibrations, i.e. 3-manifolds which are bundles over the unit interval regarded as an orbifold with two exceptional points.

Suppose M is a Seifert fibered space with hyperbolic base and a fiberwise orientation. Does $\pi_1 M$ act on the plane preserving the (singular) foliation by rays through the origin and inducing some action of the orbifold fundamental group of the base orbifold on the circle of directions?

For Seifert fibered spaces with hyperbolic base for which the Euler class is not too large and satisfies additional conditions when the base orbifold has cone points, there is an action of the fundamental group on $\mathbb{R}$. This result of R. Naimi [825,1994,Comment. Math. Helv.] is the culmination of a series of papers by J. Milnor, J. Wood, D. Eisenbud, U. Hirsch, M. Jankins and W.D. Neumann. Evidently a group that acts effectively on $\mathbb{R}$ also acts effectively on $\mathbb{R}^2$.

One would expect that if two groups act effectively on $\mathbb{R}^2$ their free product would also act effectively on $\mathbb{R}^2$. If so one could reduce to the case of prime 3-manifolds.

Problem 2.20 (Lima) *Are there commuting homeomorphisms of the 2-ball B^2 without a common fixed point ?*

Remarks: Two commuting C^1 vector fields on a surface of non-zero Euler characteristic have a common singularity [657,Lima,1963,Bull. Amer. Math. Soc.], [658,Lima,1964a,Comment. Math. Helv.], [659, Lima,1964b,Proc. Amer. Math. Soc.]. The equivalent problem for diffeomorphisms has a negative answer. One can easily see that, for example, two rotations by π on orthogonal axes of the sphere S^2 commute and have no common fixed points. But if two diffeomorphisms of S^2 are C^1-close enough to the identity map then they have a common fixed point [133,Bonatti,1989,Ann. of Math.], answering a question posed in [938,Rosenberg,1974]. Handel [437,1992,Topology], following a suggestion of Mather, defined the winding number of two commuting homeomorphisms $\phi, \psi : S^2 \longmapsto S^2$ as the class:

$$W(\phi, \psi) = [H_t] \in \pi_1(Homeo_+(S^2)) \cong \mathbb{Z}/2\mathbb{Z},$$

where ϕ_t and ψ_t are isotopies between ϕ and ψ and the identity map, and H_t is the closed path in $Homeo_+(S^2)$ defined by the isotopy $H_t = \phi_t \psi_t \phi_t^{-1} \psi_t^{-1}$. Then he proved that two commuting diffeomorphisms with $W(\phi, \psi) = 0$ have a common fixed point, extending Bonatti's result. If ϕ and ψ are only homeomorphisms, Handel's result can still guarantee the common fixed point under the additional hypothesis that the fixed point set of each map is finite.

The problem is still open for homeomorphisms of S^2 in general, as well as the original question posed by Rosenberg, [938,1974; page 305]:

Question: *Let M be a closed manifold with non-zero Euler characteristic and ϕ and ψ commuting diffeomorphisms close to the identity map; must they have a common fixed point?*

Chapter 3

3-Manifolds

- Problems 3.1–3.44 (1977), 3.45–3.113 (new).
- Classification 3.1–3.10, 3.45, 3.46, 3.52, 3.82, 3.84.
- Surfaces in 3-manifolds, 3.10–3.12, 3.17–3.19, 3.50, 3.66–3.69, 3.83, 3.96–3.99, 5.6.
- Covers and branched covers, 3.24–3.26, 3.50, 3.51.
- Links of singularities, 3.27–3.31.
- 3-Manifold groups, 3.32, 3.33, 3.74–3.79, 3.104, 3.105.
- Group actions, 3.37–3.44, 3.70–3.73.
- Hyperbolic and other geometries, 3.14, 3.15, 3.45–3.69.
- Diffeomorphism and isometry groups, 3.34–3.39, 3.47–3.49.
- Algorithms, 3.80–3.83.
- Heegaard splittings, 3.85–3.94 (also 1.72, 1.73).
- Surgery and framed links, 3.101–3.104, (also 1.77–1.82).
- Flows, 3.110–3.113.
- Topological quantum field theories (TQFT's), 3.106–3.108, (also, 1.88–1.92, 5.19–5.20).

Introduction

A 3-manifold M is *irreducible* if every smoothly imbedded S^2 bounds a B^3 in M, and is P^2*-irreducible* if it is irreducible and contains no 2-sided real projective plane $\mathbb{RP}^2$. M is *prime* if $M = M_1 \# M_2$ implies either M_1 or M_2 is S^3.

M is said to be *Haken* if it is P^2-irreducible and contains a smooth, properly imbedded, 2-sided surface F which does *not* satisfy any of the following three conditions:

(1) there exists a compressing disk D for F, i.e. a smoothly imbedded B^2 such that $D \cap F = \partial D$ and ∂D is not contractible in F;

(2) $F = S^2$ and bounds a homotopy 3-ball in M;

(3) $F = B^2$ and there is a homotopy 3-ball B_0 such that $\partial B_0 \subset F \cup \partial M$.

Such a surface F is called *incompressible*, and a 3-manifold with an incompressible surface was called *sufficiently large*, but Haken is the modern term used for 3-manifolds which are irreducible and sufficiently large.

M^3 is called *atoroidal* if:

- (geometric definition) M contains no essential, properly imbedded, nonperipheral annulus or torus;
- (algebraic definition) each $\mathbb{Z} \oplus \mathbb{Z}$ in $\pi_1(M)$ is conjugate to a subgroup in $\pi_1(\partial M)$.

The algebraic definition implies the geometric definition (clearly M contains no torus, but if M contains a nonperipheral annulus, then there is a $\mathbb{Z}$ in the center of $\pi_1(M)$, so (see Problem 3.5) M is a Seifert fibered space and these contain $\mathbb{Z} \oplus \mathbb{Z}$'s). However, the problems often do not specify which kink of atoroidal is meant; in this case, the intent of the author may be clear, or there are two versions of the problem.

Suppose M^3 is compact, orientable, irreducible and ∂M is incompressible. Then [538, Jaco & Shalen, 1979], [542, Johannson, 1979], M

has a finite family of disjoint, incompressible, 2-sided, imbedded tori which split M into pieces which are either Seifert fibered spaces or algebraically atoroidal; furthermore, a minimal such family is unique up to isotopy. For further details, the non-orientable case, and relations with the Geometrization Conjecture 3.45, see Scott's survey [979, 1983a, Bull. London Math. Soc.].

The definitions of *geometric* 3-manifold and of *orbifold* are given in Problems 3.45 and 3.46.

Problem 3.1 Poincaré Conjecture: *Every homotopy 3-sphere Σ^3 is homeomorphic to S^3.*

Here are some presumably easier conjectures:

(A) *The suspension of Σ^3 is homeomorphic to S^4.*

Remarks: The double suspension is S^5 [995,Siebenmann,1970] .

(B) *$(\Sigma^3$ – point$) \times \mathbb{R}$ is diffeomorphic to $\mathbb{R}^4$.*

(C) **(Poénaru)** *There exist two contractible open subsets, U_1 and U_2, of Σ^3 such that $\Sigma^3 = U_1 \cup U_2$ and U_1 and U_2 are imbeddable in $\mathbb{R}^3$ (see [881,Poénaru,1974,Bull. Amer. Math. Soc.]). (Added in proof. This is equivalent to the Poincaré conjecture [738,McMillan, Jr.,1970,Bull. Amer. Math. Soc.].)*

(D) *Σ^3 imbeds (smoothly?) in S^4. This implies the next two:*

(D′) *Σ^3 bounds a contractible 4-manifold, and, if the imbedding is smooth,*

(D″) *Σ^3 does not bound a smooth, almost parallelizable 4-manifold of signature 8 mod 16.*

(E) *If Σ^3 admits an involution (and is irreducible), then Σ^3 is S^3.*

(F) *The connected binding of some open book decomposition (Problem 3.13) of Σ^3 lies in a 3-ball. (This is equivalent to the Poincaré conjecture.)*

Furthermore, the Poincaré Conjecture generalizes to

(Ω) Simple homotopy equivalent, closed, orientable, 3-manifolds are homeomorphic.

Remarks: This is known for irreducible, sufficiently large manifolds [1088,Waldhausen,1968b,Ann. of Math.], and for lens spaces [220,Cohen,1973].

Update: The Poincaré conjecture is still open (presumably, as there exist manuscripts which could contain a proof) (also see Problem 4.88).

(A) True by [329,Freedman,1982,J. Differential Geom.], for $\Sigma \times \mathbb{R}$ is homeomorphic to $S^3 \times \mathbb{R}$, so the suspension is a manifold homotopy equivalent to S^4, and thus homeomorphic to it.

(B) Still open, but (Σ^3-point) $\times \mathbb{R}$ is homeomorphic to $\mathbb{R}^4$ [329,Freedman,1982,J. Differential Geom.]. If (B) is true, then no exotic $\mathbb{R}^4$ is diffeomorphic to a 3-manifold $\times \mathbb{R}$; McMillan [736,1961,Bull. Amer. Math. Soc.] (also [386,Glimm,1960,Bull. Soc. Math. France]) proves that a Whitehead manifold $\times \mathbb{R}$ is diffeomorphic to $\mathbb{R}^4$.

(C) No progress, but compare McMillan's result with Problem 4.89, last remark.

(D) Σ^3 imbeds (locally flat) in S^4 and thus bounds a TOP contractible 4-manifold (as do all homology 3-spheres [Freedman, ibid.]. (D″) was proven by Casson who showed that a homotopy 3-sphere has Rohlin invariant zero [30,Akbulut & McCarthy,1990].

(E) True if the fixed point set is 1-dimensional and hence a knot, using the proof of the Smith Conjecture (see Problem 3.38). A 2-dimensional fixed point set, a 2-sphere, is not possible because of irreducibility, and the cases of 0-dimensional or empty fixed point set are open.

(F) No progress.

The Poincaré conjecture is a special case of the Geometrization Conjecture (see Problem 3.45).

Problem 3.2 *Let M^3 be a closed, orientable, irreducible 3-manifold with infinite fundamental group. If M^3 is not sufficiently large, must it have a finite cover which is sufficiently large?*

Update: The question is still open. Note that covers of irreducible 3-manifolds are irreducible [744,Meeks, III & Yau,1982,Topology]. If the cover of M^3 is Haken (equals irreducible *and* sufficiently large), then it must be hyperbolic (or one of the rare cases in which a Seifert fibered 3-manifold covers a small non-Haken Seifert fibered 3-manifold); this follows from Thurston's geometrization theorem for Haken manifolds plus the fact that there are no characteristic tori (for otherwise they could be chosen equivariantly and would then descend to the quotient).

If M has a finite cover N with rank $H_1(N;\mathbb{Z}) > 0$, then M is called virtually $\mathbb{Z}$-representable or has virtually positive first Betti number; with irreducibility this implies M is virtually Haken. In the following list, M (as above) is virtually $\mathbb{Z}$-representable if it satisfies:

- M is Seifert fibered, for then it has a finite cover which is a circle bundle over an orientable surface.
- M is hyperbolic and has an immersed, totally geodesic surface [674,Long,1987,Bull. London Math. Soc.].
- M is obtained by certain gluings of regular hyperbolic polyhedra [848,Nicas & Stark,1984,Math. Proc. Cambridge Philos. Soc.] and [849,Nicas & Stark,1987].
- M is a quotient of H^3 by certain cocompact, arithmetic subgroups of $SL(2,\mathbb{C})$ [206,Clozel,1987,Duke Math. J.].
- M has an orientation reversing involution and also some finite cover [1098,Wang,1990,Proc. Amer. Math. Soc.].
- M is obtained by certain surgeries on the figure-8 knot [847,Nicas, 1986,Math. Proc. Cambridge Philos. Soc.], [801,Morita,1986, Sci. Papers College Arts Sci. Univ. Tokyo], [601,Kojima & Long, 1988], and [60,Baker,1991,Pacific J. Math.].
- M is one of certain branched covers over links in S^3 [477,Hempel, 1990,Topology].
- M is a certain Dehn filling of a punctured surface bundle [473, Hempel,1986,Topology Appl.], [474,Hempel,1987a,Topology Appl.], [58,Baker,1989,Proc. Amer. Math. Soc.], [59,Baker,1990,Proc. Amer. Math. Soc.], and unpublished work of Brunner, Neumann & Rubinstein mentioned in [14,Aitchison & Rubinstein,1993,J. Austral. Math. Soc. Ser. A].

Related to the above is the preprint [953,Rubinstein & Wang,1995] which gives the first example of an immersed incompressible surface which does not lift to an imbedding in any finite sheeted covering.

Problem 3.3 (Jaco) *Let M^3 be a compact 3-manifold. Are there at most a finite number of homotopy equivalent 3-manifolds? What if we* *restrict to the case where M is orientable and sufficiently large? (See [537, Jaco & Shalen, 1976, Invent. Math.].)*

Update: M^3 should be assumed to be irreducible to avoid possible fake 3-balls. Yes, for Haken 3-manifolds [1027, Swarup, 1980, Bull. London Math. Soc.], for aspherical Seifert fibered spaces [980, Scott, 1983b, Ann. of Math.], and for all closed, hyperbolic 3-manifolds satisfying Gabai's thick geodesic condition [361, Gabai, 1995a]. The question is still open for closed non-Haken 3-manifolds (see Problem 3.45). Johannson has classified homotopy equivalences of 3-manifolds with non-empty boundary [542, Johannson, 1979].

Problem 3.4 (Stallings) *Is an irreducible h-cobordism from $\mathbb{RP}^2$ to itself a product?*

Remarks: This is known if its double cover is $S^2 \times I$ [669, Livesay, 1963, Ann. of Math.] and [947, Rubinstein, 1976, Proc. Amer. Math. Soc.]. The open case is also unknown: Is a proper homotopy $\mathbb{RP}^2 \times \mathbb{R}$ homeomorphic to $\mathbb{RP}^2 \times \mathbb{R}$ (assume irreducibility of the manifold)? An affirmative answer implies that every topological involution of S^3 with two fixed points is standard.

Update: Still open. But note that a smooth or PL involution of S^3 is standard [Livesay, ibid.] and the topological case follows using a result in [492, Hirsch & Smale, 1959, Amer. J. Math.],

Problem 3.5 *Suppose M^3 is irreducible and $\pi_1 M^3$ is infinite with non-trivial center.*

(A) (Thurston) *Is M^3 a Seifert fibered space?*

Remarks: Yes if M is sufficiently large [1086, Waldhausen, 1967c, Topology]. If yes in general, then we get an affirmative answer to Problem 2.6 for finite groups. For let G be a finite subgroup of $\pi_0(\text{Homeo}F^2)$. G acts freely on the unit tangent space TF^2 with quotient space M^3 satisfying the conditions above. If M^3 is a Seifert fibered space, there is a natural, regular, branched covering of its base space with G as the group of deck transformations; the branched covering space is F^2.

(B) (Jaco) *Is the center of $\pi_1 M^3$ finitely generated?*

Remarks: Yes, if $\pi_1 M^3$ contains a (closed) surface group.

Update: P. Scott [980, 1983b, Ann. of Math.] reduced 3.5A to showing that M^3 is homotopy equivalent to a Seifert fibered space. Mess [753, 1995a] showed that either

- M was homotopy equivalent to a Seifert fibered space with Euclidean base orbifold,

  or

- the quotient of $\pi_1(M)$ by the center acted on a circle at infinity as a *convergence group*; in this case it sufficed to show that the convergence group was Fuchsian.

As a byproduct, Mess showed (B), that the center of $\pi_1(M)$ was finitely generated and therefore (by a theorem of D. B. A. Epstein [283, 1961, Proc. London Math. Soc.]) was free abelian of rank 1, 2, or 3 (except if M is non-orientable when $\mathbb{Z} \oplus \mathbb{Z}/2\mathbb{Z}$ is possible). (If ∂M contains no 2-spheres and M is compact with no fake 3-balls, then in the rank 3 case, $M = T^3$, and for rank 2, M is an I-bundle over T^2 or one of 3 different quotients of T^3 (see [472, Hempel, 1976])).

Tukia [1067, 1988, J. Reine Angew. Math.] showed that convergence groups on the circle are conjugate to Fuchsian groups except perhaps in a special case, which is in fact the heart of the problem. The special case was established by Gabai [358, 1992, Ann. of Math.] and by Casson & Jungreis [190, 1994, Invent. Math.]. Thus (A) is true.

Problem 3.6 (A) *Is the homology 3-sphere obtained by ± 1-surgery on a knot always prime?*

(B) *If so, do they all have finite covers which are either sufficiently large or S^3?*

Remarks: This is a special case of Problem 3.2 if π_1 is infinite.

(C) *Find a prime homology 3-sphere which is not obtained by ± 1-surgery on a knot. Who knows a nonprime example which is not?*

(D) **(B. Clark)** *Is there a homology 3-sphere (or any 3-manifold) which can be obtained by n-surgery on an infinite number of distinct knots?*

Remarks: Lickorish gave an example obtained from two knots [652,1977,Proc. Amer. Math. Soc.]. Problems 1.15 and 1.16 ask whether this is possible for S^3 (Property P) or $S^1 \times S^2$ (Property R). It seems likely that the Poincaré homology sphere can be obtained only from +1-surgery on the right-handed trefoil knot (or, reversing orientation, from -1-surgery on the left-handed trefoil knot).

Update:

(A) Any surgery on a knot giving a homology 3-sphere always gives a prime one [405,Gordon & Luecke,1989,J. Amer. Math. Soc.].

(B) See Update for Problem 3.2.

(C) Auckly [52,1993] has given examples using Taubes' periodic ends theorem.

(D) Still open (but see Problem 3.102).

Problem 3.7 Let M^3 be a closed, irreducible 3-manifold with infinite fundamental group (hence a $K(\pi, 1)$).

(A) *Is the universal cover always $\mathbb{R}^3$?*

(B) *Does the universal cover always imbed in $\mathbb{R}^3$?*

(C) *Is the universal cover always simply connected at infinity?*

Remarks: Yes for (A) if M or some finite cover is irreducible and sufficiently large [1088,Waldhausen,1968b,Ann. of Math.]. Yes for (B) and (C) implies yes for (A).

Update: Still open. (The hypothesis should include P^2-irreducible or orientable.) But yes for (C) alone implies yes for (A) because the universal cover is irreducible [742,Meeks, III, Simon, & Yau,1982,Ann. of Math.].

(A) is true if

- M has a foliation without Reeb components [868,Palmeira,1978, Ann. of Math.], or

- M has an essential lamination [364,Gabai & Oertel,1989,Ann. of Math.], or

- the universal cover of M is end 1-movable [148,Brin & Thickstun,1987,Topology], where *end 1-movable* means the following: an exhaustion of M is a nested sequence $K_0 \subset \text{int}(K_1) \subset K_1 \subset \text{int}(K_2) \cdots$ of compact sets (which may as well be connected submanifolds) whose union is M; M is end 1-movable if for each i there is a $j > i$ such that for all $k > j$, every loop in $M - K_j$ homotops freely through $M - K_i$ into $M - K_k$ (this is a proper homotopy invariant), or

- the universal cover of M is eventually end irreducible [1119,Wright, 1992,Topology], where *eventually end irreducible* means that M has an exhaustion (as above) for which each ∂K_i, $i > 0$, is incompressible in $M - K_0$.

Whether or not (A) is true depends only on $\pi_1(M)$ (this well known fact can be found in [Brin & Thickstun, ibid.], whereas the fact that (C) only depends on $\pi_1(M)$ is due to [533,Jackson,1981,Topology]). Then (C) is true when

- $\pi_1(M)$ is negatively curved [84,Bestvina & Mess,1991,J. Amer. Math. Soc.] and, somewhat more generally,[372,Gersten & Stallings, 1991,Internat. J. Algebra Comput.] (with an improvement in [884,Poénaru,1994,Topology]),

- contains the fundamental group of a closed, hyperbolic surface, or torus, or Klein bottle [459,Hass, Rubinstein, & Scott,1989,J. Differential Geom.],

- has non-trivial center [754,Mess,1995b] ((C) is true in this case by [Jackson, ibid.])

- is almost convex [883,Poénaru,1992,J. Differential Geom.] (also see [882,Poénaru,1991,Duke Math. J.] and [885,Poénaru & Tanasi, 1993,Geom. Dedicata]),

- is *quasi-simply-filtered* [147, Brick & Mihalik, 1995] and [760, Mihalik & Tschantz, 1995],
- or is a subgroup of finite index in a Coxeter group [758, Mihalik, 1995a].

In a different direction, define a *Whitehead manifold* to be an irreducible, contractible, open 3-manifold which is not $\mathbb{R}^3$; these are all monotone unions of *handlebodies* (0− and 1−handles) [736, McMillan, Jr., 1961, Bull. Amer. Math. Soc.]. Then no Whitehead manifold can non-trivially cover any 3-manifold when the handlebodies all have genus one [824, Myers, 1988, Topology], or have bounded genus [Wright, ibid.] (which implies eventually end irreducible).

Problem 3.8 (Jaco) A manifold M is said to have a manifold compactification if there exist a compact manifold N and an imbedding $\varphi : M \to N$ with $\varphi(\text{int}M) = \text{int}N$. Let $M(G)$ be the unique covering space determined by the conjugacy class of a subgroup G of $\pi_1(M)$.

Question: *If M is a compact, P^2-irreducible, 3-manifold and G is a finitely generated subgroup, then does $M(G)$ admit a manifold compactification?*

Remarks: The answer is yes for compact, irreducible, orientable, sufficiently large 3-manifolds iff it is true for a surface bundle over S^1, [287, Evans & Jaco, 1977]. It is yes if G is the fundamental group of an incompressible surface [ibid.] or if G is peripheral (i.e., a subgroup of the image of $\pi_1(\partial M)$) [1001, Simon, 1976, Michigan Math. J.].

Update: Still open. But $M(G)$ has a manifold compactification if M is hyperbolic and G is finitely generated and not a free product [130, Bonahon, 1986, Ann. of Math.], or if M has an essential lamination and $G = \mathbb{Z}$ [363, Gabai & Kazez, 1993, J. London Math. Soc.]. It is also true if ∂M has a component of genus ≥ 2 and M has no incompressible tori, where G is finitely generated and not a free product; this is a corollary of Thurston's hyperbolization theorem (Bonahon). Also, the answer is yes if M is a compact P^2-irreducible 3-manifold and $\pi_1(M)$ has an asynchronously bounded, almost prefix closed combing and G is quasiconvex with respect to this combing; this includes the case that $\pi_1(M)$ is automatic and G is a regular subgroup [759, Mihalik, 1995b].

Note that whether $M(G)$ has a manifold compactification depends only on the pair of groups $(\pi_1(M), G)$ [149,Brin & Thickstun,1989].

Problem 3.9 *What closed, orientable* 3*-manifolds admit a round handle decomposition?*

Remarks: A round k-handle is $S^1 \times B^k \times B^{2-k}$ attached along $S^1 \times (\partial B^k) \times B^{2-k}$, $k = 0, 1, 2$. Morgan has shown that a 3-manifold M with a round handle decomposition must be a Seifert fibered space or $\pi_1(M)$ must contain a subgroup isomorphic to $\mathbb{Z} \oplus \mathbb{Z}$. Partial converse: Seifert fibered spaces obviously have such decompositions. In dimensions > 3, round handle decompositions exist iff the Euler characteristic is zero [49,Asimov,1975,Ann. of Math.].

Update: Morgan proves [785,1979,Topology] that an orientable, closed (or with torus boundary components) 3-manifold, M^3, has a round handle decomposition iff it is a graph manifold, i.e. it decomposes along tori into a union of Seifert fibered 3-manifolds (they can be glued together along the boundary tori in any way).

Problem 3.10 (Hilden & Montesinos) Every closed, orientable 3-manifold can be constructed as follows; let F_1 and F_2 be disjoint, compact surfaces (not necessarily orientable) in S^3. Take three copies of $(S^3; F_1, F_2)$, called $S^3_a, S^3_b, S^3_c, F_{1a}, \ldots$, etc. Split S^3_a along F_1, S^3_b along F_1 and F_2, and S^3_c along F_2. Then glue one side of F_1 in S^3_a to the other side in S^3_b, and one side of F_2 in S^3_c to the other side in S^3_b.

Question: *Can the surfaces be chosen to be orientable?*

Update: The answer is yes [481,Hilden, Lozano, & Montesinos,1983, Trans. Amer. Math. Soc.].

Problem 3.11 *Classify imbeddings of orientable surfaces F_g in S^3.*

Remarks: For S^2, the Schoenflies theorem classifies. For T^2, any imbedding bounds an $S^1 \times B^2$ ([32,Alexander,1924,Proc. Nat. Acad. Sci. U.S.A.] and [321,Fox,1948,Ann. of Math.]), so the classification *reduces* to knot theory.

Any imbedding of F_g into $\mathbb{R}^3$ is unknotted as soon as some projection to a coordinate axis has only a single local maximum (or minimum) [806, Morton,1979]; (also [804,Morton,1977,Notices Amer. Math. Soc.].

Any imbedding is ε-isotopic to a real algebraic variety in $\mathbb{R}^3$ [987, Seifert,1936,Math. Zeit.]; this is true in general in codimension one.

Call a pair (S^3, F) prime if there is no pairwise connected sum $(S^3, F) \cong (S^3, F_1)\#(S^3, F_2)$. Are prime decompositions unique? Yes for genus 2 [1065,Tsukui,1970,Yokohama Math. J.], [1066,Tsukui,1975, Yokohama Math. J.]; also [1022,Suzuki,1975b,Math. Japon.], [1021, Suzuki,1975a,Hokkaido Math. J.].

Update: It is worth mentioning that a 3-manifold M imbeds in S^3 iff M imbeds in S^3 with a handlebody complement ([321,Fox,1948,Ann. of Math.] and [751,Menasco & Thompson,1989,Topology]).

Prime decompositions are not unique in general [815,Motto,1992, Trans. Amer. Math. Soc.].

Problem 3.12 Given M^3, with a Riemannian metric, consider all smooth maps $f : S^2 \to M^3$ such that $f \not\simeq 0$. There exists one of least area [962,Sacks & Uhlenbeck,1981,Ann. of Math.] which is immersed (R. Gulliver).

Question: *Is it imbedded?*

(*Added in proof*: yes, or it double covers an imbedded $\mathbb{RP}^2$ (Meeks & Yau)).

Remarks: This would give a differential geometric proof of the sphere theorem.

Update: The map of least area is imbedded unless it double covers an imbedded $\mathbb{RP}^2$ [743,Meeks, III & Yau,1980,Ann. of Math.].

Problem 3.13 Theorem: Every closed orientable M^3 contains a fibered knot K, i.e., there exists a fibration $f : M - K \to S^1$ and f is standard on a deleted tubular neighborhood of K [821,Myers,1975, Notices Amer. Math. Soc.].

This is Winkelkemper's open book decomposition but with connected binding. Note that K is homologically trivial in M.

Question (Rolfsen): *What elements of $\pi_1(M)$ are represented by fibered knots? Links? Note that if an element is represented, so is any nonzero power.*

Update: Observe that a fibered knot K in M^3 bounds a surface so K represents an element in the commutator subgroup $[\pi_1 M, \pi_1 M]$. Harer proved [441,1982b,Math. Proc. Cambridge Philos. Soc.] that exactly those elements which are in $[\pi_1 M, \pi_1 M]$ are represented by fibered knots.

Problem 3.14 (Thurston) (A) Conjecture: *Every irreducible, closed 3-manifold, with infinite fundamental group which contains no subgroup isomorphic to $\mathbb{Z} \oplus \mathbb{Z}$, has a (see definitions in the Introduction). Assume, if necessary, that some finite cover is sufficiently large.*

(Added in proof, March 1, 1977: Thurston has a proof if, in addition, the 3-manifold has an incompressible surface which is not a fiber of a fibration over S^1.)

(B) Conjecture: *Suppose G acts properly and discontinuously on a contractible 3-manifold with compact quotient. Suppose also that G has no subgroup isomorphic to $\mathbb{Z} \oplus \mathbb{Z}$. Then G is conjugate to a discrete group of isometries of hyperbolic 3-space.*

Remarks: Clearly, (B) $\Rightarrow$ (A). Furthermore, (B) holds iff ((A) and every such G is residually finite) holds. The condition $\mathbb{Z} \oplus \mathbb{Z}$ is not in G, is necessary.

(C) Conjecture (case when $\partial M \neq \emptyset$): *Suppose (i) M^3 is irreducible, (ii) $\pi_1(M)$ is infinite and every $\mathbb{Z} \oplus \mathbb{Z}$ in $\pi_1 M$ is peripheral, (iii) $\partial M = \partial_1 M \cup \partial_2 M$ where each component of $\partial_1 M$ is a torus, and each component of $\partial_2 M$ has negative Euler characteristic, (iv) ∂M is incompressible and every annulus A with $\partial A \subset \partial M$ can be deformed,* rel ∂, *into ∂M. Then $M - \partial_1 M$ has a complete hyperbolic structure, with finite volume, such that $\partial_2 M$ is totally geodesic.*

Update: This is now a part of Thurston's Geometrization Conjecture (see Problem 3.45).

(A) was proven in 1977 under the assumption that the manifold is Haken. Thurston's lectures in the late 1970's outlined the proof, and, together with his Princeton notes [1050,1977], gave proofs of some of the steps. Morgan gave a long detailed outline with proofs of some steps [786,1984], and most of the remaining steps can now be found in the literature (there is no doubt among the experts of the validity of the theorem).

The proof for the case when M^3 fibers over S^1 is exposed in Sullivan's Bourbaki lecture [1018,1981] and Thurston gave a detailed proof [1054, 1986c]. Otal [862,1995] gives a new approach to the proof using $\mathbb{R}$-trees.

When M^3 does not fiber over S^1, the main step is the existence of a fixed point for the *skinning map*, which was first proved via a compactness theorem (there are several absolute and relative versions) by Thurston [1053,1986b,Ann. of Math.]; the compactness theorem was also proved by Morgan & Shalen in a series of papers [790,1984,Ann. of Math.], [791,1988a,Ann. of Math.], [792,1988b,Ann. of Math.].

To obtain a fixed point for the skinning map (defined in [Morgan, ibid.] or [739,McMullen,1989,Invent. Math.]), Thurston studies its contraction with respect to the Teichmüller metric. Uniform contraction is established using the compactness theorems (cited above) plus control on the geometry of geometrically infinite ends (Chapters 8 and 9 of [1050,Thurston,1977], [130,Bonahon,1986,Ann. of Math.], [172,Canary,1994], and [173,Canary, Epstein, & Green,1987].

A different proof of the main step, avoiding the compactness theorems, was given by McMullen [739,1989,Invent. Math.]. Otal & Paulin are preparing a book [863,Otal & Paulin,1995] on the proof of the non-fibered case following McMullen's approach.

Further papers are planned by Thurston and at least two preprints exist [1054,Thurston,1986c] and [1055,Thurston,1986d].

(B) is known when the compact quotient is Haken (here, conjugate means topologically conjugate in that there is a homeomorphism from the contractible 3-manifold to H^3 which carries G to the discrete group of isometries).

(C) is true because M is Haken with the same references as for (A).

Problem 3.15 (Thurston) Let M^3 be orientable and let $G \subset \pi_1(M^3)$ be the homomorphic image of $\pi_1(F)$ where genus $F = 2$, e.g., any 2-generator subgroup of $\pi_1(M)$.

(A) *Does G have 2-fold symmetry (involution) which comes from $180°$ rotation about the axis below?*

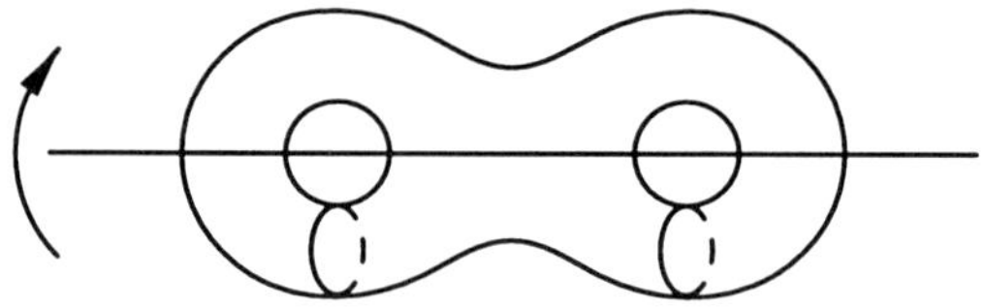

This is true for subgroups of $PSL(2,\mathbb{C})$ (which is the group of isometries of hyperbolic 3-space, the universal cover of any 3-manifold with negative curvature).

(B) *Suppose $G \subset \pi_1(M^3)$ has generators a, b with a conjugate to b in $\pi_1(M^3)$. Does G have an additional 2-fold symmetry t where $t(b) = a$? True for 2-bridge knot complements and 3-manifolds with Heegaard decompositions of genus 2.*

Update: (Mess)

(A) (i) The picture indicates that what is required is a homomorphism from $\pi_1(F)$ to G such that the kernel is invariant under the hyperelliptic involution of F, and the hyperelliptic involution induces an involution of G.

(ii) Some explanation is required for the claims in (A). For a 2 generator subgroup and G a subgroup of $PSL(2,\mathbb{C})$, the result is due to Jorgensen; 5.4.1 and 5.4.2 of [1050, Thurston, 1977] is a suitable reference. The argument of 5.4.1 and 5.4.2 extends to the case that the homomorphism factors through a free product of two free abelian groups, provided the image is not a solvable group.

More generally suppose G is a subgroup of $PSL(2,\mathbb{C})$. If G is a quasifuchsian group representing two Riemann surfaces of genus 2, then (by standard quasiconformal deformation theory) G admits a hyperelliptic involution, i.e. G is normalized in $PSL(2,\mathbb{C})$ by an involution which acts as the hyperelliptic involution on each of the boundary components at infinity. Although the Teichmüller theory predicts the involution, there is a more elementary description generalizing that of 5.4.1 and 5.4.2, and one deduces, given standard generators a_1, b_1, a_2, b_2 for the quasifuchsian group, that the two fixed points of the involution can be expressed algebraically in terms of the fixed points of a_1, b_1, a_2, b_2. (In other words the normalizing involution is defined by a rational map from the representation variety $R(G, PSL(2,\mathbb{C}))$ to $PSL(2,\mathbb{C})$. One can then deduce that there is a normalizing involution except possibly when G is a solvable subgroup of $PSL(2,\mathbb{C})$. For G a solvable discrete subgroup of $PSL(2,\mathbb{C})$ it is easy to verify that the normalizing involution exists. This confirms the claims made in the statement of part (A)).

(iii) Suppose G is a two generator subgroup. If the covering of M corresponding to $\langle a, b\rangle$ admits a Solv geometric structure, then the required involution exists.

(iv) There are counterexamples to (A) whenever M is a spherical space form, with generalized quaternion or binary polyhedral fundamental group, e.g. consider a homomorphism from the free group on 2 generators onto the quaternion group of order 8 which sends each generator to an element of order 4.

(v) Suppose G is a 2 generator group. Consider the covering M' of M corresponding to the subgroup G. If G is freely decomposable then the desired involution of G exists. If not, one should assume that M has no prime factor with finite fundamental group, because of (iv). Reduce to the case that M is prime. Either G is a finite index subgroup of M, or G is free, or G is the fundamental group of a compact Haken manifold with one or two toral boundary components.

(vi) There are examples of 3-manifolds for which the rank of the fundamental group is 2 but the Heegaard genus is 3. See [127,Boileau & Zieschang,1984,Invent. Math.]. So it is not possible to construct the involution by using the fact that every 3-manifold of Heegaard genus two is a double branched cover of the 3-sphere. In these examples, if a and b are two generators for the fundamental group G, there is nonetheless an involution of the 3-manifold which takes a and b to their respective inverses. (The involution is homotopic to the identity; it is a rotation half way around every generic fiber.)

(vii) If the simple loop conjecture is true for maps into 3-manifolds, then either the homomorphism from $\pi_1(F)$ onto G factors through a free group of rank 2, or else G contains a rank 2 abelian subgroup, or else the homomorphism from $\pi_1(F)$ onto G is an isomorphism and in that case the desired involution exists.

(viii) (Folklore.) If there is no essential simple loop on F which is mapped to a null homotopic loop in M then the map from F to M can be realized as the inclusion of an area minimizing minimal surface. If the surface was unique in its homotopy class the hyperelliptic involution would act on it. But area minimizing minimal surfaces are not always unique, so this doesn't lead to a proof.

(ix) Assuming the Geometrization Conjecture and assuming M aspherical one would hope to answer the question by extending the results of Kobayashi mentioned below.

(x) Perhaps the question was originally intended as an advertisement to topologists of the importance of geometric ideas. From this point of view one would want to forget about the original question and concentrate on the Geometrization and Orbifold Geometrization conjectures, (see Problems 3.45, 3.46).

(B) The remark, *True for 2-bridge knot complements and 3-manifolds with Heegaard decompositions of genus 2*, should probably be interpreted as follows: if a, b are free generators for the fundamental group of one of the handlebodies, and a, b are conjugate in $\pi_1(M)$ then the additional involution t exists. For the 2-bridge knot complements see e.g. [163,Burde & Zieschang,1985].

For 3-manifolds of Heegaard genus 2 and with the interpretation above, the claim is true if M is hyperbolic (see 5.4.2 in [Thurston, ibid.]). If M is not prime then M is the connected sum of two 3-manifolds of Heegaard genus 1 and the conjecture holds. Suppose M is prime. The result can probably be obtained (if it is true) using the results of [590,Kobayashi,1984,Osaka J. Math.], if M has nontrivial characteristic decomposition, although no one may have actually done so. Since M has Heegaard genus 2, M is a branched cover of S^3 over a 3-bridge knot or link, so according to a special case of the Orbifold Geometrization Conjecture, either M has a nontrivial characteristic decomposition, or M has a spherical or hyperbolic geometric structure with respect to which the covering involution is isometric. Assuming the Geometrization Conjecture one might expect to prove (A) for aspherical manifolds by extending Kobayashi's results.

Consider the standard (3,5) torus knot $K_{3,5} = \{z_1^5 = z_2^3\}$ in $S^3 \subseteq C^2$. Its double branched cover is the Poincaré homology sphere P. $S = \{\Re z_1 = 0\}$ is a 2-sphere in S^3, and the preimage F of S is a genus two Heegaard surface in P. Let B be the orbifold with underlying space the 3-ball $\{\Re z_1 \geq 0\}$ in S^3 and singular locus given by $K_{3,5} \cap B$. Then $\pi_1^{orb} B = \langle e_1, e_2, e_3 \| e_1^2, e_2^2, e_3^2 \rangle$ where each e_i generates the local group of one of the three arcs of $Sing(B)$. Let H be the preimage of B in P. Then $\pi_1 H = \langle e_1 e_2, e_2 e_3 \rangle$. There is an order 3 symmetry τ of P which projects to a symmetry $(z_1, z_2) \mapsto (z_1, \omega z_2)$ of B, where ω is a cube root of 1. This order 3 symmetry permutes e_1, e_2, e_3. Choosing a base point on one of the 2 arcs in H fixed by τ, it follows that $\tau_* e_1 e_2 = e_2 e_3$. So $e_1 e_2$ and $e_2 e_3$ are conjugate in $\pi_1 H$ and their images in $\pi_1 P$ are conjugate. Now fix a basepoint on the midpoint of the arc stabilized by e_2. We have that $i_* e_1 e_2, i_* e_2 e_3$ generate $\pi_1 P$, where $i : H \to P$ is the inclusion. The covering transformation is an involution j which inverts $e_1 e_2, e_2 e_3$ and also $a = i_* e_1 e_2, b = i_* e_2 e_3$. We may ask

(i) *Is there an involution t of P which preserves the handlebody H and satisfies $t_*(b) = a$?*

(ii) *Is there an automorphism t of $\pi_1 P$ which is an involution and commutes with j and satisfies $t_*(b) = a$?*

(iii) *Is there an automorphism of $\pi_1 P$ which is an involution and satisfies $t_*(b) = a$?*

(B(ii)) appears to be the natural interpretation of *additional* but some might think (B(iii)) is the more natural interpretation.

(B(i)) and (B(ii)) have negative answers but (B(iii)) has a positive answer. Suppose a symmetry of P preserves the Heegaard surface F. Then the symmetry is conjugate to one which commutes with the covering involution, so we will assume it does so. The symmetry then induces a symmetry of $K_{3,5}$. Up to conjugacy by a diffeomorphism isotopic to the identity, a symmetry of a torus knots is a rotation along the fibers or else an orientation preserving involution which takes fibers to fibers reversing the fiberwise orientation and fixing a great circle in S^3. So a symmetry of P which preserves the Heegaard surface F either is the covering transformation j or else exchanges H with the complementary handlebody.

Given any basepoint $*$ in P, the stabilizer of $*$ in the isometry group of P is a copy of the alternating group A_5 and this realizes the inner automorphisms of $\pi_1(P,*)$. Having chosen our basepoint, if a symmetry of $\pi_1(P,*)$ commutes with j it is realized by an isometry fixing $*$. It projects to a symmetry of $K_{3,5}$. (This symmetry will not preserve the standard metric on S^3.) As discussed above this is impossible.

It turns out that a, b necessarily project to two elements in the quotient A_5 of order 5. Considering the isometries of an icosahedron it is not hard to see that any two conjugate isometries of order 5 are conjugate by an involution, and from this one can show that the same holds for the preimages in $\pi_1 P$. So (B(iii)) has a positive answer.

Problem 3.16 (Thurston) *Is there a reasonable real-valued function C on the set of 3-manifolds which measures the complexity of $\pi_1(M^3)$? C should have the following properties:*

(A) *if M_1 is a k-fold cover of M_2, then $C(M_1) = kC(M_2)$;*

(B) $C(M_1 \# M_2) = C(M_1) + C(M_2)$;

and perhaps

(C) *if $f : M_1 \to M_2$ has positive degree, then $C(M_1) \geq C(M_2)$;*

(D) *$C(M) = volume(M)$ if M is either $\widetilde{SL}(2,\mathbb{R})$ or hyperbolic 3-space modulo a discrete group.*

Note that (A) and (B) imply that $C(M) = 0$ if M fibers over S^1 with monodromy of finite order, or if M fibers over a torus, or if M is covered by S^3.

Remarks: Via the axiom of choice, there appears to be an existence theorem for a function C satisfying (a), (b) and (d). First consider the property of a 3-manifold M,

(*) If M_1 and M_2 are homeomorphic finite covering spaces of M, then the degrees of the coverings are equal.

Question: What 3-manifolds satisfy (*)? In particular, suppose M^3 is not commensurable with $F_g \times S^1$ or a T^2 bundle over S^1; then does M^3 satisfy (*)?

Next, call M_1 and M_2 commensurable if they have homeomorphic finite covers.

Define C as follows: C must obviously be zero on M if M does not satisfy (*); C depends only on the prime components of M; if M is prime and satisfies (*), then we define C on the commensurability class of M by first defining C on M by (d) or else arbitrarily (axiom of choice here), and then if M' is commensurable with M, let $C(M') = (k/l)C(M)$ where there exists N, a k-fold cover of M and an l-fold cover of M'.

Also, see a forthcoming paper of Milnor & Thurston.

Update: See [770,Milnor & Thurston,1977,L'Enseign. Math.] for a discussion of the possibilities, as well as [413,Gromov,1982,Inst. Hautes Études Sci. Publ. Math.] and [1052,Thurston,1986a,Mem. Amer. Math. Soc.]. Also see Problem 3.84.

Problem 3.17 (Jaco) *Suppose M^3 is compact, orientable, and irreducible, and that K is a positive integer. Does there exist at most a finite number (up to homeomorphism of M) of isotopy classes of incompressible surfaces in M having Euler characteristic $\geq -K$?*

Remarks: Without the restriction on the Euler characteristic, the answer is no. Also, even with the restriction it is necessary to allow equivalence up to homeomorphism of M [534,Jaco,1970,Canad. J. Math].

Possibly an affirmative answer is contained in [428, Haken, 1961, Acta Math.].

Update: Yes, and a proof exists in Haken's paper as the remark above suggests (see also [536, Jaco & Oertel, 1984, Topology]).

Problem 3.18 (Jaco) *Up to homotopy, does there exist at most a finite number of essential maps of F_g into M^3 where M^3 is atoroidal?*

Remarks: Recall that a map is essential if it induces an injection $\pi_1(F_g) \to \pi_1(M)$ (F_g = closed, orientable surface), and M is atoroidal if M contains no essential, properly imbedded, nonperipheral annuli or tori. The answer is no without the atoroidal assumption [534, Jaco, 1970, Canad. J. Math], but even then it seems one should be able to describe the mappings.

Update: The answer is yes if M is geometric, or if $\pi_1(M)$ is hyperbolic in the sense of Gromov [417, 1993].

Problem 3.19 *Which immersed 2-spheres in $\mathbb{R}^3$ bound immersed 3-balls?*

Remarks: This is solved for one less dimension by S. Blank (see [880, Poénaru, 1969] and [326, Francis, 1970, Michigan Math. J.]).

Update: No progress.

Problem 3.20 *Under what conditions does a closed, orientable 3-manifold M smoothly imbed in S^4?*

Remarks: The torsion of $H_1(M;\mathbb{Z})$ must be of the form $T \oplus T$ [438, Hantzch, 1938, Math. Zeit.]. If $H_1(M;\mathbb{Z}) \cong \mathbb{Z}$, then the quadratic form of M is null-concordant; in particular the signature is zero and the Alexander polynomial $A(t)$ is of the form $f(t)f(t^{-1})$ [766, Milnor, 1968a], [565, Kawauchi, 1976, Osaka J. Math.]. If M is a $\mathbb{Z}/2\mathbb{Z}$-homology sphere, then the μ-invariant must be zero (see [496, Hirzebruch, Neumann, & Koh, 1971]).

Update: If M is a homology 3-sphere, then it imbeds topologically in S^4 [329,Freedman,1982,J. Differential Geom.], but in the smooth case obstructions arise from gauge theory because M would bound a smooth acyclic 4-manifold.

For non-homology spheres, topological obstructions arise from signature invariants [383,Gilmer & Livingston,1983,Topology], and in the smooth case further obstructions arise from gauge theoretic methods [311,Fintushel & Stern,1987,Topology].

Problem 3.21 *Let X be an acyclic 2-complex and M_0^3 an abstract regular neighborhood of X; $\partial M_0 = S^2$, so cap off to get a homology 3-sphere M^3. Find an effective way to compute the Rohlin invariant of M^3 in terms of X and its regular neighborhood.*

Update: No progress.

Problem 3.22 (Birman & Montesinos) *Do lens spaces admit minimal Heegaard splittings of genus > 1?* Conjecture: *Yes.*

Update: The answer is no. Bonahon & Otal [131,Bonahon & Otal, 1983,Ann. Sci. École Norm. Sup. (4)] proved that every Heegaard splitting of a lens space is a stablization of the genus 1 splitting. Every Heegaard splitting is a stabilization of the standard one for T^3 [120, Boileau & Otal,1990,J. Differential Geom.] and for orientable $F^2 \times S^1$ [974,Schultens,1993,Proc. London Math. Soc.]. Genus 2 Heegaard splittings of Seifert fibered 3-manifolds with base S^2 and 3 exceptional fibers were classified in [121,Boileau & Otal,1991,Invent. Math.].

Problem 3.23 (Jaco) A sufficiently large 3-manifold is atoroidal if it contains no essential, nonperipheral annuli or tori.

What groups appear as π_1 of an atoroidal manifold?

Remarks: Such manifolds are determined by their fundamental groups (Johannson).

Update: Assuming the 3-manifold M is orientable and irreducible, then M is hyperbolic (Thurston) (see Update to Problem 3.14) and

$\pi_1(M)$ is a finitely generated discrete subgroup of $PSL(2,\mathbb{C})$, or M is one of a special class of Seifert fibered spaces over S^2 with 3 exceptional fibers for which $\pi_1(M)$ is a central $\mathbb{Z}$-extension of a triangle group. Without irreducibility one gets free products of such groups.

Problem 3.24 (Hilden & Montesinos) *Is every homology 3-sphere the double branched covering of a knot in S^3?*

Remarks: It is known that $S^1 \times S^1 \times S^1$ is not a double branched covering (Fox) and that $S^1 \times F_g$ (F_g = surface of genus g) is not a double branched covering (Montesinos), but the arguments depend on a nontrivial first homology group.

Update: No; the first example, a suitable union of two knot spaces, appears in [823,Myers,1981b,Pacific J. Math.]. Most hyperbolic 3-manifolds, including many homology 3-spheres, have no symmetry. For more specific information, see [318,Flapan,1985,Pacific J. Math.], [996, Siebenmann,1980] and [118,Boileau, Gonzalez-Acuña, & Montesinos, 1987,Math. Ann.].

Problem 3.25 (Birman) *Let K be a knot in S^3 and $M(K)$ its 2-fold branched covering space. To what extent do topological properties of M determine K? More generally, describe the equivalence class $[K]$ of K under the relation $K_1 \approx K_2$ if $M(K_1)$ is homeomorphic to $M(K_2)$.*

Remarks:

1. If K is a 2-bridge knot, then $M(K)$ determines K among 2-bridge knots.
2. If $M(K)$ is composite, then K is composite [574,Kim & Tollefson, 1980,Michigan Math. J.].
3. The bridge index of $K \leq$ Heegaard genus of M [93,Birman & Hilden,1975,Trans. Amer. Math. Soc.].
4. There are examples of distinct prime 3-bridge knots which have homeomorphic 2-fold covering spaces [92,Birman, Gonzalez-Acuña, & Montesinos,1976,Michigan Math. J.]. In particular, the Brieskorn manifold $\Sigma(2,3,11)$ is the 2-fold branched cover of the $(3,11)$-torus knot and the knot in Figure 3.25.1 (Akbulut).

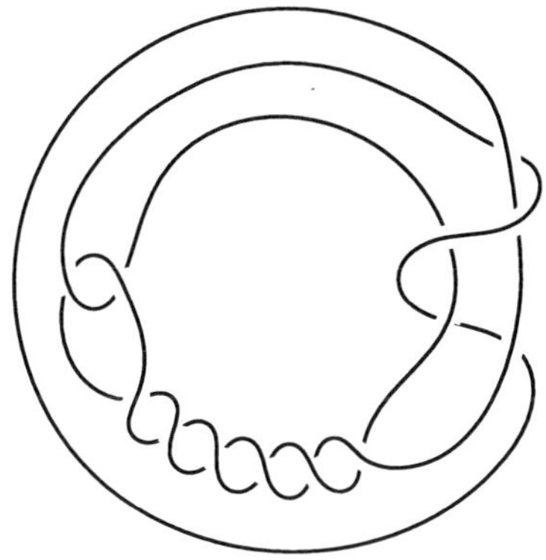

Figure 3.25.1.

Update: Montesinos & Whitten [781,1986,Pacific J. Math.] prove that if K and K' are knots in S^3, K is prime, $D(K)$ and $D(K')$ are twisted doubles of K and K' (with the same clasps), and $M(D(K))$ and $M(D(K'))$ are the double branched covers of S^3 along the doubled knots, then, K and K' are ambient isotopic iff $M(D(K))$ and $M(D(K'))$ are diffeomorphic.

If K is a 2-bridge knot, then $M(K)$ is a lens space; it is a 2-fold branched cover of a unique knot which must be K [499,Hodgson & Rubinstein, 1985].

It would follow from the Orbifold Geometrization Conjecture (see Problem 3.46) that there are only finitely many knots in the equivalence class $[K]$.

Problem 3.26 (Birman & Montesinos) *Every lens space is a 2-fold cover of S^3 branched over a unique 2-bridge knot or link. Can it be a 2-fold cover of S^3 branched over any other knot or link?* Conjecture: *No.*

Update: No, the only link is the 2-bridge one [499,Hodgson & Rubinstein, 1985].

For the following five problems of A. Durfee, P. Orlik and R. Randell, let $\mathcal{L}_k$ be the closed, orientable 3-manifolds which occur as links of isolated singularities of complex, analytic surfaces of complex codimension k in $\mathbb{C}^{2+k}$; in codimension > 1 the singularity should be normal. (Note that any analytic singularity is equivalent after an analytic change of coordinates to an algebraic singularity.) To fix notation when $k = 1$, let $f : \mathbb{C}^3 \to \mathbb{C}$ be a polynomial with an isolated singularity at the origin, and let $K = f^{-1}(0) \cap \varepsilon S^5$ for small ε. Then the map $f/|f| : S^5 - K \to S^1$ is a bundle map with fiber called F^4.

Problem 3.27 *Which closed, orientable 3-manifolds belong to $\mathcal{L}_1$? to $\mathcal{L}_k$, $k > 1$?*

Remarks: It is a classical result that a 3-manifold belongs to $\mathcal{L}_k$ iff it is the boundary of a plumbing on a finite, connected graph, not necessarily simple, with orientable surfaces at the vertices and negative definite intersection form [493,Hirzebruch,1966]. These are a special case of Waldhausen's *Graphenmannigfaltigkeiten* (see [1084,1967a,Invent. Math.], [1085,1967b,Invent. Math.]).

$S^1 \times S^1 \times S^1$ is not a link in any codimension [1016,Sullivan,1975, Topology]. Other 3-manifolds, such as the lens space $L(3,1)$, occur as links only in codimension greater than one [262,Durfee,1975].

Question: *What can be said about groups which occur as fundamental groups in $\mathcal{L}_1$?*

Update: Still open.

Problem 3.28 *Is every 3-manifold K in $\mathcal{L}_1$ irreducible?*

Remarks: Mumford [816,1961,Inst. Hautes Études Sci. Publ. Math.] (also [493,Hirzebruch,1966]) proved that if $\pi_1(K) = 0$, then $K = S^3$. If V admits an S^1 action, then K is a Seifert fibered space and Waldhausen [1084,1967a,Invent. Math.], [1085,1967b,Invent. Math.] proved that K is irreducible. The simplest class to attack next is provided by Wagreich [1082,1972,Topology] where $\pi_1(K)$ is solvable and K has a circular plumbing graph.

Update: Every 3-manifold in $\mathcal{L}_k$ for all $k \geq 1$ is irreducible: furthermore it is determined by its fundamental group, except for lens spaces [834,Neumann,1981,Trans. Amer. Math. Soc.].

Problem 3.29 *To what extent does the Seifert matrix on F^4 determine the topology of the singularity? That is, does it determine K up to diffeomorphism? F up to diffeomorphism? Up to isotopy?*

Remarks: For $f : \mathbb{C}^n \to \mathbb{C}$, $n > 3$, the Seifert matrix determines the topology of the singularity completely [261,Durfee,1974,Topology].

Update: The conjecture fails for $n = 2$; P. DuBois & F. Michel [254, 1994,J. Algebraic Geom.] give two topologically distinct plane curve germs g_1 and g_2 which have isomorphic Seifert forms. Define

$$f_{r,s}(x,y) = ((y^2-x^3)^2-x^{s+6}-4yx^{(s+9)/2})((x^2-y^5)^2-y^{r+10}-4xy^{(r+15)/2})$$

and let $s \geq 11$ and $s \neq r+8$; then g_1 and g_2 can be any pair $f_{r,s}$ and $f_{s-8,r+8}$.

Then the conjecture for $n = 3$ fails [44,Artal-Bartolo,1991,C.R. Acad. Sci. Paris Sér. I Math.] using $f_{r,s}(x,y) - z^2$ and $f_{s-8,r+8}(x,y) - z^2$, for then the corresponding K are not homeomorphic.

Problem 3.30 A polynomial $G(t, z_0, \ldots, z_n)$ is called a μ-homotopy between G_0 and G_1 if $G_t(z_0, \ldots, z_n) = G(t, z_0, \ldots, z_n)$ has an isolated singularity with constant Milnor number μ for all t in a connected open set in C containing 0 and 1. Lê and Ramanujan ([642,1976,Amer. J. Math.], and also [641,Lê Dũng Tráng,1973]) show that for $n \neq 2$, the topology of G_0 and G_1 is identical, i.e., $(\varepsilon S^4, K_{G_0})$ is pairwise diffeomorphic to $(\varepsilon S^4, K_{G_1})$ and the bundles over S^1 are isomorphic.

Prove this for $n = 2$ (what is missing is the 4-dimensional h-cobordism theorem).

Update: Still open.

Problem 3.31 *Prove that $\sigma(F^4) \leq 0$ with equality only for a nonsingular point. (See [263,Durfee,1978,Math. Ann.].)*

Update: The inequality has been proved when f is a weighted homogeneous polynomial [1124,Xu & Yau,1993,J. Differential Geom.]. f is such a polynomial with weights (positive, rational numbers) (w_0, w_1, w_2)

if f can be expressed as a linear combination of monomials $z_0^{i_0} z_1^{i_1} z_2^{i_2}$ for which $i_0/w_0 + i_1/w_1 + i_2/w_2 = 1$. Then Xu & Yau prove the inequality

$$\sigma(F) \leq -\frac{\mu}{3} - \frac{2}{3}(\nu - 1),$$

where $\mu = b_2(F)$ and ν is the multiplicity of the singularity, which is $inf\{n \in \mathbb{Z}_+ : n \geq inf(w_0, w_1, w_2)\}$.

The higher dimensional version of Durfee's conjecture is that $(-1)^n \sigma(F^{2n}) \geq 0$ where $2n$ is the complex dimension of F (in our case $n = 1$).

Problem 3.32 *What is the Whitehead group $Wh(G)$ of a 3-manifold group $G = \pi_1(M^3)$ for M compact and irreducible?*

Remarks:

(1) Suppose G is infinite. Then $Wh(G) = 0$ if M^3 is sufficiently large, but is unknown otherwise, e.g., if M^3 has a finite cover which is sufficiently large. In fact, infinite 3-manifold groups belong to the class of finitely presented, torsion free groups, and there is no such group known with nonzero Whitehead group or reduced projective class group $\widetilde{K}_0(\mathbb{Z}(G))$.

(2) If G is finite, then $Wh(G)$, when known, is cyclic. (Incorrect, see below.)

Update:

(0) The Whitehead group of a free product is the direct sum of the Whitehead groups of the factors.

(1) For infinite G, $Wh(G) = 0$ modulo the Geometrization Conjecture. For, $Wh(G) = 0$ if M is hyperbolic [292,Farrell & Jones,1986, Ann. of Math.], or if M is an aspherical Seifert fibered space ([1091,Waldhausen,1978a,Ann. of Math.] for the Haken cases, [291,Farrell & Hsiang,1981b,J. London Math. Soc.] for cases with Euclidean base, and [877,Plotnick,1980,Comment. Math. Helv.] for the remaining cases). Also, the projective class group $\widetilde{K}_0$ vanishes, the second Whitehead group vanishes, and the higher Whitehead groups are torsion groups [293,Farrell & Jones,1987].

(2) If G is any finite periodic group (including finite 3-manifold groups), then

$$Wh(G) \cong \mathbb{Z}^{r-q} \times (\mathbb{Z}/2\mathbb{Z})^k$$

where:

r = number of real representations of G = number of conjugacy classes of unordered pairs (g, g^{-1});

q = number of rational representations of G = number of conjugacy classes of cyclic subgroups; and

k = number of conjugacy classes of cyclic subgroups $C < G$ such that

(1) $|C|$ is odd,
(2) the centralizer of C has non-abelian Sylow 2-subgroup, and
(3) there is no g in normalizer(C) which acts by $x \to x^{-1}$ (which implies that $|C| > 1$).

In particular, $Wh(G)$ is torsion free for all the known finite fundamental groups of 3-manifolds (see the list of subgroups of $SO(4)$ in Problem 3.37) except for $\mathbb{Z}/n\mathbb{Z} \times T_v$ of order $n \cdot 8 \cdot 3^v$ where $(n, 6) = 1$, in which case the torsion of $Wh(G)$ is $(\mathbb{Z}/2\mathbb{Z})^k$ for $k + 1$ equal to the number of divisors of $n \cdot 3^{v-1}$; and except for $\mathbb{Z}/n\mathbb{Z} \times H$, where H is a finite subgroup of $SU(2)$ of order coprime to n with non-abelian Sylow 2-subgroup. For these calculations, see [855, Oliver, 1988].

Problem 3.33 (A) (Thurston) *Does every finitely generated 3-manifold group G have a faithful representation in $GL(4, \mathbb{R})$?*

Remarks: If so, G is residually finite (part (B)). Since $PSL(2, \mathbb{C}) \hookrightarrow GL(4, \mathbb{R})$, this is true for hyperbolic 3-manifolds (see the Introduction).

(B) *Is G residually finite?*

Remarks: See [472, Hempel, 1976] for an extensive discussion. (Residually finite means that for each $g \neq 1$, there exists a representation λ of G to a finite group for which $\lambda(g) \neq 1$.) (Added in proof, March 1, 1977: Thurston has probably shown that $G = \pi_1(M^3)$ is residually finite if M or a finite cover is sufficiently large.)

(C) *Is G Hopfian?*

Remarks: Residually finite implies Hopfian (for finitely generated groups). Hopfian means that every epimorphism $G \to G$ is monic.

(D) *Is the Frattini subgroup of G trivial?*

Remarks: Yes, if G is a knot group; if the Frattini subgroup is nonzero for an orientable, compact, irreducible, sufficiently large 3-manifold, then the 3-manifold must be a Seifert fibered space (in which case the Frattini subgroup is cyclic) [33, Allenby, Boler, Evans, Moser, & Tang, 1979, Trans. Amer. Math. Soc.]. The Frattini subgroup F is the intersection of the maximal subgroups; equivalently, $g \in F$ if every set of generators of G containing g does not need g.

Update:

(A) Still open.

(B) G is residually finite if the 3-manifold is virtually geometric (implied by virtually Haken) [1051, Thurston, 1982, Bull. Amer. Math. Soc.] and [475, Hempel, 1987b].

(C) Residually finite implies Hopfian, so see (B).

(D) Yes when G is residually finite, so see (B).

Problem 3.34 (Smale) Conjecture: *Diff*$^{+}(S^3)$ *is homotopy equivalent to* $SO(4)$.

Remarks: $\pi_0(\mathrm{Diff}^{+}(S^3)/SO(4)) = 0$ [191, Cerf, 1968]. (Added in proof, March 1, 1977: A. Hatcher has announced a proof of the conjecture.)

Update: True, as proved by Hatcher [464, 1983, Ann. of Math.].

Problem 3.35 (Hatcher) *Compute* $\pi_0(\mathit{Diff}(L^3))$*, the space of diffeomorphisms of a lens space.*

Remarks: $\pi_0(\text{Diff}(\mathbb{RP}^3)) = \mathbb{Z}/2\mathbb{Z}$.

Update: Solved by Bonahon [129,1983b,Topology] and by Hodgson and Rubinstein [499,1985]. $\pi_0(\text{Diff}(L(p,q)))$ is isomorphic to

a) $\mathbb{Z}/2\mathbb{Z}$ if $p = 2$ or if $p \neq 2$ and $q \equiv \pm 1 \bmod p$,

and when $p \neq 2$ and $q \not\equiv \pm 1 \bmod p$, one has 3 cases.

b) $\mathbb{Z}/2\mathbb{Z} \oplus \mathbb{Z}/2\mathbb{Z}$ if $q^2 \equiv 1 \bmod p$.

c) $\mathbb{Z}/4\mathbb{Z}$ if $q^2 \equiv -1 \bmod p$.

d) $\mathbb{Z}/2\mathbb{Z}$ if $q^2 \not\equiv \pm 1 \bmod p$.

Moreover, homotopy implies isotopy for lens spaces, and, in fact, for geometric 3-manifolds other than hyperbolic ones (see the Remarks to Problem 3.68).

Problem 3.36 (W.-C. Hsiang) Conjecture*: $rank(\pi_1 Diff(L(p,q)) \otimes Q) \geq \frac{1}{2}(p-1)$.*

Remarks: For lens spaces of dim ≥ 5, the corresponding statement is true [512,Hsiang & Jahren,1983,Pacific J. Math.]. These elements are detected by Atiyah–Singer invariants.

Note that $\text{rank}(\pi_1 \text{Iso}(L(p,q)) \otimes \mathbb{Q}) \leq 2$ (W. C. Hsiang & B. Jahren). If the conjecture is true, it means that the diffeomorphism group of lens spaces is generally very different from the group of isometries (for any Riemannian metric) and the H-space of homotopy equivalences. Compare the case of 3-manifolds whose universal cover is hyperbolic 3-space (Problem 3.14).

Update: The conjecture is false by the following work of Ivanov. An oriented 3-manifold with finite fundamental group which contains a Klein bottle has a fundamental group of the form $\langle a, b \mid abab^{-1} = 1, a^{2m}b^{2n} = 1\rangle$, $(m,n) = 1$; this group determines the 3-manifold which can be denoted $Q(m,n)$. If $\text{Diff}_0(M)$ denotes the identity component of $\text{Diff}(M)$, then Ivanov shows that

- for $m, n \neq 1$, $\text{Diff}_0(Q(m,n)) \cong S^1$,
- for $n \neq 1$, $\text{Diff}_0(Q(1,n)) \cong S^1 \times S^1$,

in [522,Ivanov,1979a,Докл. Акад. Наук СССР], [523,Ivanov,1979b, Докл. Акад. Наук СССР], [528,Ivanov,1984,J. Soviet Math.] and in his Ph.D. thesis (Leningrad, 1980). Since $Q(1,n)$ is the lens space $L(4n,2n-1)$, this disproves the conjecture. Also see Problem 3.47.

Problem 3.37 (C. B. Thomas) *Classify free actions of finite groups on S^3.*

(A) Existence: If a finite group Γ acts freely on S^3, then Γ is, up to direct product with a cyclic group of coprime order, one of the following types:

(1) $\mathbb{Z}/r\mathbb{Z}$,

(2) an extension of a cyclic group by one of order 2^k,

(3) a generalized binary tetrahedral group T_v^* of order $8 \cdot 3^v$,

(4) the binary octahedral group O^* of order 48, or the binary icosahedral group of order 120,

(5) a certain split extension of $\mathbb{Z}/(2n+1)\mathbb{Z}$ by the binary dihedral group of order 8 (see [764,Milnor,1957a,Amer. J. Math.], and [644,Lee,1973,Topology]).

All these groups except for type (A–5) admit faithful representations in $SO(4)$, i.e., have free, linear actions on S^3 [1118,Wolf, 1967]. No nonlinear actions are known.

Question: *What about groups of type (A–5)?*

Remarks: If $SO(4) \simeq \mathrm{Diff}^+ S^3$ (see Problem 3.34), then these groups do not act freely on S^3 (C. B. Thomas).

(B) Uniqueness: The absence of nonlinear examples suggests:

(1) If Γ acts freely on S^3, is S^3/Γ homotopy equivalent to the quotient of a linear action?

Remarks: Yes for $\mathbb{Z}/r\mathbb{Z}$, the binary dihedral groups D_{8k}^* (in (5) above), and T_v^*, $v \geq 2$. Yes for the remaining groups if a sequence of obstructions in $\pi_i(\mathrm{Diff}^+ S^3/SO(4))$ vanish; so again there is a reduction to the Smale conjecture (Problem 3.34) [645,Lee & Thomas,1973,Bull. Amer. Math. Soc.], and [1046,Thomas,1977,Math. Ann.].

We can ask if S^3/Γ is simple homotopy equivalent or even homeomorphic to the quotient of a linear action, but this looks hard for arbitrary Γ. It may be easier for the following special cases:

(2) Compute the Reidemeister torsion of an arbitrary free $\mathbb{Z}/r\mathbb{Z}$-action; in particular, is every cyclic quotient simple homotopy equivalent to a lens space $L(r,q)$?

Remarks: Yes, trivially, for $r = 2,3,4,6$. At the Poincaré complex level, one can realize geometrically Reidemeister torsions distinct from those of the $L(r,q)$.

(3) Let $\mathbb{Z}/2^k\mathbb{Z}$ act freely on S^3. Is the quotient homeomorphic to a lens space?

Remarks: Yes for $k = 1,2,3$ [668, Livesay, 1960, Ann. of Math.], [926, Rice, 1968, Duke Math. J.], [928, Ritter, 1975, Trans. Amer. Math. Soc.].

(4) If D_8^* acts freely on S^3, is the quotient homeomorphic to the unique linear quotient?

Remarks: By (B–1) and the vanishing of $Wh(D_8^*)$ [569, Keating, 1973, Mathematika], the quotient is simple homotopy equivalent to the linear quotient, and its double cover is homeomorphic to $L(4,\pm 1)$ [Rice, ibid.].

Update: (A) The groups in (A–5) do not act freely on S^3 [950, Rubinstein, 1994]. Regarding the remark above, Hatcher proved that $SO(4) \simeq \mathrm{Diff}^+ S^3$, but Thomas' claim has not been proved.

The groups in (A–5) can be better described as follows: the binary dihedral group of order 8 is usually called the quaternionic group, Q_8, (which consists of $\pm 1, \pm i, \pm j, \pm k$). Let $2n+1 = a\cdot b\cdot c$ be a factorization of $2n+1$ into coprime integers, and let $G = Q(8,a,b,c) = \mathbb{Z}/a\mathbb{Z}\times\mathbb{Z}/b\mathbb{Z}\times \mathbb{Z}/c\mathbb{Z}\tilde{\times}Q_8$ where Q_8 acts by letting i,k act by inversion on the first factor while j acts trivially, j,k act by inversion on the second factor while i acts trivially, and i,j act by inversion on the third factor while k acts trivially. The G constructed from ordered, odd, coprime triples $\{a,b,c\}$ form the groups in (A–5). Each acts linearly on S^7.

It is possible to find a complex X, homotopy equivalent to S^3, on which G acts. Then there is an obstruction [1024, Swan, 1960, Ann. of Math.] to X being a finite complex, which may be non-zero (e.g. $Q(8,3,7,1)$) or zero (e.g. $Q(8,3,13,1)$) [762, Milgram, 1985].

When Swan's obstruction is zero, one can attempt to use the surgery program to find a 3-manifold which is a homology 3-sphere (in dimension 3 surgery can't control the fundamental group, only homology) on which G acts. There is a second obstruction in $L_3^p(G)$ which measures the obstruction to the existence of a free action on an open 4-manifold of the form $\mathbb{R} \times M^3$ with M^3 a homology 3-sphere. When the second obstruction vanishes, there is a third obstruction in $L_3^h(G)$ which measures the obstruction to a free action on an actual homology 3-sphere.

The first case where the second obstruction vanishes but the third does not is the group $Q(8,7,29,1)$, and the first example where the third obstruction (in $L_3^h(G)$) vanishes is $Q(8,41,223,1)$ [Milgram, ibid]. Thus $Q(8,41,223,1)$ acts on a homology 3-sphere M, but it is unknown what restrictions exist on $\pi_1(M)$. (For further number theory, see [77, Bentzen,1987,Proc. London Math. Soc.].) Thus, Rubinstein has shown that none of these groups can act on S^3, but Milgram has shown further that many of them cannot act on a homology 3-sphere, or even a finite 3-complex.

(B) Rubinstein [ibid.] has shown that the following groups act only via their faithful representations in $SO(4)$: any cyclic group of order $p2^k3^n$ for p prime; the groups listed in (A–2) and (A–3), and O^*, as well as their products with a cyclic group of coprime order. Thus the only remaining possibilities for groups which act non-linearly on S^3 are the remaining cyclic groups and direct products of a cyclic group of order prime to $2, 3, 5$ with the binary icosahedral group I^*.

The actions of the remaining cyclic groups could all be classified if one could prove the Smale conjecture for lens spaces, that is, the group of diffeomorphisms of $L(p,q)$ is homotopy equivalent to the group of isometries (see Problem 3.47. It is true for $L(2p,q)$, $q \neq 1$, [732, McCullough & Rubinstein,1989].

Earlier work on uniqueness includes: [949,Rubinstein,1979b,Trans. Amer. Math. Soc.], [948,Rubinstein,1979a,Math. Ann.], [822,Myers, 1981a,Topology].

Problem 3.38 (C. Giffen) The Smith Conjecture (K,r): *If K is a smooth, nontrivial knot in S^3, then K is not the fixed point set of a homeomorphism $h : S^3 \to S^3$ of least period r $(r > 1)$.*

Remarks: The conjecture first appeared in [1007,Smith,1939,Ann. of Math.]. Note that K must be smooth, for there are wild knots which are fixed point sets [783,Montgomery & Zippin,1954,Proc. Amer. Math. Soc.] and [88,Bing,1964,Ann. of Math.]; however, they bound wild disks. The conjecture is known for (K,r) if r is even [1090,Waldhausen, 1969,Topology], or if K is a torus knot [376,Giffen,1964], and [325, Fox,1967,Michigan Math. J.], or if K is a 2-bridge knot (and others) (Cappell & Shaneson), or if K is a cable knot, a cable braid, or a double of a knot (R. Myers), or other special cases of (K,r).

(A) Covering Conjecture (K,r): *If K is a smooth, nontrivial knot in a homotopy sphere Σ^3, then the r-fold cyclic branched covering $\Sigma_r(K)$ over K is not simply connected if $r > 1$.*

Remarks: This is equivalent to the conjecture that $\pi_1(\Sigma^3 - K)/\langle\mu^r\rangle \ncong Z/rZ$ for μ = meridian of K. Also the Covering Conjecture (K,r) for all K, r implies the Smith Conjecture (K,r) for all K,r and the converse holds if the Poincaré Conjecture is true. The Covering Conjecture is known for doubles of any nontrivial knot in Σ^3 [377,Giffen,1967,Illinois J. Math.], see also [401, Gordon,1977,Quart. J. Math. Oxford Ser. (2)] and cases covered in [576,Kinoshita,1958,Osaka Math. J.] and [322,Fox,1958,Osaka Math. J.].

(B) Let $F(S^3,K) = \{h : S^3 \to S^3 \mid h = \mathrm{id} \text{ on} K\}$ with compact open topology. The path component $F(S^3,K)_0$ of the identity contains no elements of finite order if K is nontrivial [377,Giffen, 1967,Illinois J. Math.]. Thus a counterexample to the Smith Conjecture implies $\pi_0(F(S^3,K))$ ($\cong F(S^3,K)/F(S^3,K)_0$) has an element of order $r > 1$. Let $\Gamma(S^3,K) \equiv \mathrm{Aut}(\pi_1(S^3 - K),\mu,l)/I_\mu$ be the group of automorphisms of $\pi_1(S^3 - K)$ which fix a meridian-longitude pair μ, λ, divided by the normal subgroup, I_μ, generated by the inner automorphism *conjugation-by-μ*. Then $\Gamma(S^3,K) \cong \pi_0 F(S^3,K)$ (Giffen).

Algebraic Conjecture (K,r): *$\Gamma(S^3,K)$ contains no element of least order r, $r > 1$.*

Remarks: Clearly this conjecture implies the Smith Conjecture (K,r), and the reverse is true for fibered knots; also there is some information about $\pi_0(F(S^3,K))$ (Giffen).

Update: The Smith Conjecture (for homotopy 3-spheres) was proved (and (A) was proved) in a joint effort by Thurston, and Meeks and Yau, with help from Bass, Shalen, Gordon and Litherland [69, Bass & Morgan, 1984a]. Furthermore, (B) follows from (A) using [1141, Zimmermann, 1982, Math. Zeit.].

Problem 3.39 Let $PL(M^3)$ be the group of PL homeomorphisms of a compact 3-manifold.

(A) **(Tollefson)** *Does $PL(M)$ have only finitely many conjugacy classes of finite cyclic subgroups of given order? (If yes, for S^3, then the Smith Conjecture holds.) What about finite subgroups?*

(B) **(Giffen)** *Suppose M admits no S^1 action. Is it possible for $PL(M)$ to contain an infinite torsion subgroup?*

(C) **(Thurston)** *Is there a bound to the order of finite subgroups of $PL(M)$? Note that this is true for surfaces F_g since a finite subgroup represents faithfully in $GL(2g-2, \mathbb{Z}/3\mathbb{Z}) = Aut(H_1(F_g; \mathbb{Z}/3\mathbb{Z}))$.*

Update:

(A) Still open, even for S^3 where the solution to the Smith conjecture does not cover free actions.

(B) (Mess) Still open. Even in the simple case $M = S^2 \times S^1 \# S^2 \times S^1$, if $PL(M)$ contained a finitely generated infinite torsion group, then it would have a finite index subgroup acting homotopically trivially, but apparently there is no way to exclude this possibility.

If M is aspherical and not a counterexample to the Geometrization Conjecture (see Problem 3.45), then M does not admit an action of a finitely generated, infinite torsion group unless M admits an S^1 action *in which case it is not known whether M admits an effective action of a finitely generated torsion group.* The proof of this statement requires standard methods and the Seifert fibered space conjecture (see Problem 3.5).

It can be asked instead whether $PL(M)$ can contain an infinite, locally finite, group of homeomorphisms (a group is locally finite if every finitely generated subgroup is finite). It can be shown, using the equivariant sphere theorem [744,Meeks, III & Yau,1982, Topology], that the problem reduces to the case of prime M. In that case, if M has finite fundamental group, then no more can be said as in (A). If M is closed, aspherical, but not atoroidal, then using the equivariant torus theorem together with rigidity theorems of Mostow [814,1973] and Prasad [889,1973,Invent. Math.], or the related topological theorems of Johannson on the mapping class groups of Haken manifolds [542,Johannson,1979], one can show that there is no action of an infinite locally finite group unless there is an S^1-action. One obtains the same conclusion if M is hyperbolic.

(C) It should be assumed here, as in (B), that M admits no S^1 action. If there is a counterexample, there is a prime counterexample, so we may assume that M is prime. Then there is a bound unless either M has finite fundamental group or is aspherical, atoroidal and a counterexample to the Geometrization Conjecture . (If M is oriented and has finite fundamental group and admits no S^1 action, then M is a counterexample to the Geometrization Conjecture (Problem 3.45).)

Problem 3.40 (Nielsen) *Let h be a homeomorphism of a 3-manifold M such that the nth iterate h^n is homotopic to the identity. When does there exist a map g homotopic to h such that $g^n =$ identity?*

Remarks: For $n = 2$ and M a closed, orientable 3-manifold fibered over S^1 with fiber F, there exist examples of such h which are not homotopic to any involution (for $F =$ torus, [919,Raymond & Scott,1977,Archiv. Math. (Basel)] and for genus $F > 1$, J. L. Tollefson).

For n a prime and M an orientable, closed 3-manifold fibered over S^1 with genus $F > 1$, there always exists a periodic g (homotopic to h) whenever either $H_1(M;\mathbb{Q}) = \mathbb{Q}$ or $n > 2$ and M is a Seifert fibered space [1058,Tollefson,1976,Trans. Amer. Math. Soc.]. (Added in proof, April 1, 1977): Consider the natural map

$$\mathrm{Diff}(M^3) \xrightarrow{\pi_0} \pi_0(\mathrm{Diff}(M^3)).$$

Thurston has shown that there exists an inverse ρ such that $\pi_0\rho = \mathrm{id}$ if M^3 is closed, irreducible, sufficiently large and $\pi_1(M^3)$ is infinite and contains no $\mathbb{Z} \oplus \mathbb{Z}$, i.e., if M^3 is hyperbolic (see Problem 3.14); furthermore, $\pi_0(\text{Diff}M^3)$ is finite under the same assumptions.

Update: To begin with, M should be assumed to be irreducible. Otherwise there are straightforward counterexamples such as: let $h : S^1 \times S^2 \to S^1 \times S^2$ rotate the 2-sphere once while traversing the circle; h^2 is homotopic (indeed isotopic) to the identity, but h is not isotopic to an involution (Mess).

The comment which was added *in proof* above, should simply state that the answer to the Problem is yes for hyperbolic manifolds (they do not need to be Haken).

However it was not known in 1977 that the kernel of the map $\pi_0(\text{Diff}M^3) \to Out(\pi_1 M)$ was trivial for all hyperbolic manifolds, or even that $\pi_0(\text{Diff}M^3)$ was finite; it is still not known although Gabai [360,1994b,Bull. Amer. Math. Soc.] has proved this for *almost all* hyperbolic 3-manifolds and it is expected that computations by N. Thurston will show that in fact there are no exceptions.

If M is elliptic, i.e. a quotient of S^3 by a finite linear action, then the answer is yes as a corollary of work of [499,Hodgson & Rubinstein,1985] and [129,Bonahon,1983b,Topology] for lens spaces, and [121,Boileau & Otal,1991,Invent. Math.].

If M is Haken, then there is an obstruction in $H^3(\mathbb{Z}/n\mathbb{Z};\mathcal{G})$ which vanishes iff h is realized by a periodic diffeomorphism g (here, $\mathcal{G}$ are twisted coefficients in the center of $\pi_1(M)$). The obstruction was introduced in [222,Conner & Raymond,1972].See [1140,Zieschang,1981] for a survey of work on this problem. See also [1141,Zimmermann,1982, Math. Zeit.]; Zimmermann's work allows for the case of a finite, not necessarily cyclic, group.

The obstruction vanishes if M is prime and aspherical, but not a Seifert fibered space by an application of the Seifert fibered space conjecture (see Problem 3.5), because the obstruction is trivial when the center of the fundamental group is trivial.

Problem 3.41 (Montesinos) *Is there a 3-manifold with an infinite number of nonequivalent involutions with S^3 as orbit space?*

Remarks: Given N, there exists a manifold with more than N such involutions (Montesinos).

Update: A negative answer would follow from the Orbifold Geometrization Conjecture (see Problem 3.46).

Problem 3.42 (J. L. Tollefson) *Is every periodic homeomorphism of a 3-manifold homotopic to a periodic PL homeomorphism of the same period?*

Remarks: Yes for homeomorphisms which are locally nice so that the quotient is locally triangulable, for then we can triangulate the 3-manifold so that the original homeomorphism is PL. Tameness of the fixed point set often implies this niceness [776,Moise,1979,Trans. Amer. Math. Soc.], [777,Moise,1980,Trans. Amer. Math. Soc.].

Update: (Mess) The answer is yes for a compact manifold with (possibly empty) boundary if the 3-manifold is not a counterexample to the Geometrization Conjecture. If the manifold is hyperbolic, use Mostow rigidity. If the manifold is Haken the result of [1141,Zimmermann,1982, Math. Zeit.] applies. If the manifold is aspherical but neither hyperbolic nor Haken, it must be a Seifert fibered space with only 3 exceptional fibers. It is an exercise in homotopy theory to show that the periodic homeomorphism must be homotopic to a periodic homeomorphism which preserves the geometric structure. If the manifold has finite fundamental group and is not a counterexample to geometrization, it is an exercise in homotopy theory to show that the periodic homeomorphism is homotopic to an isometry of the spherical structure.

If the manifold is not prime, an exercise in homotopy theory shows that the periodic homeomorphism is homotopic to one which permutes a certain collection of spheres, and so the problem reduces to the previous cases.

There is, however, a more satisfying point of view towards the question. Claim: an action of a finite group G of homeomorphisms on a 3-manifold (possibly noncompact) can be approximated by a PL action of the same group.

Sketch of proof of claim: First assume that the action preserves orientation. Triangulate the complement of the singular locus in the quotient space. Use this triangulation to construct a G invariant handlebody (perhaps of infinite genus in the noncompact case) around the upstairs singular set. Then use the equivariant loop theorem to cut the handlebody up into little pieces, in each of which a small alteration makes the action PL (in a new PL structure). By Moise's approximation theorems, the new PL structure is the pullback of the old one by a homeomorphism arbitrarily close to the identity, so the action of G can be approximated by a G action which is PL in the original structure. With some more work, the nonorientable case can be dealt with too. This settles this problem in the affirmative.

Problem 3.43 (Casson) *Does every homology 3-sphere H with an orientation reversing diffeomorphism have Rohlin invariant zero?*

Remarks: If not, then $H\#H \cong H\#(-H)$, which bounds an acyclic manifold, giving an element of order two in θ^3_H, the homology bordism classes of homology 3-spheres; this triangulates higher dimensional manifolds (see Problem 4.4). Brieskorn homology 3-spheres $\Sigma(p,q,r)$ do not even admit orientation reversing homotopy equivalences. Probably the homology 3-spheres arising as irregular branched covers of amphicheiral knots have Rohlin invariant zero; cyclic covers do have invariant zero.

Update: Yes. Casson's invariant [30,Akbulut & McCarthy,1990] is zero on such homology 3-spheres H and is also an integral lift of the Rohlin invariant.

Problem 3.44 (Meeks) Let M^3 be a closed, orientable 3-manifold with universal cover $\mathbb{R}^3$; then $G = \pi_1(M)$ acts on $\mathbb{R}^3$ with quotient M^3. Let Γ_1 and Γ_2 be graphs in $\mathbb{R}^3$ which are invariant under G, with $\pi_1(\mathbb{R}^3 - \Gamma_i)$ free, $i = 1, 2$. Let N_1 and N_2 be equivariant regular neighborhoods of Γ_1 and Γ_2.

Conjecture: *∂N_1 is isotopic to ∂N_2.*

Update: The conjecture is true and was proved by Frohman and Meeks [348,Frohman & Meeks, III,1990,Bull. Amer. Math. Soc.]

NEW PROBLEMS

Problem 3.45 Geometrization Conjecture (Thurston): *An orientable, compact, irreducible 3-manifold M with incompressible boundary can be decomposed canonically along incompressible tori into geometric pieces.*

Remarks: There are eight possible model geometries:

- S^3 with isometry group $SO(4)$,
- $\mathbb{R}^3$ with isometry group consisting of translations and rotations,
- H^3 with isometry group $PSL(2, \mathbb{C})$,
- and five others. (A good general reference is [979,Scott,1983a, Bull. London Math. Soc.]).

A closed M^3 is geometric if its universal cover is one of the eight models and $\pi_1(M^3)$, as the deck transformations of the model, is isomorphic to a discrete subgroup of the corresponding group of isometries.

If ∂M is non-empty, then M is geometric means that int$M$ has universal cover equal to one of the models and $\pi_1(M^3)$ is again a discrete group of isometries.

M^3 has a canonical collection (possibly empty) of disjoint, imbedded, incompressible tori; when non-empty, canonical is with respect to the condition that each component of $M - tori$ is either a Seifert fibered space or is atoroidal (which means that each $\mathbb{Z} \oplus \mathbb{Z}$ in $\pi_1(N)$ is conjugate into $\pi_1(\partial N)$) [538,Jaco & Shalen,1979], [542,Johannson,1979], and [978,Scott,1980,Amer. J. Math.].

The Geometrization Conjecture then states that the components are geometric, or if the collection of tori is empty, then M itself is geometric.

The Geometrization Conjecture naturally breaks into three parts according to the structure of $\pi_1(N)$:

(I) Conjecture: *If $\pi_1(N)$ is finite, then $N = B^3$ or $N = S^3/\pi_1(N)$ where $\pi_1(N)$ acts on S^3 as a subgroup of $SO(4)$. (Compare Problem 3.37 and see its Update.)*

Remarks: The geometric 3-manifolds of the form $S^3/\pi_1(N)$ are either S^3, lens spaces, or Seifert fibered spaces over S^2 with 3 exceptional fibers of type (2,2,n), (2,3,3), (2,3,4), or (2,3,5) (i.e. $1/p+1/q+1/r > 1$). (Of course, S^3 is a Seifert fibered space with no exceptional fibers, and lens space have two (or less) exceptional fibers, so (I) conjectures that finite π_1 implies N is Seifert fibered.)

(II) Theorem: $\pi_1(N)$ contains $\mathbb{Z} \oplus \mathbb{Z}$ iff either N contains an incompressible torus or N is a Seifert fibered space over S^2 with 3 exceptional fibers. A discussion of this case can be found in the Update to Problem 3.5.

(III) Conjecture: *If $\pi_1(N)$ is infinite and does not contain a $\mathbb{Z} \oplus \mathbb{Z}$, then N is hyperbolic.*

For a statement of the non-orientable case of the conjecture, see [979, Scott,1983a,Bull. London Math. Soc.].

Problem 3.46 (Mess) Definition: a smooth 3-dimensional *orbifold* is a paracompact, Hausdorff, stratified space which is covered by open sets, each homeomorphic to $\mathbb{R}^3/G$ or else to a quotient $\mathbb{R}^3_{x_3 \geq 0}/G$ of a half space, where G is a finite group acting linearly on $\mathbb{R}^3$ ($G \subset O(3)$) (and preserving the plane $\{x_3 = 0\}$ in the half space case) with the obvious gluing conditions satisfied; moreover each subset $\mathbb{R}^3/G$ must be a stratified space in which each stratum is labeled by the conjugacy class in $O(3)$ of the stabilizer of a point in $\mathbb{R}^3$ which projects to a point of the stratum, and similarly for the case with boundary. In the orientable case, G can be a cyclic or dihedral group, or can be a subgroup of the full group of isometries of one of the five Platonic solids; in the non-orientable case, G can also be certain $\mathbb{Z}/2\mathbb{Z}$-extensions of the above groups.

There is a (now standard) notion of covering space for orbifolds. A *bad orbifold* is one for which the universal orbifold covering space is *not* a manifold. An orbifold that is not bad is *good* and a good orbifold is *very good* if there is a finite cover which is a manifold. All compact 2-orbifolds which are good are very good, but, e.g. S^2 with one singular point of order $p > 1$ is bad.

Following Thurston, and Bonahon & Siebenmann in [132,1987,Math. Ann.], we make further definitions.

A 2-orbifold is respectively *spherical, discal, toric* or *annular* if it is the quotient of a sphere, disc, torus or annulus by a finite group of isometries. A compact 3-orbifold is *irreducible* if it contains no bad 2-suborbifold and in addition every spherical 2-suborbifold bounds a discal 3-suborbifold, which by definition is the quotient of a 3-disc by a finite group of isometries of the standard metric. The prime decomposition theorems of Kneser and Schubert for 3-manifolds and links generalize to show that every compact 3-orbifold which contains no bad 2-orbifold can be decomposed by splitting along spherical 2-orbifolds and capping the boundaries by discal 3-orbifolds (see page 445 of [Bonahon & Siebenmann, ibid.]). The collection of spherical 2-orbifolds is not canonical but the resulting collection of irreducible 3-orbifolds is.

The Orbifold Geometrization Conjecture: *[1051, Thurston, 1982, Bull. Amer. Math. Soc.]*

(i) Any compact, irreducible 3-orbifold without boundary can be cut along incompressible, π_1-injective, toric 2-orbifolds to get pieces which are geometric.

(ii) Any compact, irreducible 3-orbifold with incompressible boundary can be decomposed along a canonical family of incompressible toric and annular 2-orbifolds to get pieces which are geometric.

(iii) Any compact irreducible 3-orbifold with boundary can be decomposed along a family of incompressible discal, toric and annular 2-orbifolds into pieces which are geometric. (In this case the collection of pieces is uniquely determined by the orbifold but the family of 2-orbifolds is not canonical nor even unique up to isotopy. The collection of discal orbifolds is any collection minimal with respect to the property that cutting along the discal orbifolds leads to a 3-orbifold with incompressible boundary.)

Preliminary Remarks: In fact this is a generalization of the Geometrization Conjecture, and those with truly logical minds might wish the two problems to be stated as a single problem. (ii) is apparently more general than (i) but follows from (i) by a doubling trick due to Thurston. Similarly, by the doubling trick, (iii) is not essentially more general than the conjunction of (i) and the theorem asserting the *prime decomposition* of 3-orbifolds without bad 2-suborbifolds along spherical 2-orbifolds.

The collection of incompressible 2-orbifolds is canonical, and its existence is not conjectural: it is established in [Bonahon & Siebenmann, ibid.] The pieces are orbifolds with boundary. A piece is either

(a) a spherical 3-orbifold without boundary, or

(b) the quotient of $S^2 \times \mathbb{R}$ by a cocompact group of isometries, or

(c) the quotient of Nil by a cocompact group of isometries, or

(d) the quotient of $\mathbb{R}^3$ by a cocompact group of isometries, or the quotient of $\mathbb{R}^2 \times I$ by a cocompact group of isometries, or

(e) the quotient of a subset of $H^2 \times \mathbb{R}$ which has totally geodesic boundary by a cocompact group of isometries, or

(f) the quotient of the universal cover of $SL(2, \mathbb{R})$ by a cocompact group of isometries of a left invariant Riemannian metric with maximal symmetry, or

(g) the quotient of *Solv* by a cocompact group of isometries, or

(h) the orbifold associated with a non-elementary, geometrically finite group acting on hyperbolic 3-space. (In (h), let Γ be the group, and let $\Omega(\Gamma)$ be its discontinuity domain, a subset of the sphere at infinity. The orbifold is obtained from the quotient $(H^3 \cup \Omega(\Gamma))/(\Gamma)$ by attaching toric orbifolds to the cusps so as to obtain a compact orbifold with boundary.)

Note that the geometry on a geometric piece is not necessarily unique: a manifold with boundary which admits a $H^2 \times \mathbb{R}$ structure with totally geodesic boundary also admits a geometric structure modeled on the universal cover of $SL(2, \mathbb{R})$, and it is a matter of convention whether one should use an $H^2 \times \mathbb{R}$ structure with geodesic boundary or a non-compact structure with finite volume. Further, the geometry is not canonical: there is a Teichmüller space of possibilities in the case where the geometry is modeled on $H^2 \times \mathbb{R}$ and there is a finite dimensional moduli space in the case where the geometry is Euclidean.

(Editor's note: A proof of this conjecture when the singular set is non-empty and 1-dimensional was announced in Thurston's Bulletin article above. However, in the intervening 14 years, no proof has appeared nor is the editor aware of anyone else who claims to know a proof. Despite Thurston's reputation for accuracy, it seems healthy for this subject to list the conjecture as an open problem and to encourage the development of an up-to-date, elegant proof.)

Remarks: The main application of the conjecture would be to show that a compact orientable 3-manifold with a finite group of symmetries could be decomposed along surfaces to give pieces which are geometric. In fact (using the result of McCullough & Miller mentioned below) every compact good 3-orbifold with a decomposition into geometric pieces is the quotient of a compact 3-manifold by a finite group action which preserves a family of surfaces which decompose the 3-manifold into pieces on which the group acts geometrically.

There are related results: [1034,Takeuchi,1988,Trans. Amer. Math. Soc.] and [1033,Takeuchi,1991,Trans. Amer. Math. Soc.]; [132,Bonahon & Siebenmann,1987,Math. Ann.] in which the torus decomposition is achieved; [255,Dunbar,1988,Topology Appl.].

Thurston's original approach was to start with a hyperbolic structure on the complement of the singular set and then deform it. A detailed outline of this approach is given in notes of Hodgson, and in the case that the singular set is a 1-manifold, additional details appear in [1139, Zhou,1990].

Outside the framework of Thurston's program, but related to the Conjecture, are [157,Buchdahl, Kwasik, & Schultz,1990,Invent. Math.] and [634,Kwasik & Schultz,1992a,Invent. Math.] which show that an action of the icosahedral group on $\mathbb{R}^3$ has a fixed point (this is an important complement to results in Chapters X, XI of [69,Bass & Morgan, 1984a]).

A major result on the Orbifold Conjecture is [741,Meeks, III & Scott, 1986,Invent. Math.] which shows that every action of a finite group on a Seifert fibered space with base orbifold having infinite fundamental group, and every action of a finite group on a product $M^2 \times I$, where M is a surface, respects the geometry. In addition, see [297,Feighn,1989, Trans. Amer. Math. Soc.] [499,Hodgson & Rubinstein,1985] in which involutions of lens spaces with one dimensional fixed point set are shown to be standard.

Further papers (this is surely not an exhaustive list of papers on the Orbifold Conjecture) are [714,Matsumoto & Montesinos-Amilibia,1991, Tokyo J. Math.], [731,McCullough & Miller,1989,Topology Appl.] and [729,McCullough & Miller,1986a] which shows that an orbifold with a geometric decomposition is very good and has residually finite fundamental group (the orbifold analogue of [475,Hempel,1987b]).

Problem 3.47 (Mess & Rubinstein (Generalized Smale Conj.)) A strong version of the Geometrization Conjecture would give information about the diffeomorphism group of a 3-orbifold. For comparison, instead of just proving that each compact 2 manifold admits a metric of constant curvature, one can also show that the inclusion of the identity component of the isometry group (of any one of its constant curvature metrics) into the identity component of the diffeomorphism group is a homotopy equivalence.

It is folk knowledge that the conjectures below would follow naturally if there was a natural flow on the space of metrics, given by a parabolic differential equation or otherwise, which deformed the space of all metrics on a given manifold, which admits one of the standard geometries, to the space of metrics locally isometric to one of the eight model geometries. Although there are partial results on conjectures A and B below, a uniform approach is desirable.

(A) *Suppose that M is a compact, boundaryless, 3-orbifold, modeled on one of Thurston's eight geometries.*

(A1) Conjecture: *if M is modeled on S^3, H^3, or Solv, then the inclusion of the isometry group $I(M)$ of M into Diff(M) is a homotopy equivalence.*

Remarks: This conjecture is not true for $M = S^2 \times S^1$ (because the diffeomorphism which spins $S^2 \times \theta$ by θ is not isotopic to an isometry ($\theta \in S^1$)). In fact, Diff($S^2 \times S^1$) is homotopy equivalent to $O(2) \times O(3) \times \Omega O(3)$ [463,Hatcher,1981,Proc. Amer. Math. Soc.].

In the case that M is modeled on S^3, the conjecture is true for S^3 [464,Hatcher,1983,Ann. of Math.], Ivanov in [522,Ivanov,1979a, Докл. Акад. Наук СССР], [523,Ivanov,1979b,Докл. Акад. Наук СССР], [528,Ivanov,1984,J. Soviet Math.] proved it is true for most of the cases when M contains a Klein bottle (see Problem 3.36), and McCullough & Rubinstein have claimed the conjecture for the remaining cases (when M contains a Klein bottle), and for the 3-manifolds with fundamental group equal to a direct sum of a binary dihedral group with a coprime cyclic group (see Problem 3.37).

If M is modeled on H^3 and has nonempty 1 dimensional singular set, it is known that the identity component of Diff(M) is contractible, but the corresponding statement is not known if M has isolated singular points. Mostow rigidity ([814,Mostow,1973], and for the non-compact case, [889,Prasad,1973,Invent. Math.]) shows that π_0Diff(M) maps onto $Out(\pi_1(M))$ for M a finite volume hyperbolic manifold, and in fact $Out(\pi_1(M))$ is a retract of π_0Diff(M).

If M is a Haken (or more generally, a non-compact, finite volume) hyperbolic 3-manifold then [1089,Waldhausen,1968c,Ann. of Math.], together with [Mostow, ibid.] and [Prasad, ibid.], implies that $\pi_0(\text{Diff}(M)) = Out(\pi_1(M))$. If M is a hyperbolic 3-manifold then $\pi_0(\text{Diff}(M)) = Out(\pi_1(M))$ holds provided that a closed geodesic satisfies a conjecturally true geometric condition, e.g. is the core of a (log3)/2 tube [360,Gabai,1994b,Bull. Amer. Math. Soc.], [361,Gabai,1995a]. This condition is always true after passing to a finite sheeted covering space [359,Gabai, 1994a,J. Amer. Math. Soc.]. Mostow [ibid.] and Prasad [ibid.] show that a hyperbolic metric on a finite volume (not necessarily closed) hyperbolic 3-manifold is unique up to an isometry homotopic to id_M. The additional fact $\pi_0(\text{Diff}(M)) = Out(\pi_1(M))$ implies the stronger statement that hyperbolic metrics are unique up to isotopy or equivalently the space of hyperbolic metrics is path connected.

The conjecture is known if M is a Solv manifold or Solv orbifold, by the combination of the fact that a Solv manifold is necessarily Haken, the work of Meeks & Scott [741,1986,Invent. Math.] and the work of Hatcher and Ivanov on spaces of diffeomorphisms, *except* if M is an orbifold with isolated singular points, in which case it is not known, although the induced map on components is a bijection by the work of Meeks & Scott [ibid.].

For an aspherical space X, the space $Eq(X)$ of homotopy self-equivalences of X is homotopy equivalent (as an H-space) to the discrete group $Out(\pi_1(X))$, except when $\pi_1(X)$ has a non-trivial center in which case each component has the homotopy type of a $K(\mathbb{Z}(\pi_1(X),1))$ and the group of components is $Out(\pi_1(X))$. For a hyperbolic manifold of finite volume (which is not necessarily closed) and dimension > 2, $Out(\pi_1(X))$ is canonically isomorphic to the isometry group of the manifold by Mostow–Prasad rigidity [ibid.]. Hatcher showed that the PL homeomorphism group of a Haken manifold M is naturally homotopy equivalent to the H-space $Eq(M)$; Ivanov proved the same for the diffeomorphism group modulo the Smale conjecture, (see [461,Hatcher, 1976,Topology], [521,Ivanov,1976,Issledovaniya po Topologii 2, Zapiski Nauchnykh Seminarov Leningradskogo Otdeleniya Matematicheskiĭ in-ta. im. Steklova]).

(A2) Conjecture: *if M is modeled on $\mathbb{R}^3$ or Nil then the inclusion of the group Aff(M) of affine maps from M to itself into the group Diff(M) is a homotopy equivalence.*

Remarks: This conjecture naturally breaks into two parts: that the induced map on components is an isomorphism (which is known by the work of Meeks & Scott [ibid.]), and that the inclusion of identity components, $\text{Aff}_0 M \hookrightarrow \text{Diff}_0(M)$, is a homotopy equivalence. Here the essential difficulty occurs when M is a non-Haken manifold or an orbifold with isolated singular points.

(A3) Conjecture: *if M is modeled on $H^2 \times \mathbb{R}$ or $\widetilde{SL}(2,\mathbb{R})$, then the space of Seifert fibered space structures on M is contractible. Equivalently, the inclusion of the fiber preserving diffeomorphisms into Diff(M) is a homotopy equivalence. Equivalently, the inclusion of the identity component of the isometry group of M into the identity component of the diffeomorphism group of M is a homotopy equivalence.*

Remarks: The conjecture is true if M is a Haken manifold, by reformulations of results of Hatcher and Ivanov [ibid.].

The conjecture that the space of Seifert fibered space structures is contractible is made only for these two geometries because it is not true in general. For example, the SFS structures on T^3 are not connected; the the space of SFS structures on S^3 which are isomorphic to the Hopf fibration has the homotopy type of S^2; on some Nil manifolds or Euclidean manifolds the SFS structure is not unique up to homotopy.

(B) *This is a reformulation of the conjecture: let G be a finite group acting isometrically on a geometric manifold M. Then the inclusion of the {identity component of the centralizer of G in the isometry group of M} into the {identity component of the centralizer of G in the diffeomorphism group of M} is a homotopy equivalence.*

Remarks: There are good reasons why one should only look at the identity component. One reason is related to the fact that for certain hyperbolic 2 dimensional orbifolds the mapping class group has a nontrivial finite kernel in its action on the Teichmüller space of the orbifold, and the other is the existence of Dehn twists (in the sense of Johannson) on vertical tori in Seifert fibered spaces which have hyperbolic bases with first Betti number positive.

Conjecture (A) is equivalent to the conjunction of (B) and the statement that any smooth action of a finite group on a geometric manifold M is conjugate by a diffeomorphism isotopic to the identity to an action which is geometric with respect to a geometry which lies in the same component of the space of geometric structures as the original geometry on M. (The awkwardness of the statement is inevitable given the existence of Teichmüller space and our incomplete knowledge on the problem of whether homotopic diffeomorphisms are isotopic (see Problem 3.95).)

Problem 3.48 (Kontsevich) Let M^3 be a connected 3-manifold with non-empty boundary, and let BDiff(M rel ∂M) be the classifying space for the group of diffeomorphisms of M which are the identity on ∂M.

Conjecture: *BDiff(M rel ∂M) has the homotopy type of a finite CW-complex.*

A particularly interesting case occurs when M is equal to the complement of the unlink of n components in S^3, $n > 1$, that is, the connected sum of n solid tori.

Remarks: (McCullough) The conjecture does not hold for closed 3-manifolds. Suppose M is a closed Haken manifold. By [461,Hatcher, 1976,Topology], each component of Diff(M) is contractible, hence BDiff(M) is a $K(\pi_0(\text{Diff}(M)), 1)$. But whenever $\pi_0(\text{Diff}(M))$ contains torsion (e. g. is a nontrivial finite group, as happens in many hyperbolic examples), there can be no finite-dimensional complex which is a

$K(\pi_0(\text{Diff}(M)), 1)$. However, the conjecture is true for $M \# B^3$ (M punctured) when M is Haken (to see this, use the homotopy exact sequence of the fiber bundle with base space $Emb(B^3, M^3)$ and total space $\text{Diff}(M)$).

Problem 3.49 (McCullough) Generalizing the construction of Dehn twist homeomorphisms of 2-manifolds, define a *Dehn homeomorphism* as follows: Let $(F^{n-1} \times I, \partial F^{n-1} \times I) \subset (M^n, \partial M^n)$, where F is a connected codimension-1 submanifold, and $F \times I \cap \partial M = \partial F \times I$. Let $\langle \phi_t \rangle$ be an element of $\pi_1(Homeo(F), 1_F)$, i.e. for $0 \le t \le 1$, ϕ_t is a continuous family of homeomorphisms of F such that $\phi_0 = \phi_1 = 1_F$. Define $h \in \pi_0(Homeo(M)) = \mathcal{H}(M)$ by

$$h = \begin{cases} h(x,t) = (\phi_t(x), t) & \textit{if } (x,t) \in F \times I \\ h(m) = m & \textit{if } m \notin F \times I \end{cases}$$

Note that when $\pi_1(Homeo(F))$ is trivial, a Dehn homeomorphism must be isotopic to the identity. Define the *Dehn subgroup* $\mathcal{D}(M)$ of $\mathcal{H}(M)$ to be the subgroup generated by Dehn homeomorphisms.

The following table lists $\pi_1(Homeo(F))$ for connected 2-manifolds, and the names of the corresponding Dehn homeomorphisms of 3-manifolds.

F	$\pi_1(Homeo(F))$	*Dehn homeomorphism*
$S^1 \times S^1$	$\mathbb{Z} \times \mathbb{Z}$	*Dehn twist about a torus*
$S^1 \times I$	$\mathbb{Z}$	*Dehn twist about an annulus*
D^2	$\mathbb{Z}$	*twist*
S^2	$\mathbb{Z}/2\mathbb{Z}$	*rotation about a sphere*
$\mathbb{RP}^2$	$\mathbb{Z}/2\mathbb{Z}$	*rotation about a projective plane*
Klein bottle	$\mathbb{Z}$	*Dehn twist about a Klein bottle*
Möbius band	$\mathbb{Z}$	*Dehn twist about a Möbius band*
$\chi(F) < 0$	*0*	

(A) Dehn Subgroup Conjecture: *Let M be a compact 3-manifold. Then $\mathcal{D}(M)$ has finite index in $\mathcal{H}(M)$.*

Remarks: For M orientable, (A) is true if it is true for irreducible manifolds [726,McCullough,1986]. Johannson (Corollary 27.6 in [542,Johannson,1979]) proved (A) for boundary-irreducible Haken manifolds, and this was extended to all Haken manifolds in [729, McCullough & Miller, 1986a].

Denote by $\mathcal{D}_{>0}(M)$ the subgroup of $\mathcal{D}(M)$ generated by Dehn homeomorphisms using D^2, S^2, and $\mathbb{RP}^2$ (the surfaces of positive Euler characteristic).

By an argument similar to the proof of Proposition 1.2 of [727, McCullough, 1990], one can prove that if ∂M is incompressible, then $\mathcal{D}_{>0}(M)$ is a finite abelian group.

When the boundary of M is compressible, the following results were proved in [725, McCullough, 1985, Topology]:

- If ∂M is almost incompressible, then $\mathcal{D}_{>0}(M)$ is a finitely generated abelian group (almost incompressible means that in each boundary component F of M, there is at most one simple closed curve up to isotopy that bounds a disc in M but does not bound a disc or Möbius band in F);
- If ∂M is not almost incompressible, then $\mathcal{D}_{>0}(M)$ is infinitely generated and nonabelian.

(B) Kernel Conjecture: *$\mathcal{D}_{>0}(M)$ has finite index in the kernel of $\mathcal{H}(M) \to Out(\pi_1(M))$.*

Remarks: In general, $\mathcal{D}_{>0}(M)$ need not equal the kernel, as shown by the example of reflection in the fibers of an I-bundle. For orientable M containing no fake 3-cells, (B) is true if it is true for irreducible M [727, McCullough, 1990]. The main case in which (B) is unknown is when M is irreducible, aspherical and not sufficiently large, although even here some cases are known [360, Gabai, 1994b, Bull. Amer. Math. Soc.], [361, Gabai, 1995a].

Define $Out_\partial(\pi_1(M))$ to be the subgroup of $Out(\pi_1(M))$ consisting of the automorphisms ϕ such that for every boundary component F of M, there exists a boundary component G so that $\phi(i_\#(\pi_1(F)))$ is conjugate in $\pi_1(M)$ to $j_\#(\pi_1(G))$, where $i: F \to M$ and $j: G \to M$ are the inclusions. This subgroup contains the image of $\mathcal{H}(M) \to Out(\pi_1(M))$.

(C) Image Conjecture: *The homomorphism $\mathcal{H}(M) \to Out_{\partial M}(\pi_1(M))$ has image of finite index.*

Remarks: In general, the image is not all of $Out_{\partial M}(\pi_1(M))$. Again, (C) is true if it is true for irreducible manifolds [727,McCullough,1990].

(B) and (C) combine to give the following conjecture, where *almost exact* means that images have finite indexes in kernels (rather than equaling kernels as in exactness).

(D) Almost Exactness Conjecture: *Let M be a compact 3-manifold. Then the sequence*

$$1 \to \mathcal{D}_{>0}(M) \to \mathcal{H}(M) \to Out_{\partial}(\pi_1(M)) \to 1$$

is almost exact.

(E) Finiteness Conjecture: *Let M be closed, irreducible, but not sufficiently large. Then $\mathcal{H}(M)$ is finite.*

Remarks: Note that (E) follows from the Dehn Subgroup Conjecture (A). (E) has been proven by Gabai for many aspherical but not sufficiently large manifolds [360,Gabai,1994b,Bull. Amer. Math. Soc.], [361,Gabai,1995a]. Also, $\mathcal{H}(M)$ should be finite when $M = S^3/G$ for $G \subset SO(4)$ for then it is conjectured that $\mathcal{H}(M) = \pi_0(Isom(M))$ (Problem 3.47).

(F) Finite Presentation Conjecture: *$\mathcal{H}(M)$ is finitely presented.*

Remarks: For orientable M, (F) is true if it is true for irreducible manifolds [465,Hatcher & McCullough,1990], and is known in many cases, for example lens spaces [129,Bonahon,1983b,Topology] and Haken manifolds [408,Grasse,1989,Topology Appl.], [1092, Waldhausen,1978b].

Recall that a group is said to have a property virtually if some finite-index subgroup has the property.

(G) Virtual Geometric Finiteness Conjecture: *Let M be a compact 3-manifold. Then*

(i) $\mathcal{H}(M)$ is virtually torsion free.

(ii) $\mathcal{H}(M)$ is virtually of finite cohomological dimension.

(iii) $\mathcal{H}(M)$ *is virtually geometrically finite (a group is geometrically finite if it is the fundamental group of a finite aspherical complex).*

Remarks: Since (iii) implies (ii) and (ii) implies (i), this is really a sequence of three successively stronger conjectures. All hold for compact 2-manifolds ([442,Harer,1986,Invent. Math.], [443, Harer,1988], [452,Harvey,1981], [451,Harvey,1979]), for Haken manifolds [728,McCullough,1991,J. Differential Geom.], and hold trivially in the cases where the mapping class group is known to be finite. For non-irreducible 3-manifolds, the following is a preliminary question. Define the *rotation subgroup* $\mathcal{R}(M)$ to be the subgroup generated by rotations about 2-spheres and 2-sided projective planes in M; it is a finite normal abelian subgroup of $\mathcal{H}(M)$. Is there a finite-index subgroup of $\mathcal{H}(M)$ that intersects $\mathcal{R}(M)$ trivially? If not, replace $\mathcal{H}(M)$ by $\mathcal{H}(M)/\mathcal{R}(M)$ in the conjecture.

Problem 3.50 (Mess) Let M^3 be a closed, hyperbolic 3-manifold (or just a closed, atoroidal, irreducible, aspherical 3-manifold). Here are some variants of a question of Waldhausen, namely:

(A1) *Does* M^3 *have a finite cover which is Haken?*

Remarks: This is Problem 3.2 for hyperbolic 3-manifolds, and the Update lists positive results.

(A2) *If* M^3 *is Haken, does* M *have a finite cover* $\widetilde{M}$ *with positive first betti number* $b_1 = \mathrm{rank} H_1(\widetilde{M};\mathbb{Q})$*?*

(A3) *If* $b_1(M) > 0$*, does* M *have finite covers with arbitrarily large* b_1*?*

(B1) *Does* $\pi_1(M)$ *contain a subgroup* G *isomorphic to the fundamental group of a (necessarily) hyperbolic closed surface* F_g*? (Equivalently, does* M *contain an incompressible immersed closed surface?)*

Remarks: Yes if there exists an immersed surface with principal curvatures everywhere ≤ 1 in absolute value.

(B2) *If* $G = \pi_1(F_g) \subset \pi_1(M)$*, does* M *have a finite cover which is Haken? has* $b_1 > 0$*? has* b_1 *arbitrarily large?*

(B3) *If $G = \pi_1(F_g) \subset \pi_1(M)$, does M have a finite cover $\widetilde{M}$ in which G is the fundamental group of an imbedded surface? or at least a finite index subgroup of G is the fundamental group of an imbedded surface?*

(C) *Does $\pi_1(M)$ contain a subgroup $G \cong \pi_1(F_g)$ such that G is separable in $\pi_1(M)$?*

Remarks: If so, then M has finite covers with arbitrarily large collections of disjoint, homologically independent surfaces whose fundamental groups are finite index subgroups of G. (G *separable in* $\pi_1(M)$ means that for any $x \in \pi_1(M) - G$, there exists a finite index subgroup of $\pi_1(M)$ which contains G but not x.)

Problem 3.51 (Thurston) *Does a hyperbolic 3-manifold M have a finite cover which is a surface bundle over S^1?*

Remarks: A Seifert fibered space has a finite cover which is a circle bundle over a surface. But not all closed 3-manifolds have finite covers which are bundles.

A closed Seifert fibered space has a finite cover which fibers over S^1 iff its Euler number or its orbifold characteristic is zero (the two cases can be found in [837,Neumann & Raymond,1978] and [354,Gabai, 1986]).

Luecke & Wu [682,1995] give conditions under which a graph manifold (that is, the union of Seifert fibered spaces along torus boundary components) admits a finite cover which is a bundle; necessary and sufficient conditions in terms of a kind of incidence matrix associated to the manifold can be found in [836,Neumann,1995].

Reid gives the first example of a non-Haken, hyperbolic 3-manifold with a finite cover which fibers over S^1 in [922,Reid,1995,Pacific J. Math.]; the methods are arithmetic and give other examples.

On the other hand, such a cover must be complicated because Boileau and S. Wang show [125,Boileau & Wang,1995] that for any integer $n > 0$ there exist infinitely many closed, orientable, hyperbolic 3-manifolds M with first Betti number n, such that no tower of abelian coverings over M contains a surface bundle over S^1.

Problem 3.52 (Canary) (A) Conjecture (Marden): *If M^3 is a complete hyperbolic 3-manifold and $\pi_1(M)$ is finitely generated, then its ends are topologically tame, i.e. they are all products $F^2 \times \mathbb{R}$ for F^2 a closed, orientable surface. (This is equivalent to conjecturing that M is the interior of a compact 3-manifold (see Problem 3.8).)*

Remarks: The conjecture first appears in [693,Marden,1974, Ann. of Math.]. It is true if $\pi_1(M)$ is not a free product [130, Bonahon,1986,Ann. of Math.], and in other cases, [174,Canary & Minsky,1995], [853,Ohshika,1995a]. A good general reference is [171,Canary,1993b].

(B) Ahlfors' Measure Conjecture: *If Γ is a finitely generated Kleinian group, then its limit set either has measure zero or is the entire sphere at infinity. Moreover, if its limit set is the entire sphere at infinity, then Γ acts ergodically on the sphere at infinity.*

Remarks: Conjecture (A) implies Ahlfors' Conjecture which was first made, and proved for geometrically finite Kleinian groups, in [11,Ahlfors,1966,Proc. Nat. Acad. Sci. U.S.A.]. Thurston proved Ahlfors' Conjecture for geometrically tame hyperbolic 3-manifolds, and D. Canary [170,1993a,J. Amer. Math. Soc.] proved it for topologically tame Kleinian groups, so (A) implies (B).

Conjecture (A) would also imply that the geodesic flow on a hyperbolic 3-manifold associated to a finitely generated Kleinian group is *ergodic* iff its limit set is the entire sphere at infinity (this is a natural ergodic theoretic extension of the Ahlfors' measure conjecture).

(Recall that a *Kleinian group* is a discrete subgroup of $PSL(2, \mathbb{C})$, whereas a *Fuchsian group* is a discrete subgroup of $PSL(2, \mathbb{R})$.)

(C) Conjecture: *The fundamental group of a finite volume hyperbolic 3-manifold has the finitely generated intersection property (FGIP) iff it does not have a finite cover which fibers over the circle (see Problem 3.51).*

Remarks: (A) also implies (C) (for a discussion of this, see [172, Canary,1994]). Every co-infinite volume Kleinian group has the FGIP [476,Hempel,1987c], [36,Anderson,1991,Complex Variables Theory Appl.], whereas, for example, the fundamental group of a surface (of negative Euler characteristic) bundle over S^1 does not have the FGIP [535,Jaco,1980]. (Definition: a group G has the FGIP if the intersection of every pair of finitely generated subgroups of G is again finitely generated.)

(D) Conjecture: *A Kleinian group has the FGIP iff it has infinite covolume.*

Remarks: (D) would follow from (C) and Thurston's conjecture (Problem 3.51) that every hyperbolic 3-manifold has a finite cover which fibers over the circle.

(E) Conjecture: *Two isomorphic Kleinian groups, G_1 and G_2, have topologically conjugate actions on the 2-sphere at infinity iff they are quasiconformally conjugate.*

Remarks: The actions are topologically conjugate if there exists a homeomorphism $h : S^2 \to S^2$ such that the following diagram commutes (where $\iota : G_1 \to G_2$ is the isomorphism):

$$\begin{array}{ccc} G_1 \times S^2 & \longrightarrow & S^2 \\ \iota \Big\downarrow h & & \Big\downarrow h \\ G_2 \times S^2 & \longrightarrow & S^2 \end{array}$$

The actions are quasiconformally conjugate if h is quasiconformal (so the implication $\Leftarrow$ is obvious).

The conjecture is true for geometrically finite Kleinian groups [693, Marden,1974,Ann. of Math.]. It is also true if the quotient manifold has a lower bound on the injectivity radius and the group is a closed surface group [771,Minsky,1994,J. Amer. Math. Soc.] or if the group is topologically tame [854,Ohshika,1995b].

Problem 3.53 (Christy) (A) *Which hyperbolic 3-manifolds have Anosov flows?*

Remarks: M^3 has an Anosov flow if the tangent bundle T_M splits as a Whitney sum of line bundles $E_S \oplus F \oplus E_U$ where F is always tangent to the flow ϕ_t, the splitting is preserved by $D\phi_t$, points in E_S converge exponentially to 0 as t increases, and points in E_U

converge exponentially to 0 as t decreases, (see [39,Anosov,1969, Proc. Steklov Inst. Math.] for a precise and detailed definition).

If M^3 has an Anosov flow, then it is irreducible, [851,Novikov, 1963,Trans. Moscow Math. Soc.], [169,Camacho & Lins Neto, 1985], [364,Gabai & Oertel,1989,Ann. of Math.], and $\pi_1(M)$ grows exponentially [694,Margulis,1967,Успехи Мат. Наук]. The latter condition rules out Euclidean and Nil manifolds.

A Seifert fibered 3-manifold has an Anosov flow iff it is a finite cover of the unit tangent bundle of a hyperbolic orbifold [374, Ghys,1984,Ergod. Th. & Dynam. Sys.], [66,Barbot,1995].

If $\pi_1(M)$ is solvable, then an Anosov flow is conjugate to a suspension of an Anosov diffeomorphism of a surface, [876,Plante,1981, J. London Math. Soc.], [64,Barbot,1992].

Given an Anosov flow on M and a closed orbit C, then there is a line of Dehn surgeries on C which also have Anosov flows [399,Goodman,1983] and [342,Fried,1983,Topology]. This gives examples of hyperbolic 3-manifolds with Anosov flows; there are no known hyperbolic 3-manifolds without Anosov flows.

(B) *Does every irreducible, atoroidal 3-manifold with an Anosov flow have a hyperbolic metric?*

(C) *Given an integer N, does there exist a hyperbolic 3-manifold with at least N Anosov flows which are topologically inequivalent.?*

Remarks: Two Anosov flows are topologically equivalent if there exists a homeomorphism of M taking oriented orbits to oriented orbits. There exist T^2-bundles over S^1 for which there exists an Anosov flow which is not topologically equivalent to the same flow with time reversed; these bundles are characterized by the fact that their monodromies are not conjugate to their inverses. Geodesic flows on tangent circle bundles to surfaces are topologically equivalent to their time reversals. Barbot [65,1994] elucidated the construction of Bonatti & Langevin [134,1994,Ergod. Th. & Dynam. Sys.] and showed that their example lives on a 3-manifold N which admits two topologically inequivalent Anosov flows which are not related by time reversal.

Problem 3.54 (Christy) The geodesic flow (which is Anosov) on the tangent S^1 bundle of a closed, hyperbolic surface without boundary can be constructed by Dehn surgery on closed orbits of the suspension flow of a mapping torus (a T^2 bundle over S^1) [342,Fried,1983,Topology], [399,Goodman,1983].

Question: *Does every Anosov flow on a closed compact 3-manifold arise in this way? If so, can it even arise from the mapping torus with monodromy* $\left(\begin{smallmatrix}2&1\\1&1\end{smallmatrix}\right)$ *?*

Problem 3.55 (Thurston) *Does every closed hyperbolic 3-manifold admit a 2-dimensional foliation without Reeb components?*

Remarks: A possible generalization, if the hyperbolic part of the Geometrization Conjecture fails (see Problem 3.45), is to replace the hypothesis of hyperbolicity by prime, atoroidal, infinite π_1. Also, one can ask in general if a Reebless foliation can by used to construct a geometric structure on a 3-manifold.

Problem 3.56 (Thurston) Given a hyperbolic 3-manifold with a foliation without Reeb components, then all leaves lift to topological planes in $\mathbb{H}^3$ which are conformally equivalent to the interior of the unit disk D in $\mathbb{C}$ (see [175,Candel,1993,Ann. Sci. École Norm. Sup. (4)], which produces a metric on M which is smooth along the leaves and continuous in the transverse direction, and for which all leaves have constant curvature -1).

Question: *Does the inverse map of intD into* $\mathbb{H}^3$ *extend continuously over* ∂D *to the sphere at infinity?*

Remarks: This is true for surface bundles over S^1 [177,Cannon & Thurston,1995] (with published proof in [771,Minsky,1994,J. Amer. Math. Soc.]), and for depth one foliations [298,Fenley,1992,J. Differential Geom.].

Problem 3.57 (Mess) *Given a compact, hyperbolic 3-manifold* M*, show that a finite iteration of the following operation results in a link complement in* S^3*: remove the shortest geodesic and put a complete hyperbolic structure on the resulting manifold.*

Remarks: The complement of the geodesic is atoroidal so it has a unique (up to diffeomorphism) hyperbolic structure. However it is not known that the hyperbolic structure on the filled manifold is obtained from the hyperbolic structure on the complement by a deformation through cone manifolds.

Problem 3.58 (Cooper) *Given $K > 0$, is there a hyperbolic, rational homology 3-sphere M such that the minimum over all points $p \in M$ of the injectivity radius at p is bigger than K?*

Remarks: Presumably the answer is yes. If N is a closed hyperbolic 3-manifold then N has finitely many geodesics shorter than a given constant, so N has a finite cover with arbitrarily large injectivity radius, which, using a *no answer* to the problem, is not a rational homology sphere and therefore has $b_1 > 0$. Long & Reid [676,1995] use the fact that the group of a hyperbolic 3-manifold is residually simple to show that a *no answer* implies that a hyperbolic rational homology sphere has infinite virtual betti number. (A group G is residually P, where P is a property of a group, if the following statement holds for any element $g \in G$: there exists a group H with property P and an epimorphism $h : G \to H$ which is non-trivial on g.) The Long & Reid proof is that there are many surjections of $\pi_1(N)$ onto simple groups of the form $SL(2, F_p)$ with F_p a finite field, giving covers N_p with rational homology. One now uses simplicity plus the fact that N has no rational homology to argue that the common cover of these covers has large b_1.

Problem 3.59 *If a complete, orientable, hyperbolic, 3-manifold with finite volume has cusps, show it has an immersed, closed, orientable, incompressible surface with no accidental parabolics.*

Remarks: The hyperbolic structure corresponds to a representation of π_1 to $PSL(2, \mathbb{C})$ and all elements of a $\mathbb{Z} \oplus \mathbb{Z}$ must go to parabolic elements. Any other such elements called *accidental parabolic.*

Problem 3.60 (C. C. Adams) Hyperbolic 3-manifolds come in many flavors: orientable or not, closed or with totally geodesic boundaries (which are hyperbolic surfaces) or with one or more cusps (homeomorphic to $S^1 \times S^1 \times \mathbb{R}$ or to $K \times \mathbb{R}$ in the non-orientable case where K

is the Klein bottle); one can also consider orbifolds (see definition in Problem 3.46).

The map taking volumes of hyperbolic 3-orbifolds is a finite-to-one map whose image is a well ordered set of the reals $\mathcal{S}$ (Jørgensen and Thurston for manifolds, [256,Dunbar & Meyerhoff,1994,Indiana Univ. Math. J.] for orbifolds). The limit points of the set of volumes of the closed, orientable, hyperbolic 3-manifolds all correspond to volumes of hyperbolic 3-manifolds, each with a single cusp. Similarly, the limit points of the set of volumes of the orientable, hyperbolic 3-manifolds with n cusps correspond to volumes of orientable, hyperbolic 3-manifolds with $n+1$ cusps. Filling in an orientable cusp by a Dehn surgery lowers volume (when the surgery results in a hyperbolic manifold, as it does for all but a finite number of Dehn surgeries), whereas a non-orientable cusp can only be filled in in one way, and then the hyperbolic structure may or may not extend. The volume of a non-orientable 3-manifold is half that of its oriented cover, so the smallest volumes in the various cases could be non-orientable.

(A) *Determine the closed, orientable, hyperbolic 3-manifold of least volume.*

Remarks: In 1983, Przytycki conjectured that the punctured torus bundle with monodromy $\left(\begin{smallmatrix}-2 & -1\\ -1 & -1\end{smallmatrix}\right)$ (see Problem 1.77) and Dehn filling (3,1) was hyperbolic and had smallest volume. This manifold can be also described by (5,1), (5,2) surgery on the right handed Whitehead link (drawn below in Figure 3.60.1; with crossings which are right handed with respect to orientations of the components of the link), and Weeks (independently) has shown it is hyperbolic with volume 0.9247... [1100,Weeks,1985]. This manifold is arithmetic and has recently been shown to be the smallest such [202,Chinburg, Friedman, Jones, & Reid,1995].

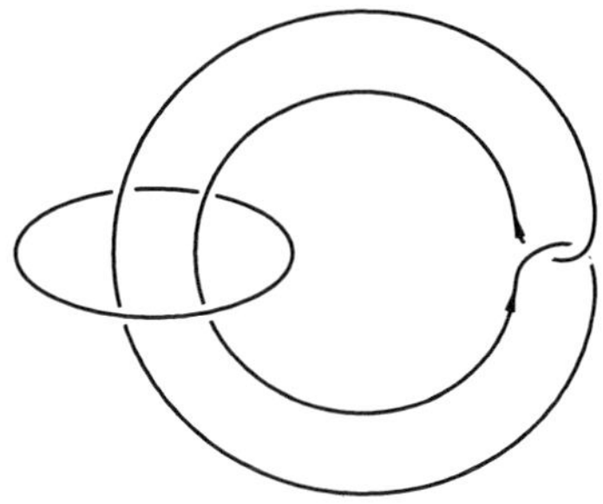

Figure 3.60.1.

In the closed, orientable case, the least volume is $\geq .00115$ [368, Gehring & Martin,1991,J. Reine Angew. Math.], and if the first Betti number is ≥ 3, then the least volume is ≥ 0.92 [230,Culler & Shalen,1992,J. Amer. Math. Soc.].

(B) *Determine the non-orientable, closed, hyperbolic 3-manifold of least volume.*

Remarks: The smallest so far has volume 2.02988... (J. Weeks). It is constructed from a single surgery on any of several 3-manifolds obtained by gluing together faces of a small number of ideal tetrahedra. Surprisingly, its volume is precisely twice that of the Gieseking manifold below, which is non-orientable with one cusp.

(C) *Determine the cusped, orientable, hyperbolic 3-manifold of least volume. A special case is to determine the hyperbolic knot in S^3 of least volume.*

Remarks: The conjectured answer in both cases is the figure-8 knot complement and also, in the first case, the manifold obtained by (5,1) surgery on one component of the right handed Whitehead link.

(D) *Determine the n-cusped hyperbolic 3-manifold of least volume for $n \geq 3$.*

Remarks: For $n = 1$, the least volume is $v_0 = 1.01494...$, (which equals the volume of the ideal tetrahedron in hyperbolic 3-space), for the non-orientable Gieseking 3-manifold [2,Adams,1987,Proc. Amer. Math. Soc.], [375,Gieseking,1912]. This manifold is constructed from the ideal tetrahedron by identifying A with A' and B with B' in a way which preserves the orientations on the edges in Figure 3.60.2; its double cover is the figure-8 knot complement. For $n = 2$, the least volume is $2v_0$ and for $n \geq 3$, the least volume is known to be at least nv_0 [3, Adams, 1988, J. London Math. Soc.].

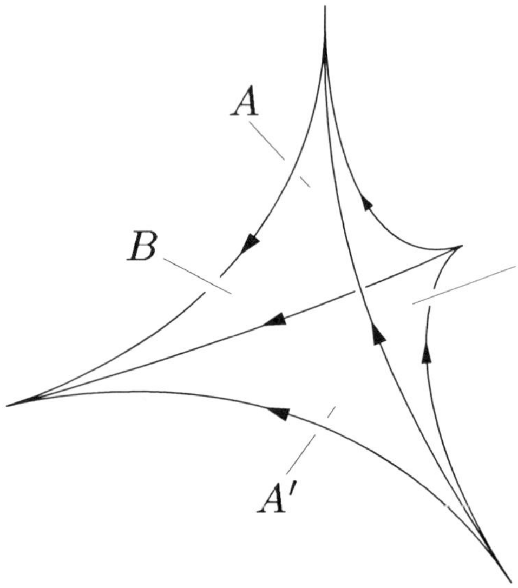

Figure 3.60.2. Gieseking manifold

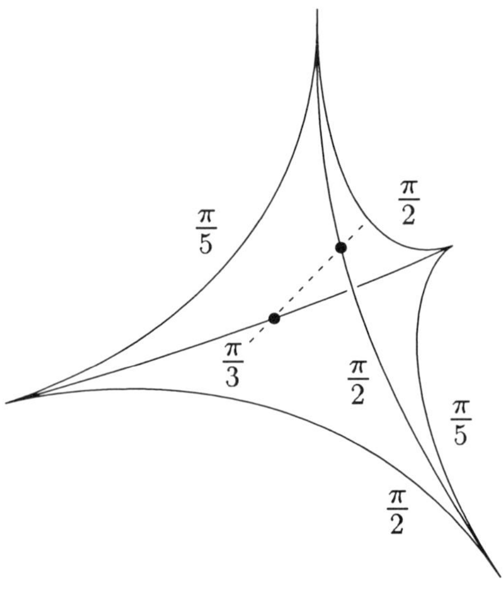

Figure 3.60.3.

(E) *For each n, determine the hyperbolic 3-manifold of least volume with n orientable cusps.*

Remarks: It is expected that such a manifold is orientable. (The volume of such a manifold will be the smallest n-fold limit point of the set of all volumes of hyperbolic 3-manifolds.)

(F) *Determine the smallest volume orientable and non-orientable hyperbolic 3-orbifolds.*

Remarks: The smallest known orientable example is the following: consider the tetrahedron T imbedded in H^3 so that its dihedral angles are $\pi/5, \pi/5, \pi/3, \pi/2, \pi/2, \pi/2$, as drawn in Figure 3.60.4. Tesselate H^3 by reflections of T through its faces. Take the quotient of H^3 by the orientation preserving symmetries of this tesselation (including the rotation by π about the line L). This quotient is S^3 with orbifold singular set drawn in Figure 3.60.4. This is the smallest example among all orbifolds with at least one edge with orbifold degree ≥ 4 [369, Gehring & Martin, 1994, Math. Res. Lett.] and the smallest among all orbifolds arising from symmetries of tetrahedral tesselations.

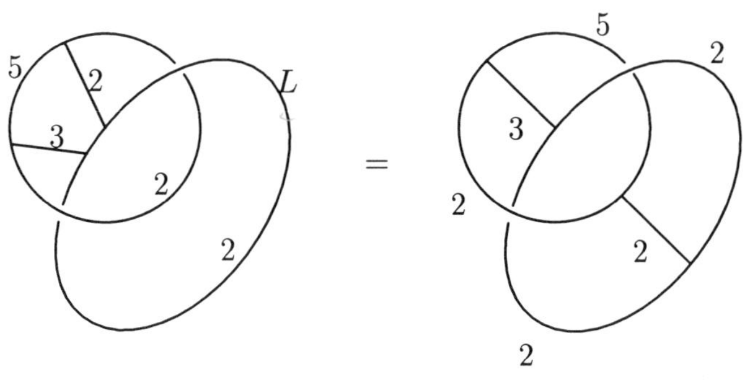

Figure 3.60.4.

(G) Consider the punctured torus bundle P with monodromy $\binom{3\ 2}{1\ 1}$

The Dehn fillings $(3, 1)$ and $(5, -1)$ of P both have the same volume [85, Betley, Przytycki, & Zukowski, 1986, Kobe J. Math.].

Twins conjecture*: These are the smallest volume hyperbolic twins.*

(H) **(Przytycki)** *The Dehn fillings (μ, λ) and $(\mu + 2\lambda, -\lambda)$ of P are hyperbolic twins for all μ and λ [ibid.]. Is each pair of these twins* *commensurable (that is, do they have a common finite cover)?*

Further remarks: The six smallest cusped, hyperbolic 3-orbifolds are known and they are all orientable ([757,Meyerhoff,1985,Bull. Amer. Math. Soc.] for the first of the six, and [5,Adams,1992a] for the rest). All six are arithmetic [839,Neumann & Reid,1992b]. The smallest orientable one is $Q = H^3/\Gamma$ where $\Gamma = PGL(2, \mathcal{O}_3)$ and $\mathcal{O}_3$ is the ring of integers in $\mathbb{Q}(\sqrt{-3})$; Q is the double cover of the smallest, cusped, non-orientable orbifold whose volume is $v_0/24$.

The three smallest limit volumes for hyperbolic 3-orbifolds are known [4,Adams,1991,J. Differential Geom.]. The four smallest orientable (and non-orientable), multiply cusped, hyperbolic, 3-orbifolds are known [6, Adams,1992b,Indiana Univ. Math. J.]. The smallest arithmetic, hyperbolic, 3-orbifold [203,Chinburg & Friedman,1986,Invent. Math.] is known.

The smallest compact, hyperbolic 3-manifold with totally geodesic boundary [602,Kojima & Miyamoto,1991,J. Differential Geom.], and the smallest hyperbolic 3-manifold with closed totally geodesic boundary of any given genus ≥ 2 [772,Miyamoto,1994,Topology] are known.

- The following notions are useful in the next few problems.

A subgroup of $PSL(2,\mathbb{C})$ is *arithmetic* (or a hyperbolic 3-manifold is arithmetic) if the subgroup arises by the following construction.

Start with a field k which imbeds in $\mathbb{C}$ in only one way (plus its conjugate). This happens iff $k = \mathbb{Q}(\lambda)$ where λ is a complex root of an irreducible $\mathbb{Q}$-polynomial P with exactly two complex roots, $\lambda, \bar{\lambda}$, and all others real. Now take a quaternion algebra A over k (a semi-simple algebra over k^4) which is ramified at all real imbeddings (i.e. each real root of P gives a real imbedding $\tau : k \to \mathbb{R}$ (e.g. $\mathbb{Q}(\sqrt[3]{2}) = \mathbb{Q}[x]/(x^3 - 2)$) and then $A \otimes_\tau \mathbb{R}$ = Hamiltonian quaternions). Then take an order (a subring O of A which is, for O_k = ring of integers of k, an O_k-module of dimension 4 over O_k and O generates A over k). Then take the invertible elements O^* in O; O^* imbeds naturally as a subgroup of $SL(2,\mathbb{C})$ (using the complex imbedding $\tau : k \to \mathbb{C}$, $A \otimes_\tau \mathbb{C} = M_{\mathbb{C}}(2 \times 2)$, so $O^* \subset A \subset SL(2,\mathbb{C})$ = invertible elements of $M_{\mathbb{C}}(2 \times 2)$). Then $O^*/\pm 1$

is an arithmetic subgroup, as is any subgroup which is commensurable with an $O^*/\pm 1$.

A subgroup Γ of $PSL(2,\mathbb{C})$ is not arithmetic if the set of subgroups commensurable with Γ contains a maximal (under inclusion) element Γ_0 (Margulis); however, in the arithmetic case there are infinitely many maximal subgroups in the commensurability class of Γ [137,Borel,1981, Ann. Scuola Norm. Sup. Pisa Cl. Sci. (4)].

Problem 3.61 (Neumann) *What fields occur as invariant trace fields of hyperbolic 3-manifolds?*

Remarks: The *trace field* of $M = H^3/\Gamma$ is the field

$$K(\Gamma) := Q(\{tr(\gamma)|\gamma \in \Gamma\}).$$

The *invariant trace field* is $k(\Gamma) := K(\Gamma^{(2)})$ where $\Gamma^{(2)} \subset \Gamma$ is the subgroup generated by squares of elements of Γ. It is a commensurability invariant [920,Reid,1990,Bull. London Math. Soc.]. A number field occurs as the invariant trace field of an arithmetic hyperbolic 3-manifold iff it has exactly one conjugate pair of complex imbeddings. Thus such fields occur. But for fields with more than one conjugate pair of complex imbeddings, very little is known (other than many examples). Note that the invariant trace field comes with a specific imbedding $k(\Gamma) \subset \mathbb{C}$, so one can refine the question to ask what pairs (field, imbedding) occur.

There are additional related invariants: the invariant quaternion algebra $A(\Gamma)$ and the set $P(\Gamma)$ of primes of $k(\Gamma)$ at which traces of elements of $\Gamma^{(2)}$ are non-integral [838,Neumann & Reid,1992a]. One can ask which of these are realized; arithmetic manifolds realize precisely the cases with $k(\Gamma)$ as described above, $A(\Gamma)$ ramified at each real imbedding of $k(\Gamma)$, and $P(\Gamma) = \emptyset$.

Problem 3.62 (Neumann) Let k be a number field with exactly one conjugate pair of complex imbeddings. Then there exist arithmetic 3-manifolds M with invariant trace field k.

(A) Conjecture: *The Chern–Simons invariant $CS(M)$ is rational iff $k = \bar{k}$.*

Remarks: The *if* part is true [843, Neumann & Yang, 1995]. As an example, $k = \mathbb{Q}(\sqrt[3]{2})$ should give an irrational Chern–Simons invariant. The conjecture is a special case of the Ramakrishnan Conjecture [914, 1989], which generalizes a conjecture of Milnor [768, 1983, L'Enseign. Math.] (see [843, Neumann & Yang, 1995] for a discussion and more general conjectures). See also Problem 3.63.

(B) *What is the number theoretic significance of $CS(M)$?*

Remarks: $CS(M)$ is expected to have number theoretic meaning since it is intimately related to $vol(M)$ which does (it is a rational multiple of $\pi^2 J_k(2)$ where J_k is the Zeta-function for k) [137, Borel, 1981, Ann. Scuola Norm. Sup. Pisa Cl. Sci. (4)]. Note that $CS(M)$ is determined, up to rational multiples, by k (this uses the assumption that k has a unique pair of complex imbeddings, but not that M is arithmetic [Neumann & Yang, ibid.]).

Problem 3.63 (J. D. S. Jones) *Is the Chern–Simons invariant of a closed hyperbolic 3-manifold M rational?*

Remarks: Here closed means compact with no boundary and *rational* means a rational multiple of a fixed constant (there are constants involved in the definition of the Chern–Simons invariant). One way to phrase the question so that it is independent of these constants is as follows:

Is the ratio of the Chern–Simons invariant of any two closed hyperbolic 3-manifolds rational?

This problem is related to the calculation of the algebraic K-group $K_3(\mathbb{C})$ as follows. By definition, the group $K_3(\mathbb{C})$ is $\pi_3(BGL^+(\mathbb{C}))$. Here $BGL(\mathbb{C})$ is the classifying space of the group $GL(\mathbb{C})$ regarded as a discrete group and the superscript $+$ means Quillen's plus construction ([907, Quillen, 1970], [908, Quillen, 1975]); attach 2 and 3 cells to abelianize π_1 and preserve homology.

A closed hyperbolic 3-manifold M equipped with a spin structure α determines an element $[M, \alpha]$ in $K_3(\mathbb{C})$ by the following procedure. The spin structure gives a representation $\pi_1(M) \to SL(2, \mathbb{C})$ lifting the canonical homomorphism $\pi_1(M) \to PSL(2, \mathbb{C})$. This gives a map

$M \to BSL(2,\mathbb{C})$ and composing with

$$BSL(2,\mathbb{C}) \to BGL(\mathbb{C}) \to BGL^+(\mathbb{C})$$

gives a map

$$M \to BGL^+(\mathbb{C}).$$

Now $BGL^+(\mathbb{C})$ is an N-fold loop space for any N; say $\Omega^N Y_N \simeq BGL^+(\mathbb{C})$. Thus taking adjoints gives a map

$$\Sigma^N M \to Y_N.$$

Now since M is a closed 3-manifold, a spin structure determines a stable framing and hence, by the Pontryagin–Thom construction, a degree one map

$$S^{N+3} \to \Sigma^N M.$$

Thus, composing with the map $\Sigma^N M \to Y_N$, we get an element

$$[M, \alpha] \in \pi_{N+3}(Y_N) = K_3(\mathbb{C}).$$

There are, of course, other ways of constructing this invariant.

There is a homomorphism

$$r : K_3(\mathbb{C}) \to \mathbb{C}/\mathbb{Z}$$

which is known as a secondary Chern character, or a regulator. This homomorphism is studied in detail in [561, Karoubi, 1987] and also [550, Jones & Westbury, 1995] (where it is called the e-invariant in algebraic K-theory). It is proved in [Jones & Westbury, ibid.] that

$$r[M, \alpha] = \frac{1}{2} CS(M) - \frac{i}{4\pi^2} vol(M)$$

where CS is the Chern–Simons invariant, and vol is the volume. Thus, if the imaginary part of r is always rational then it follows that the Chern–Simons invariant of every closed hyperbolic 3-manifold is rational.

What is known about $K_3(\mathbb{C})$? Suslin has shown that $K_3(\mathbb{C})$ is isomorphic to $F \oplus \mathbb{Q}/\mathbb{Z}$ where F is a uniquely divisible group [1019, Suslin, 1984, J. Pure Appl. Algebra]. There are many well-known elements in $K_3(\mathbb{C})$. The easiest to describe are the products xyz where $x, y, z \in K_1(\mathbb{C}) = \mathbb{C}^*$. A deep theorem of Borel [136, 1974, Ann. Sci. École Norm. Sup. (4)] computes the torsion free part of $K_*(E)$ where

E is a number field. Elements in the torsion free part of $K_*(E)$ are known as Borel classes as are their images in $K_3(\mathbb{C})$. The torsion summand $\mathbb{Q}/\mathbb{Z}$ can be described in many ways. For example, in [Jones & Westbury, ibid.] it is shown that the torsion elements can be constructed as $[M,\rho]$ where M is a Seifert fibered homology 3-sphere and $\rho : \pi_1(M) \to SL(2,\mathbb{C})$ is a (possibly non-faithful) representation and $[M,\rho]$ is constructed by the above procedure.

There are various conjectures concerning the algebraic K-theory of fields described in [1020,Suslin,1987]. The strongest of these, applied to the group $K_3(\mathbb{C})$ is that the products from K_1, the Borel classes, and the torsion generate $K_3(\mathbb{C})$.

This line of thought suggests two other natural questions.

(A) Does every element of $K_3(\mathbb{C})/\text{Torsion}$ arise as $[M,\alpha]$ where M is a closed hyperbolic 3-manifold? Note that the volume term in $r[M,\alpha]$ shows that $[M,\alpha]$ has infinite order. If not, is it possible to characterize the subgroup of $K_3(\mathbb{C})$ which can be constructed in this way?

(B) Are the known values of the Chern–Simons invariant and the volume of closed hyperbolic 3-manifolds consistent with the conjecture that $K_3(\mathbb{C})$ is generated by the products from $K_1(\mathbb{C})$, the torsion summand $\mathbb{Q}/\mathbb{Z}$, and the Borel classes?

The papers [925,Reznikov,1995b] and [924,Reznikov,1995a,Ann. of Math.] contain very interesting related results.

Problem 3.64 (A) Conjecture (Neumann & Reid): *If $M^3 = H^3/\Gamma$ is a knot complement, then the commensurator of Γ, $com(\Gamma)$, is equal to the normalizer of Γ, $norm(\Gamma)$, except in the cases of the figure-8 knot and the two Aitchison–Rubinstein knots.*

Remarks: A non-arithmetic subgroup has among its commensurable subgroups a maximal one called the *commensurator*. A non-arithmetic hyperbolic 3-manifold, $M^3 = H^3/\Gamma$, is a branched cover of a minimal orbifold which corresponds to $H^3/com(\Gamma)$. The figure-8 knot is the only arithmetic knot complement [921,Reid, 1991,J. London Math. Soc.].

The two Aitchison–Rubinstein knots [13,1992] are obtained by identifying (in two ways) the faces of the ideal dodecahedron so as to get a knot complement.

The symmetry group of a hyperbolic knot equals the isometry group of H^3/Γ which equals $norm(\Gamma)/\Gamma$.

Evidence for this conjecture can be found in [838,Neumann & Reid,1992a].

(B) *Is there a knot other than one of the above three whose complement is hyperbolic with cusp parameter in* $\mathbb{Q}(\sqrt{-1})$ *or* $\mathbb{Q}(\sqrt{-3})$*?*

(C) *Are there knots other than the two dodecahedral knots for which the cusp field (field generated by the cusp parameter) is a proper subfield of the trace field?*

Problem 3.65 (Mess) *In each of the geometries Nil, Solv, and* $\widetilde{SL}(2,\mathbb{R})$*, answer the following questions for a 3-manifold modeled on the geometry:*

(A) *Are closed geodesics unique in their homotopy classes?*

(B) *Are shortest geodesics unique in their homotopy classes?*

(C) *Describe the different knots formed by simple closed geodesics in compact manifolds modeled on these spaces. In particular, under what circumstances does the complement in the universal cover of the preimage of a simple closed geodesic have free fundamental group?*

Remarks: In (A) and (B) uniqueness means up to an isometry in the identity component of the isometry group. Of course, yes to (A) implies yes to (B).

Problem 3.66 (Freedman, Luo & Hass) Conjecture: *Let* M^3 *be irreducible and non-compact, with* $\partial M = \emptyset$*, and with a homogeneously regular, complete Riemannian metric. Then an essential, proper imbedded plane is proper homotopic to a least area imbedded plane, where least area means that any Jordan curve in the plane bounds a least area disk which also lies in the plane.*

Remarks: Because of irreducibility, one does not have *bubbling off*. Without irreducibility, one could conjecture: if there exists an essential, proper imbedded plane, then there exists an essential least area plane.

Homogeneous regularity [803,Morrey Jr.,1966] is a condition which is essentially the same as having upper and lower bounds on the curvature and injectivity radius. Such a bound always holds in manifolds which cover compact manifolds. The standard method to get existence theorems for a plane is to take a limit of disks, which exist given Morrey's condition. The paper [460,Hass & Scott,1988,Trans. Amer. Math. Soc.] contains an example showing that Morrey's condition is necessary sometimes. A complete metric is put on $\mathbb{R}^3$ in which the unit circle in the xy-plane bounds no least area disk. Presumably this example can be modified to give a counterexample to the conjecture without the homogeneous regularity condition.

Problem 3.67 (Hass) (A) Let $f : F \to M^3$ be a one-sided proper map of a surface into a Riemannian 3-manifold which induces an isomorphism on π_1; also assume $f(F)$ has least area among all such surfaces.

Conjecture: *f is an imbedding.*

Remarks: The assumptions imply that M is homotopy equivalent to a non-trivial I-bundle over a surface, or to $\mathbb{RP}^3$.

The corresponding result for two-sided maps is known [332,Freedman, Hass, & Scott,1983,Invent. Math.] and [744,Meeks, III & Yau,1982,Topology], [743,Meeks, III & Yau,1980,Ann. of Math.].

Note that an affirmative answer to (A) confirms Problem 3.1 E since an irreducible 3-manifold containing an imbedded one-sided $\mathbb{RP}^2$ is $\mathbb{RP}^3$, and implies Problem 3.4 since an imbedded incompressible annulus from one $\mathbb{RP}^2$ to the other would imply that the h-cobordism is standard.

(B) Let $f : F \to M$ be a one-sided, least area (in its homotopy class), π_1-injective proper map which is homotopic to an imbedding.

Conjecture: *f is an imbedding.*

Remarks: By results in [458,Hass & Rubinstein,1986,Michigan Math. J.], an affirmative answer to conjecture (A) implies the same for (B).

(C) A simple special case of this is a conjecture of Meeks: Let M^3 be $S^1 \times B^2$ with a metric in which the boundary is convex, and let K be a $(2,1)$-torus knot in $S^1 \times S^1$. Let F be a least area Möbius band bounded by K.

Conjecture: *F is imbedded.*

Remarks: Here one is minimizing area among all mappings of Möbius bands with a given boundary. It is known from an easy extension of [973,Schoen & Yau,1979,Ann. of Math.] to the non-orientable case, that a least area Möbius band does exist and that it is immersed. It is also known that there is a least area Möbius band in the class of imbedded Möbius bands [742,Meeks, III, Simon, & Yau,1982,Ann. of Math.]. The problem thus asks if they are always the same.

Problem 3.68 (Hass) (A) *Find a smooth flow on surfaces in Riemannian manifolds that*

- *keeps imbedded surfaces imbedded, and*
- *takes an incompressible surface to a minimal surface.*

Remarks: The mean curvature flow, which works for curves on a surface [409,Grayson,1989,Ann. of Math.] doesn't work.

(B) *Find a flow on the space of metrics on a closed 3-manifold M that converges to a constant curvature metric if M is irreducible and atoroidal.*

Remarks: This would prove the Geometrization Conjecture (Problem 3.45).

(C) *Find a smooth flow on hyperbolic manifolds that takes a diffeomorphism that is homotopic to the identity, through diffeomorphisms, to the identity.*

Remarks: It is known that homotopic diffeomorphisms are isotopic for geometric 3-manifolds other than hyperbolic ones. See the list of references in Problem 3.95.

Problem 3.69 (Harnack) *What is the maximal number N of components of a non-singular, real algebraic surface in $\mathbb{RP}^3$ of degree m?*

Remarks: The first open case is $m = 5$ when $22 \leq N \leq 25$ [520, Itenberg & Kharlamov, 1993].

The known upper bound for the number $N(m)$ is

$$N(m) \leq m(5m^2 - 18m + 25)/12$$

(this inequality is not sharp for $m = 4$); it comes from combining the Harnack–Smith inequality (for the total Betti number of the real surface) and the Petrovskii inequality (for its Euler characteristic).

The known lower bound for even m is

$$N(m) \geq (m^3 - 2m^2 + 4)/4$$

and for $m \equiv 2 \pmod 4$

$$N(m) \geq (7m^3 - 24m^2 + 32m)/24$$

(the latter case can be realized by an M-surface) [1078, Viro, 1979b, Soviet Math. Dokl.].

Problem 3.70 (G. Martin) *Given a Kleinian group Γ and a triangle subgroup Δ with an invariant hemisphere Π in H^3, is it true that for all $g \in \Gamma$ either $g(\Pi) = \Pi$ or $g(\bar{\Pi}) \cap \bar{\Pi} = \emptyset$ (i.e. the 2-orbifold Π/Δ is imbedded in H^3/Γ)?*

Remarks: This is true for the triangle groups $(2,3,p)$, $(2,4,p)$, and $(2,5,p)$, all for $p \geq 7$ [698, Martin, 1995].

Problem 3.71 (G. Martin) (A) *Let G be a discrete subgroup of $Homeo(S^n)$ and suppose that G is a convergence group. When is G conjugate to a subgroup which acts conformally on S^n?*

Remarks: Recall that G is a *convergence group* if given a sequence $\{g_i\} \subset G$, there exist two points x, y (possibly $x = y$) and a subsequence $\{g_{i_k}\}$ such that $\{g_{i_k}\}$ converges locally uniformly to the constant map x in $S^n - y$ and $\{g_{i_k}^{-1}\}$ converges locally uniformly to y in $S^n - x$. This definition is equivalent to G acting properly discontinuously on $S^n \times S^n \times S^n -$ {big diagonal}. If G is a Mobius or quasiconformal group, or if G acts properly discontinuously in $S^n -$ {Cantor set}, then G is a convergence group.

If Γ is a conformal subgroup of $Homeo(S^n)$ and $f : S^n \to S^n$ is a continuous map such that $fG = \Gamma f$, then G is a convergence group; such an f is a semi-conjugacy, and (A) ask for a conjugacy, i.e. for f to be a homeomorphism.

An excellent general reference for this problem is [697,Martin, 1988,Rev. Mat. Iberoamericana.].

When $n = 1$, every discrete convergence group G extends as a convergence group to B^2 ([358,Gabai,1992,Ann. of Math.] and [190,Casson & Jungreis,1994,Invent. Math.]) and thus is topologically conjugate to a Fuchsian (i.e. conformal) group [700,Martin & Tukia,1988].

When $n = 2$, not every discrete convergence group G is topologically conjugate to a Kleinian (i.e. conformal) group. However:

(B) Conjecture: *if G is a discrete convergence group of S^2 and for each $x \in S^2$, $stab_G(x)$ is a finite extension of $\{1\}$, $\mathbb{Z}$, or $\mathbb{Z} \oplus \mathbb{Z}$, then G is topologically conjugate to a Kleinian group.*

Remarks: The conjecture is true if the limit set of G consists of points and topological circles [699,Martin & Skora,1989,Amer. J. Math.] (for a slight generalization, see [Martin & Tukia, ibid.]). Note that the stabilizer of a point in a Kleinian group has to be an extension as above, and (B) conjectures that the converse is true.

(C) Conjecture: *if G is a discrete convergence group of S^2, then there exists a Kleinian group Γ isomorphic (via ι) to G and a Γ-equivariant upper semicontinuous cellular decomposition f of S^2 such that the following diagram commutes:*

$$\begin{array}{ccc} \Gamma \times S^2 & \longrightarrow & S^2 \\ \iota \Big\downarrow f & & \Big\downarrow f \\ G \times S^2 & \longrightarrow & S^2 \end{array}$$

Remarks: This is true in the same cases as conjecture (B). (C) is motivated by the fact that such decompositions provide examples of discrete convergence groups which are not conjugate to

Kleinian groups (although they may still be isomorphic), and (C) conjectures that this is the only way such examples arise.

(D) Conjecture: *If G is a convergence group of S^2 with $(S^2 \times S^2 \times S^2 - \{big\ diagonal\})/G$ compact, then G is topologically conjugate to a Kleinian group.*

Remarks: Gabai [358,1992,Ann. of Math.] and Casson & Jungreis [190,1994,Invent. Math.] prove this for S^1 instead of S^2. If M^3 is a negatively curved closed 3-manifold, then $\pi_1(M)$ acts as a convergence group on the 2-sphere at infinity with the compactness property above, so if (D) is true then $\pi_1(M)$ is isomorphic to a Kleinian group (which implies M is hyperbolic if a suitable geometric condition holds [360,Gabai,1994b,Bull. Amer. Math. Soc.]).

(E) *Suppose that G is a convergence group acting on S^3, that the limit set $L(G)$ of G is a Cantor set in S^3, that $(S^3 - L(G))/G$ is compact, and that G is torsion free (which implies that G is free). Then, does G extend to B^4 as a convergence group?*

Remarks: The answer is yes iff 4-dimensional topological surgery works for all fundamental groups [331,Freedman,1986,Topology Appl.] (see Problems 4.6 and 5.9).

Problem 3.72 (Gromov) *Suppose that G is a convergence group acting cocompactly on the space of distinct triples of X. Is G necessarily a word hyperbolic group with boundary X?*

Remarks: Of course, the special case of $X = S^n$ is covered in Problem 3.71, where it is asked whether G is a subgroup of the isometries of hyperbolic $(n+1)$-space.

A *word hyperbolic group* G is defined as follows: choose a finite set of generators and form the Cayley graph Γ; put the word metric on Γ (the distance between two vertices is the minimal number of edges joining them); then G is word hyperbolic if there exists a universal constant K such that for any geodesic triangle, each edge is within a K-neighborhood of the other two edges (this turns out to be independent of the choice of generators). The boundary of G is then defined to be the equivalence classes of geodesics from a fixed point to infinity, where two geodesics are equivalent if they are a bounded distance apart.

Problem 3.73 (Mess) *Is there any example in any dimension of a finitely generated, infinite, torsion group acting effectively on a compact manifold? Or of an action of such a group on a noncompact manifold which fails to be properly discontinuous?*

Remarks: Certainly a finitely generated, infinite, torsion group can act as a group of deck transformations on a noncompact manifold. Also, Olshanskii has shown that a finitely generated, infinite torsion group can act effectively on a finite dimensional complex (2-dimensional in his first examples). This was generalized by Gromov. A complete proof of a stronger result is given in [857,Ol'shanskii,1992,Math. USSR.-Sb].

It is not clear that an answer in the PL (respectively smooth) category yields an answer in the smooth (respectively PL) category. It seems natural to consider the PL version, for there one can consider the dynamics of the action of the group on the space of all triangulations of the space, as well as the action on the space itself. Special case: suppose f is simplicial in a triangulation K_1 of a compact PL manifold $|K_1|$ and g is simplicial in a subdivision K_2 of the triangulation K_1. Then f and g are certainly of finite order, and the question is whether f and g can generate an infinite torsion group. The question can also be asked for a compact polyhedron.

The answer is negative in dimension 1. The answer for the 2-sphere is not clear. By application of Hurwitz's $84(g-1)$ theorem, a torsion group acting on a compact connected 2-manifold of negative Euler characteristic is a finite group. (The application of Hurwitz's theorem is indirect if the manifold is nonorientable. If the manifold has boundary, then the negative answer in dimension 1 implies that a torsion group is finite.) A torsion group which acts on the 2-torus has a subgroup of index at most 12 which acts freely, since the extended mapping class group has no finite subgroups of order greater than 12 and a finite order homeomorphism of the torus which is isotopic to the identity acts freely.

Suppose an infinite torsion group acts on the 2-sphere, preserving orientation, and all point stabilizers are of finite order. Then all point stabilizers are isomorphic to finite cyclic groups of rotations. By an inverse limit of blowing up processes, one can obtain a new action of the same group, which setwise preserves a Sierpinski curve, and for which the stabilizer of any closed complementary domain is conjugate to a group of rotations of the disc. The same holds for a group of piecewise linear homeomorphisms of the 2-sphere, but it is not clear that the pointwise stabilizers of a torsion group of homeomorphisms of the 2-sphere are necessarily abelian.

This question is related to Problem 3.39: *Suppose M admits no S^1 action. Is it possible for $PL(M)$ to contain an infinite torsion subgroup?* Assuming that $PL(M)$ doesn't contain an infinite finitely generated torsion subgroup, any infinite torsion subgroup is a direct limit of finite groups. It is then an exercise in the equivariant sphere theorem and the Smith conjecture to deduce that M has finite fundamental group, and is prime or else M is prime, aspherical, atoroidal, but not hyperbolizable, and that there are regular coverings $M \to N_i$ of arbitrarily large degree; this last assertion depends on the Orbifold Geometrization Conjecture (Problem 3.46). Furthermore, homological group theory gives severe restrictions on the possibilities for the deck groups of these coverings.

Problem 3.74 (Mess) *Let M^3 be compact, closed, and aspherical. Suppose $\pi_1(M)$ is not word hyperbolic. Then show that $\mathbb{Z} \oplus \mathbb{Z}$ occurs as a subgroup of $\pi_1(M)$.*

Remarks: In [415,Gromov,1987], Gromov shows that if $\pi_1(M)$ is not word hyperbolic then there is a conformal map of the complex plane to the universal cover of M which is area minimizing on every disc in the complex plane. Thus the canonical approach to this problem is to show that if there is such a minimal surface then there is also a minimal surface which covers a torus and is stabilized by a rank two free abelian subgroup of the fundamental group $\pi_1 M$. A partial result is that a single area minimizing surface can be improved to a lamination [812, Mosher & Oertel,1995].

Problem 3.75 (Waldhausen) *If a closed M^3 is irreducible with infinite $\pi_1(M)$, does $\pi_1(M)$ contain a surface group?*

Remarks: If not, then all infinite index, finitely generated, subgroups of $\pi_1(M)$ are free. Apparently there is no reference for this fact although some experts (e.g. Jaco and Scott) knew it in the 1970's. Here is a sketch (Scott):

First note that if M contains a P^2, then $\pi_1(M)$ contains a surface group namely $\mathbb{Z}/2\mathbb{Z}$. From now on, assume that M is P^2-irreducible so that M is aspherical. Suppose that there is a non-free subgroup H which is finitely generated and has infinite index. Then H has an indecomposable free factor, K, which is not $\mathbb{Z}$. Now in the cover M_K corresponding to K there is a core whose boundary is π_1-injective in M_K. This boundary cannot be empty because the cover corresponding to K is non-compact, as K is of infinite index in $\pi_1(M)$. As M is aspherical, so is M_K. Hence any S^2 component of the boundary of the core must bound a fake ball which we can add to the core. By repeating this addition, we construct a core of M_K, whose boundary has no S^2 components. Any component of this boundary must be a π_1-injective surface, as required.

Problem 3.76 (Hass) *Determine which finitely generated, 3-manifold groups are LERF. In particular, is the figure-8 knot group LERF? What about hyperbolic 3-manifold groups?*

Remarks: G is LERF (locally extended residually finite) if for any finitely generated subgroup $S \subset G$ and any element $g \in G - S$, there exists a finite index subgroup H such that H contains S but not g. All surface groups are LERF and all Seifert fibered 3-manifold groups are LERF ([977,Scott,1978,J. London Math. Soc.] and correction [982, Scott,1985b,J. London Math. Soc.]), but many graph manifold groups are not ([165,Burns, Karrass, & Solitar,1987,Bull. Austral. Math. Soc.] and [675,Long & Niblo,1991,Math. Zeit.]).

Nothing has been proved about closed or finite volume hyperbolic 3-manifold groups; it is possible that all are LERF, or that none are. This problem is closely related to the question (Problem 3.52) of the tameness of the ends of hyperbolic 3-manifolds (see [981,Scott,1985a, Topology] for a discussion of this).

If the fundamental group of M has a surface subgroup and is LERF, then the surface imbeds in a finite cover, and thus M is virtually Haken. Furthermore, there is a finite cover of M which has infinite first homology.

Problem 3.77 (Scott) A Poincaré duality group G of dimension n (PDn-group) is a finitely presented group whose homology and cohomology satisfy Poincaré duality over $\mathbb{Z}[G]$ with a fundamental class in dimension n. The only known PDn-groups are fundamental groups of closed aspherical n-manifolds, and every PD2-group is the fundamental group of a closed aspherical surface ([268,Eckmann & Muller,1980, Comment. Math. Helv.] and [267,Eckmann & Linnell,1983,Comment. Math. Helv.]). (See Problem 5.29.)

(A) *Is every PD3-group G the fundamental group of a closed, aspherical 3-manifold?*

Remarks: If G is a PD3-group with non-trivial center and if $H_1(G)$ is infinite, then G is the fundamental group of an aspherical Seifert fibered space [484,Hillman,1985,Math. Zeit.]. The answer is also yes if G is solvable or has a finitely presented normal subgroup of infinite index [1047,Thomas,1984,Math. Zeit.]. Thomas [1048,1995] has just published a survey on (A).

(B) Conjecture: *If G has non-trivial center, then G is the fundamental group of an aspherical Seifert fibered space.*

(B) is a special case of (A), and is motivated by the theorem that if a closed 3-manifold has fundamental group with center $\mathbb{Z}$, then it is a Seifert fibered space (see the Update to Problem 3.5).

(C) Conjecture: *There does not exist a finitely presented, infinite, torsion group.*

Remarks: This conjecture would be useful in some approaches to (B). This is because, in (B), the quotient of G by the central $\mathbb{Z}$ must be finitely presented and infinite. Mess showed that if G is a 3-manifold group, then this quotient cannot be torsion (see Problem 3.73).

Problem 3.78 (Cochran & Freedman) Given a group G, recall that the lower central series $\ldots G_{n-1} \supset G_n \supset G_{n+1} \ldots$ is defined inductively by $G_1 = G$ and $G_n = [G, G_{n-1}]$. (Note that if $G = \pi_1(X)$, then an element of G_n is a product of elements represented by maps of an nth-order, half-grope into X, where the boundary loop of the grope lies in G_n (see [330,Freedman,1984])). For the first countable ordinal ω, we can define $g \in G_\omega$ if $g \in G_n$ for all n. Then $G_{\omega+1} = [G, G_\omega]$, and so on, defining G_α for all ordinals α. (Note that $g \in G_\omega = (\pi_1(X))_\omega$ does not mean that g corresponds to an infinite half-grope.) Finally, $|G|$ is some cardinal, so there exists an ordinal α (with cardinality no more than than $|G|$) for which $G_\alpha = G_{\alpha+1} = \ldots$ and the lower central series is constant after α. Let the smallest such α be called the *length* of G.

(A) *Find a closed 3-manifold group G such that $F/F_k \cong G/G_k$ for all $k \in \mathbb{Z}_+$ and F some free group, with the property that the length of G is not ω.*

Remarks: Given such a G, then G/G_ω is a finitely generated, parafree group satisfying $H_2(G/G_\omega) \neq 0$, which would provide a counterexample to Baumslag's Parafree Conjecture: a finitely generated, parafree group satisfies $H_2(G;\mathbb{Z}) = 0$ [73,Baumslag,1967, Trans. Amer. Math. Soc.], [74,Baumslag,1969,Trans. Amer. Math. Soc.] (parafree is defined in the Problem 3.79).

(B) *For what ordinals α does there exist a compact 3-manifold M for which $\pi_1(M)$ has length α? More generally, what are the restrictions on the lower central sequence $\{G/G_\alpha\}$ of a closed 3-manifold group.*

Remarks: First note that any countable ordinal is the length of some finitely generated group [431,Hall & Hartley,1966,Proc. London Math. Soc.], but it is not known which are the lengths of finitely presented groups, let alone compact 3-manifold groups.

Second, [648,Levine,1995] contains the first examples of finitely presented groups with length larger than ω.

Third, there are closed, oriented 3-manifolds whose groups have length at least 2ω (i.e. $G_{\omega+k+1} \neq G_{\omega+k}$ for any non-negative integer k). These 3-manifolds can be taken to be the connected sum of a lens space and integral surgery on the Borromean rings or the Whitehead link (their actual length is not yet known) [217, Cochran & Orr,1995].

Furthermore, given a close, oriented, connected 3-manifold M with fundamental group G, there is a *hyperbolic* 3-manifold M' with group G' and a degree one map $f : M' \to M$ which induces an isomorphism on H_1 (and hence induces isomorphisms $f_\alpha : G'/(G')_\alpha \cong G/G_\alpha$ for each ordinal α). As a consequence, G and G' have the same length [ibid.] .

(C) *Is the length of a 3-manifold group invariant under homology bordism?*

Remarks: A yes answer would make an example for (A) very difficult to find, and would offer strong support for the *Parafree Conjecture.*

(D) *Is the property,* length bigger than ω, *generic in some sense for closed 3-manifold groups, or is it rare?*

Problem 3.79 (Freedman) Call a group G *almost parafree* if there exists a homomorphism $\varphi : F \to G$, F a finitely generated free group, which induces isomorphisms

$$\varphi_n : F/F_n \xrightarrow{\cong} G/G_n$$

for all integers $n > 0$. (Recall that these isomorphisms are implied by φ_2 = isomorphism and $\varphi_* : H_2(F) \to H_2(G)$ is an epimorphism (in the case above, $H_2(F) = 0$ so $H_2(G) = 0$ is necessary) [1009, Stallings, 1965, J. Algebra] and [212, Cochran, 1985, Math. Proc. Cambridge Philos. Soc.] for a topological proof.) An almost parafree group is called *parafree* if in addition $\cap_n G_n = e$ [73, Baumslag, 1967, Trans. Amer. Math. Soc.], [74, Baumslag, 1969, Trans. Amer. Math. Soc.].

A stronger condition, $\mathcal{E}$, which implies almost parafree is that there is an epimorphism $e : G \twoheadrightarrow F$ which induces an isomorphism $e_* : H_1(G) \to H_1(F)$. There exists a closed 3-manifold M whose $\pi_1(M)$ is almost parafree but does not satisfy the stronger condition $\mathcal{E}$, for example, 0-surgery on any ribbon link which is not a homology boundary link [483, Hillman, 1981].

Question: *Given any closed 3-manifold M with $\pi_1(M)$ almost parafree, is M homology bordant to a 3-manifold N whose $\pi_1(N)$ (which is necessarily almost parafree) satisfies the stronger condition $\mathcal{E}$?*

Remarks: This is in analogy with a link L in S^3 and its fundamental group $\pi_1(S^3 - L)$. The $\bar{\mu}$ invariants of L are all zero iff $\pi_1(S^3 - L)$ is almost parafree; L is a homology boundary link iff the stronger condition $\mathcal{E}$ holds (that is, there is an epimorphism $e : \pi_1(S^3 - L) \twoheadrightarrow F$ inducing an isomorphism on H_1); (L is a boundary link iff in addition e takes meridians to generators). Homology boundary links have all $\bar{\mu} = 0$, but the converse is not true [Hillman, ibid.]. There also exist links with $\bar{\mu} = 0$ which are not homology boundary links *or* ribbon links (fusions of boundary links) [213, Cochran, 1987, Invent. Math.].

Therefore the above open question is very closely related to the question, *is every link in* S^3 *with* $\bar{\mu} = 0$ *concordant to a homology boundary link?*. There exist links L with $\bar{\mu} = 0$ which are not concordant to any boundary link [216,Cochran & Orr,1993,Ann. of Math.].

Problem 3.80 (Hass) *Find an algorithm to determine if a 3-manifold* M^3 *is simply connected.*

Remarks: Rubinstein has outlined an algorithm (see [1049,Thompson, 1994,Math. Res. Lett.] for a moderately different version) to determine if M is S^3, but it does not detect homotopy 3-spheres which are not S^3.

Problem 3.81 (Rubinstein) *Given a loop* γ *in a compact 3-manifold* M*, is there an algorithm to decide whether* γ *is contractible?*

Remarks: There is an algorithm when M is Haken [1089,Waldhausen, 1968c,Ann. of Math.] which works for 3-manifolds with finite covers which are Haken; therefore there is an algorithm for Seifert fibered spaces. The problem is equivalent to finding an algorithm for building the universal cover of M by putting fundamental polyhedra together.

An algorithm could be found if the Geometrization Conjecture holds, or if $\pi_1(M)$ is residually finite. The problem may be easier if M has an immersed, incompressible surface (for some special cases, see [1003, Skinner,1994,Topology]).

Problem 3.82 (Rubinstein) *Given two compact 3-manifolds,* M_1 *and* M_2*, find an algorithm to decide whether they are homeomorphic.*

Remarks: An algorithm exists for Haken manifolds [428,Haken,1961, Acta Math.], [429,Haken,1962,Math. Zeit.], modulo work of Hemion [470,1979,Acta Math.] solving the conjugacy problem for the mapping class group; an improved version can be found in [536,Jaco & Oertel, 1984,Topology]. An algorithm also exists when the Heegaard genus of M_1 is ≤ 2 [951,Rubinstein,1995].

Problem 3.83 (Rubinstein) *Is there an algorithm to decide whether a 3-manifold* M *has a 2-sided, immersed, incompressible surface?*

Problem 3.84 (A) *Formulate an integer valued notion of complexity of a closed 3-manifold, say $C(M^3)$. It should have the property that there are only finitely many 3-manifolds $F(N)$ with complexity less than any given integer N. The number of tetrahedra in a minimal combinatorial triangulation of M^3 is an example.*

Let $\mathcal{S}$ be a collection of closed 3-manifolds; we say that $\mathcal{S}$ is *almost all* 3-manifolds if $\lim_{N\to\infty} \frac{[\mathcal{S}\cap N]}{F(N)} = 1$ where by definition $[\mathcal{S}\cap N]$ is the number of 3-manifolds in $\mathcal{S}$ of complexity less than N. Define equivalence of two notions of complexity, C and C', to mean that $\mathcal{S}$ is *almost all* for C iff $\mathcal{S}$ is almost all for C'.

(B) *What other notions of complexity are equivalent to the example in (A)?*

(C) *Let $\mathcal{H}$ be the set of Haken 3-manifolds. In the universe of irreducible 3-manifolds, is $\mathcal{H}$ almost all 3-manifolds? If not, what is the limit $\lim_{N\to\infty} \frac{[H\cap N]}{F(N)}$?*

Problem 3.85 (Boileau) Conjecture: *A closed, orientable 3-manifold M^3 has a unique unstabilized Heegaard splitting if it is modeled on one of the four geometries,*

(1) S^3

(2) E^3

(3) $S^2 \times \mathbb{R}$

(4) *Nil.*

Remarks: This conjecture is true for S^3 and $S^2 \times S^1$ [1087,Waldhausen, 1968a,Topology], for lens spaces [131,Bonahon & Otal,1983,Ann. Sci. École Norm. Sup. (4)], for T^3 [120,Boileau & Otal,1990,J. Differential Geom.], and for Seifert fibered 3-manifolds with three exceptional fibers and whose base is spherical or Euclidean [121,Boileau & Otal, 1991,Invent. Math.], [122,Boileau & Otal,1994]. Thus the conjecture is known for the geometries S^3, E^3, and $S^2 \times \mathbb{R}$, except for two Euclidean manifolds containing an incompressible Klein bottle.

A positive answer to the conjecture would prove that a closed, minimal (mean curvature zero) surface with a given genus in a 3-manifold with Ricci curvature ≥ 0 is unique up to isotopy. This follows because Ricci curvature ≥ 0 implies that M^3 has one of the geometries S^3, E^3 or $S^2 \times \mathbb{R}$ [436,Hamilton,1986,J. Differential Geom.] and because a closed, minimal surface is a Heegaard splitting surface of genus > 1 (or a flat T^2 in T^3).

Definitions: Suppose S is a Heegaard surface, splitting M into two compression bodies, H and H'. If essential disks D and D' can be found in H and H' respectively, then,

(a) the splitting is *reducible* if $\partial D = \partial D'$,

(b) the splitting is *stabilized* if ∂D and $\partial D'$ intersect in exactly one point, (and is unstabilized if no such D and D' exist),

(c) the splitting is *weakly reducible* if $\partial D \cap \partial D' = \emptyset$, and

(d) the splitting is *strongly irreducible* if there are no D and D' satisfying (a) or (b) or (c).

Note that (a) implies (c), just by pushing one boundary off the other. Also, (b) implies (a) unless $M = S^3$ and S is the genus one splitting. If M is irreducible, then (a) implies (b) [1087,Waldhausen, 1968a,Topology]. And if M is reducible (is a connected sum), then any Heegaard splitting is reducible [430,Haken,1968].

Problem 3.86 *What can be said about Heegaard splittings of closed, orientable 3-manifolds M^3 with one of the other four geometries,*

(1) $H^2 \times \mathbb{R}$

(2) $\widetilde{PSL}(2, \mathbb{R})$

(3) *Solv*

(4) H^3.

Remarks: In (1) uniqueness holds for $S^1 \times F_g^2$, $g > 1$, [974,Schultens, 1993,Proc. London Math. Soc.]. 3-Manifolds in (1)–(3) are all Seifert fibered spaces, and their irreducible Heegaard splittings are either horizontal or vertical [797,Moriah & Schultens,1995], [975,Schultens,1995].

There are examples with non-unique splittings of the same genus [92, Birman, Gonzalez-Acuña, & Montesinos,1976,Michigan Math. J.], [795, Moriah,1988,Invent. Math.], [115,Boileau, Collins, & Zieschang,1986, C.R. Acad. Sci. Paris Sér. I Math.] and [686,Lustig & Moriah,1991, Topology]. In (4) there exist non-unique examples [781,Montesinos & Whitten,1986,Pacific J. Math.].

Problem 3.87 (Moriah) *Is there a general theory describing Heegaard splittings of arbitrary 3-manifolds, as opposed to many special cases?*

Remarks: Moriah & Rubinstein proved that Heegaard splittings of bounded genus of manifolds obtained by Dehn surgery on a hyperbolic knot or link are, for large enough surgery, Heegaard splittings of the knot or link complement (i.e. the Heegaard surface can be moved off the link). Furthermore, either the link components are cores of the handlebodies, i.e. the Heegaard splittings are tunnel constructions and can be viewed as *vertical*, or there is a simple closed curve on the Heegaard surface which is isotopic to a component of the link, and therefore can be viewed as *horizontal*. This is similar to the known Heegaard splittings of Seifert fibered spaces in general.

Problem 3.88 (Boileau) *Does every closed, oriented, irreducible, 3-manifold have a finite cover which has a unique Heegaard splitting?*

Remarks: Schultens [975,1995] has shown that every irreducible splitting of a surface bundle over S^1 with finite monodromy is vertical, but these are not necessarily unique [686,Lustig & Moriah,1991,Topology]. But such a bundle has a finite cover which is a product, and these have unique irreducible splittings [974,Schultens,1993,Proc. London Math. Soc.]. Compare this problem with Problem 3.51.

Problem 3.89 (Moriah) *Any two Heegaard splittings of a given 3-manifold M become equivalent (isotopic) after stabilizing sufficiently many times by connected summing with the genus one Heegaard splitting of S^3. Is there a universal bound N such that two splittings of M*

become equivalent when the higher genus splitting is stabilized N times and the other is stabilized appropriately?

Remarks: There is no known example needing more than one stabilization. $N = 1$ suffices for the case of 2-bridge knot or link complements and non-simple tunnel number one knot complements [427,Hagiwara, 1994,Kobe J. Math.], for vertical Heegaard splittings of Seifert fibered spaces [975,Schultens,1995] and (E. Sedgwick), and for the examples of Casson & Gordon (described in [591,Kobayashi,1988,Osaka J. Math.]), (E. Sedgwick). Rubinstein & Scharlemann [952,1995] have shown in the non-Haken case that there is a bound which is linear in the genera of the two splittings. Johannson [544,Johannson,1995] has shown that there is a polynomial bound in the Haken case, and Rubinstein & Scharlemann give a quadratic bound.

Problem 3.90 (Johannson) (A) Conjecture: *A closed, irreducible, atoroidal 3-manifold has at most a finite number of Heegaard splittings of a given genus.*

(B) *For manifolds satisfying (A), does the number of Heegaard splittings of genus g grow polynomially in g?*

Remarks: An atoroidal, Haken manifold has at most a finite number of Heegaard splittings of a given genus [543,Johannson,1990,Bull. Amer. Math. Soc.], [544,Johannson,1995].

A closed hyperbolic manifold has at most a finite number of genus 2 Heegaard splittings [456,Hass,1992,Proc. Amer. Math. Soc.].

Problem 3.91 (Gordon) *Let H_i be an unstabilized Heegaard splitting of a 3-manifold M_i, $i = 1, 2$. Is $H_1 \# H_2$ an unstabilized Heegaard splitting of $M_1 \# M_2$?*

Remarks: If H_i is a *minimal genus* Heegaard splitting of M_i, then $H_1 \# H_2$ is a minimal genus Heegaard splitting of $M_1 \# M_2$ [430,Haken, 1968]. However an unstabilized splitting may have arbitrarily higher genus than a minimal genus splitting of M [593,Kobayashi,1992,Osaka J. Math.].

Problem 3.92 (Moriah) *Find more examples of minimal genus Heegaard splittings of 3-manifolds for which the Heegaard genus is greater than the minimal number of generators in a presentation of $\pi_1(M^3)$.*

Remarks: Boileau & Zieschang [127,1984,Invent. Math.] found examples of this phenomena among Seifert fibered 3-manifolds over S^2 with four exceptional fibers, and proved that they did not exist for most other Seifert fibered spaces.

Recall that a *minimal genus* Heegaard splitting means the obvious: no other Heegaard splitting has smaller genus.

Problem 3.93 (Moriah) *Given any closed, orientable 3-manifold with any metric, show that the Heegaard surface of any Heegaard splitting is isotopic to a minimal surface.*

Remarks: Pitts & Rubinstein proved that a strongly irreducible, Heegaard surface in a negatively curved 3-manifold is isotopic to either a minimal surface or to the boundary of an I-bundle over a non-orientable surface with boundary.

Problem 3.94 (Adams) *What is the minimal volume for a closed, orientable, hyperbolic 3-manifold M of Heegaard genus g?*

Remarks: Using unpublished work of Thurston, there is a universal constant K such that $vol(M) > Kg$, so there is a linear lower bound, but no values for K are known (see the proof of Theorem 2.1 in [7, Adams,1995] for a statement of Thurston's work). The conjectured (by J. Weeks) lower bound for the volume of any M is .947 for a genus 2 manifold.

Problem 3.95 *If M^3 is closed, irreducible and $|\pi_1(M)| = \infty$, and if $f : M \to M$ is a homeomorphism with $f \simeq id$, then is f isotopic to the identity?*

Remarks: This does not follow immediately from the Geometrization Conjecture. It is false for reducible 3-manifolds [343,Friedman & Witt,

1986,Topology]. It is true for all geometric 3-manifolds except possibly hyperbolic ones. For hyperbolic 3-manifolds it is true after passing to a finite sheeted covering [359,Gabai,1994a,J. Amer. Math. Soc.] and is true for those hyperbolic 3-manifolds having a closed geodesic with a $log(3)/2$ neighborhood [361,Gabai,1995a].

An unpublished theorem of Siebenmann is relevant: If $f \simeq \mathrm{id}$, M is hyperbolic and there is a geodesic γ such that $f(\gamma)$ is isotopic to γ, then f is isotopic to the identity. (Sketch of proof: Assume $f(\gamma) = \gamma$. By Thurston's geometrization theorem, $M - \gamma$ is hyperbolic, so by [813, Mostow,1967,Inst. Hautes Études Sci. Publ. Math.] $f|(M - \gamma)$ is homotopic to an isometry g and hence isotopic to that isometry. (The latter statement uses [1088,Waldhausen,1968b,Ann. of Math.] and the fact that $M - \gamma$ is the interior of a Haken manifold.) Since the isometry group of a finite volume hyperbolic manifold is finite, g is periodic. This g extends to a periodic homeomorphism h of M which is homotopic to f and fixes γ. The lift of h to B^3 fixes S^2 (since $f \simeq \mathrm{id}$), and is periodic, but the only periodic map with this property is the identity [846,Newman,1931,Quart. J. Math. Oxford Ser. (2)].)

The answer is yes for Haken 3-manifolds [F. Waldhausen, ibid.]; for Seifert fibered 3-manifolds with infinite fundamental group from [48,Asanao,1978,Yokohama Math. J.], [98,Birman & Rubinstein,1984, Proc. London Math. Soc.], [949,Rubinstein,1979b,Trans. Amer. Math. Soc.] in special cases, and in general from [981,Scott,1985a,Topology] and [119,Boileau & Otal,1986,C.R. Acad. Sci. Paris Sér. I Math.]; and for lens spaces from [129,Bonahon,1983b,Topology] and [499,Hodgson & Rubinstein,1985].

Problem 3.96 (Hass) Let $f : F \to M^3$ be a 2-sided immersion of a surface such that $f_* : \pi_1(F) \to \pi_1(M)$ is not injective.

Simple loop conjecture for 3-manifolds: *There exists an essential, simple loop on F whose image is null homotopic in M.*

Remarks: This conjecture arises when trying to surger an immersed surface which is not π_1-injective, and in trying to characterize 3-manifold groups among Poincaré duality groups.

The conjecture is true for M a Seifert fibered space [455,Hass,1987, Proc. Amer. Math. Soc.]. The conjecture is also true if F is an imbedded surface by the loop theorem, but is false for 1-sided surfaces, e.g. a torus in $\mathbb{RP}^2 \times S^1$.

Problem 3.97 *Let $f : S^2 \to S^3$ be a topological imbedding. Suppose that for any $\epsilon > 0$, there is a continuous function $g : S^2 \to C$, where C is either complement, such that $|f(x) - g(x)| < \epsilon$ for all $x \in S^2$. Does it follow that the closure of C is homeomorphic to B^3?*

Remarks: The question is open in all higher dimensions, but a counterexample in this dimension should, by suspension, give counterexamples in all higher dimensions. The 2-dimensional Schoenflies Theorem [775, Moise, 1977] gives an affirmative answer in one lower dimension, even without the hypothesis about g.

Problem 3.98 (Haken) Let $f : S^2 \to M^3$ be a smooth immersion into an orientable 3-manifold M. Assume that the double point set in $f(S^2)$ is an immersed circle (its preimage must be two circles) with $2k$ triple points (there must be an even number). Let N be a regular neighborhood of $f(S^2)$. $\pi_1(N)$ is either trivial or is $\mathbb{Z}/3\mathbb{Z}$, and in either case ∂N is a union of 2-spheres.

Conjecture: *N is a punctured S^3 or $L(3,1)$.*

Remarks: More generally, if the double point set of $f(S^2)$ consists of n immersed circles, then $\pi_1(N)$ has n generators and each triple point gives a relator of length 3. Every orientable 3-manifold can be punctured enough so that it is then realized as the regular neighborhood of some immersion $f : S^2 \to M^3$, or just as an abstract thickening of S^2 with various pairs of immersed circles identified.

Problem 3.99 (A) *Show that a non-orientable, irreducible 3-manifold with $\pi_1 = \mathbb{Z}/2\mathbb{Z}$ is $\mathbb{RP}^2 \times I$.*

Remarks: A warm up problem would be to show that for an arbitrary metric on $\mathbb{RP}^2 \times I$, a least area, one-sided, essential, proper annulus is imbedded. (Note that this is essentially the old Problem 3.4.)

(B) *Show that an irreducible, closed, orientable 3-manifold with $\pi_1 = \mathbb{Z}/2\mathbb{Z}$ is $\mathbb{RP}^3$.*

Remarks: Again, to warm up show that for an arbitrary metric on $\mathbb{RP}^3$, a least area $\mathbb{RP}^2$ is imbedded. Both (A) and (B) are true if their covers are standard [668,Livesay,1960,Ann. of Math.], [669,Livesay,1963,Ann. of Math.], and [947,Rubinstein, 1976,Proc. Amer. Math. Soc.].

Problem 3.100 (Y. Rong) Let M^3 be closed and orientable.

(A) *Are there only finitely many irreducible 3-manifolds N such that there exists a degree one map $M \to N$?*

Remarks: There exists infinitely many if M is non-orientable (Rong). A degree one map $f : M \to N$ is always homotopic to one of the form: there is a Heegaard splitting $H \cup H'$, $H \cap H' = F_g$, of N such that $f|f^{-1}(H)$ is a homeomorphism, there exist g disjoint Seifert surfaces in $f^{-1}(H')$ for the g meridians in $f^{-1}(F_g)$, which are mapped in the obvious way to the 2-disks in H', and then the remainder of M is crushed to the remaining 3-ball in N [937,Rong & Wang,1992,Math. Proc. Cambridge Philos. Soc.].

(B) *Does there exist an integer N_M such that if*

$$M \stackrel{deg1}{\to} M_2 \stackrel{deg1}{\to} M_3 \stackrel{deg1}{\to} \dots \stackrel{deg1}{\to} M_k$$

is a sequence of degree 1 maps with $k > N_M$, then the sequence must contain a homotopy equivalence?

Remarks: Note that a yes answer to (A) implies a yes answer to (B). If $k = \infty$ and M is geometric, then after some finite stage all degree one maps are homotopy equivalences [934,Rong,1992, Trans. Amer. Math. Soc.].

One can also ask these questions when $\partial M \neq \emptyset$ and the degree one maps must be proper. In the knot complement case, there is an obvious relation with parts (C) and (D) of Problem 1.12.

Problem 3.101 (Rourke) (A) *Given two oriented, framed links, L_1 and L_2, in S^3 which define the same 3-manifold, is it possible to get from L_1 to L_2 by a sequence of coherent K-moves, where a coherent K-move is an ordinary K-move in which all strands passing through the ± 1 unknot are oriented in the same direction?*

Remarks: See [300,Fenn & Rourke,1979,Topology] for a definition of a K-move. (A) is equivalent to the following question about surgery on framed braids.

(B) *Given two (oriented), framed, closed braids, L_1 and L_2, in S^3 which define the same 3-manifold, is it possible to get from L_1 to L_2 using the Markov moves and braided K-moves, where a braided K-move is a K-move in which the ± 1 unknot lies in a plane transverse to the braid and encircles the leftmost s strands in the braid, $0 \leq s \leq n$, for an n-strand braid (of course the ± 1 unknot can be oriented and tilted to form part of the braid).*

Remarks: This seems to be the natural question combining surgery on framed links and braids. A harder question (in that an affirmative answer implies yes to (A) and (B)) is:

(C) *Since any 3-manifold can be obtained by surgery on a pure, framed closed braid, suppose two such give the same 3-manifold; is it possible to move from one braid to the other by a sequence of braided K-moves (which preserve pureness) and conjugacy of braids?*

Remarks: The two Markov moves (under which two braids are equivalent if their closures form the same link) can be replaced by one move [638,Lambropoulou & Rourke,1996].

Ko & Smolinsky [589,1992,Proc. Amer. Math. Soc.] in the circumstances of (B), show how to get from L_1 to L_2 by Markov moves, blowing up and down, braid preserving handle slides, and a certain way of reversing the orientation of a ± 1 unknotted component.

Problem 3.102 (Auckly) Define the *surgery number*, $S(M)$, of a 3-manifold to be the minimal number of components in a framed link needed to define M. In the same way, define the *Dehn surgery number*, $S_D(M)$.

(A) *Is it true that the connected sum of n non-trivial homology 3-spheres has $S_D \geq n$?*

Remarks: Neither S nor S_D is additive, as can be seen by considering surgery on cabled knots. (A) is true for $n = 2$ by [404, Gordon & Luecke,1987,Math. Proc. Cambridge Philos. Soc.].

(B) *Find an irreducible, atoroidal, homology 3-sphere, M, with $S(M) > 2$ or $S_D(M) > 2$, or even arbitrarily large.*

Remarks: There is a hyperbolic M with $S_D(M) > 1$ [54,Auckly, 1995]. Probably one can use lens spaces to find 3-manifolds with $S - S_D$ arbitrarily large.

Problem 3.103 (Hoste) Every open 3-manifold (paracompact) can be obtained by surgery on a possibly infinite link imbedded in $S^3 - X$ where X is a closed subset of a fixed tame Cantor set in S^3 [507,Hoste, 1995].

Question: *What is the calculus relating the different links that give the same 3-manifold?*

Remarks: It is known that it may take infinitely many moves to pass between different framed links for the same manifold [ibid.], but it may be possible to do a finite number of sets-of-moves each of which consists of an infinite number of simultaneous moves in different isolated locations.

Problem 3.104 (Auckly) (A) *Is every closed 3-manifold M with a representation $\varphi : \pi_1(M) \to SU(2)$ given by a framed link L which has a sublink S satisfying the following two conditions:*

1. *for any meridian m of S, $\varphi(m) = \pm I \in SU(2)$.*
2. *$L - S$ can be divided into sublinks $L_1 \cup L_2 \cup \ldots \cup L_k$, $k \geq 1$, such that each L_i lies in a 3-ball B_i and the $\{B_i\}$ are pairwise disjoint; then we require that, for each $i \in \{1, \ldots, k\}$, all meridians of L_i are taken by φ into some $U(1)$ in $SU(2)$ which may vary with i.*

Remarks: Yes if M is Seifert fibered or a graph manifold or some bundles over S^1 [53,Auckly,1994,Math. Proc. Cambridge Philos. Soc.] .

(B) *Is every closed 3-manifold M, with a representation $\varphi : \pi_1(M) \to SU(2)$, bordant via W^4 to a connected sum of lens spaces such that φ extends over the bordism to $\Phi : \pi_1(W) \to SU(2)$?*

Remarks: Representations φ correspond to flat connections A on the trivial $SU(2)$ bundle over M. An affirmative answer to either (A) or (B) would imply that the Chern–Simons invariant $CS(M, A)$ is rational [Auckly, ibid.] .

Problem 3.105 (A) *Does every compact 3-manifold M^3 have a non-trivial representation $\pi_1(M) \to SU(2)$? If $\pi_1(M)$ is non-abelian when can the representation have non-abelian image?*

Remarks: It is not always true that if $\pi_1(M)$ is non-abelian, then the representation can be chosen so that its image is not Abelian; if two trefoil knot complements are glued together so that the meridian of one matches the fiber of the other copy when it is thought of as a Seifert fibered 3-manifold (in which case the fiber is lm^{-6}), then the resulting fundamental group is non-abelian but has only Abelian representations to $SU(2)$ (Klassen). One can ask for a similar example among homology 3-spheres.

One tool for finding non-trivial representations is to place a homology 3-sphere in a 4-manifold with positive b_+^2 on both sides and non-trivial Donaldson invariants.

(There are finitely presented, infinite, simple groups with no non-trivial representations to any matrix group. Higman [480,1951,J. London Math. Soc.] gave the following example, $\langle a, b, c, d | \bar{a}ba = b^2, \bar{b}cb = c^2, \bar{c}dc = d^2, \bar{d}ad = a^2 \rangle$, which has no subgroups of finite index, whereas any finitely generated subgroup of a matrix group must have subgroups of finite index; this group is not simple (although it is clearly perfect ($b = \bar{b}\bar{a}ba$ etc)), but its quotient by any maximal proper normal subgroup is simple.)

(B) *Find an M^3 with an isolated representation $\rho : \pi_1(M) \to SU(2)$ for which $H^1(M; ad(\rho)) \neq 0$. There are examples when $\pi_1(M)$ is replaced by an arbitrary group G, e.g. [679,Lubotzky & Magid, 1985].*

Problem 3.106 (Stern) *Find an example of a Floer sphere, i.e. find an irreducible homology 3-sphere, Σ^3, other than S^3, with vanishing Floer homology groups.*

Problem 3.107 (Braam) *The Casson invariant of homology three spheres is defined as the geometric oriented intersection of two subvarieties of the space of flat connections on a surface. Can a homology theory be defined such that these subvarieties are cycles in this theory?*

Remarks: Something like this has been done for generalized Casson invariants of knots ([347,Frohman,1993,Topology], [349,Frohman & Nicas,1994,Topology]) but the case of homology 3-spheres seems to escape known theories.

Problem 3.108 (Garoufalidis) *Does there exist a closed 3-manifold* M^3 *such that* $Z(M) = Z(S^3)$ *(*M *might be called a* Witten sphere *or* quantum sphere*)?*

Remarks: The invariant Z can have several interpretations:

(i) Witten's definition of Z for $SU(N)$ [1116,1989,Comm. Math. Phys.];

(ii) Z could mean the perturbative Chern–Simons invariants [55,Axelrod & Singer,1992] and [603,Kontsevich,1994];

(iii) Z could be the invariants at r^{th} roots of unity of [923,Reshetikhin & Turaev,1991,Invent. Math.] for $sl(2,\mathbb{C})$ and [1108,Wenzl,1993, Invent. Math.] or [1071,Turaev & Wenzl,1993,Internat. J. Math.], for other Lie algebras.

It would be interesting to find an M with $Z(M) = Z(S^3)$ for just $sl(2,\mathbb{C})$ and all $r > 1$ in (iii). Such an M would have to be a rational homology sphere (using the formula for τ_3 in [582,Kirby & Melvin, 1991b,Invent. Math.]), and thus an integral homology sphere with Casson invariant zero [817,Murakami,1994,Math. Proc. Cambridge Philos. Soc.].

Note that Z for $sl(2,\mathbb{C})$ and all r does not even distinguish lens spaces, for $L(65,8)$ and $L(65,18)$ have the same invariant ([581,Kirby & Melvin,1991a,Abstracts Amer. Math. Soc.] for prime r, and [1129, Yamada,1995,J. Knot Theory Ramifications] for all r).

Let us consider the Reshetikhin–Turaev invariant $Z(M)$ of (iii) for $sl(N,\mathbb{C})$ and r^{th} roots of unity. A growth rate conjecture of Witten can be stated as follows:

Conjecture: Z(M) is asymptotic to $c(sl(N,\mathbb{C}),M)r^{\theta_{sl(N,\mathbb{C})}(M)}$ as $r \to \infty$, where $c(sl(N,\mathbb{C}),M)$ is a nonzero constant and

$$\theta_{sl(N,\mathbb{C})}(M) = max_\rho(h^1(M,ad\rho) - h^0(M,ad\rho))/2$$

where ρ runs over all generic representations of $\pi_1(M)$ in $SU(N)$ and $h^i(M,ad\rho)$ denotes the dimension of the ith cohomology of M with coefficients in ρ.

Garoufalidis [366,1995] has shown that if $Z(M) = Z(S^3)$ for all the $sl(N,\mathbb{C})$ Reshetikhin–Turaev invariants and all r then,

- either the above growth rate conjecture of Witten regarding the asymptotics of $Z(M)$ fails, or

- there is a closed 3-manifold M whose fundamental group is not residually finite, or

- the Poincaré conjecture in dimension 3 fails.

This Problem is analogous to asking for a *Jones knot*, that is, a knot with the same Jones polynomial as the unknot (see Problem 1.88). In this case, Garoufalidis [ibid.] has shown that

- either there exists a framed knot in S^3 all of whose colored $sl(N,C)$ Jones polynomials do not distinguish it from the unknot or,

- the growth rate conjecture of Witten fails.

Problem 3.109 (The Closing Lemma) *If V is a C^r vector field on a closed manifold with a non-trivial recurrent orbit (the orbit is contained in its backward or forward closure, [866,Palis, Jr. & de Melo, 1982a]) is there a C^r-perturbation V_0 of V with a closed orbit?*

Remarks: This problem originated in [886,Poincaré,1892; page 82]; Poincaré's application then was to show that for generic hamiltonian systems, the periodic trajectories are dense in the compact energy surfaces. Thom gave a half page proof in his 1960 preprint [1045,Thom, 1960; pages 5–6], but Peixoto pointed out an error and the Closing

Lemma, as it became known, turned into one of the most sought after building blocks of the theory of dynamical systems.

Among the results implied by a positive solution to the Closing Lemma are that generic dynamical systems have periodic trajectories which are dense in the set of recurrent orbits (Axiom Ab of [1005,Smale, 1967,Bull. Amer. Math. Soc.]), and generic density theorems (periodic sets are dense in the non-wandering set) for fields and diffeomorphisms in general.

Peixoto [870,1962,Topology] was able to obtain the C^r-generic density theorems, mentioned above, for compact orientable surfaces and $1 \leq r \leq \infty$, circumventing the Closing Lemma. In 1967, Pugh [905, 1967,Amer. J. Math.] proved the C^1-version of the Lemma and in 1983 Pugh & Robinson [906,1983,Ergod. Th. & Dynam. Sys.] extended it to the class of hamiltonian fields, proving a C^1-version of Poincaré's initial assertion [*ibid*]. Very little is known beyond that. For example, it is not known whether:

- *all recurrent trajectories of a C^2 differential equation can be closed by a C^2 perturbation of the equation;*
- *for non-orientable surfaces and all higher dimensional manifolds, a generic C^2 differential equation has its periodic trajectories dense in the non-wandering set.*

Problem 3.110 (Freedman) *Find a compactly supported vector field on $\mathbb{R}^3$ which generates a volume preserving flow $\psi_t : \mathbb{R}^3 \to \mathbb{R}^3$ so that for some closed loop, γ*

$$\textit{Whitney norm}(\psi_t(\gamma)) \geq c_0 e^{c_1 t}$$

for some positive constants c_0 and c_1.

Are such flows generic?

Remarks: If Whitney norm is replaced by *length*, then it is not hard to find such a flow. The Whitney norm [1113,Whitney,1957] allows one to minimize using the following idea: find a 2-chain of small area whose boundary is two nearly parallel, oppositely oriented, long strands in $\psi_t(\gamma)$ and two short arcs at either end; one then eliminates the length of the two long strands at the cost of adding the length of the short arcs plus the area of the 2-chain. The infimum over such constructions is the Whitney norm of $\psi_t(\gamma)$.

Problem 3.111 (G. Kuperberg) Conjecture: *Every closed, orientable m-manifold M^m with zero Euler characteristic has a minimal flow. In particular, S^3.*

Remarks: A flow is minimal if every orbit is dense in M^3. One can impose varying degrees of differentiability, but the conjecture is even open for the continuous case where one asks for a topological 1-dimensional foliation with each leaf dense. The n-torus T^n has an analytic, minimal flow. Fathi & Herman [295,1977] have shown that a manifold with a locally free C^∞ S^1-action has a minimal C^∞ diffeomorphism (a $\mathbb{Z}$-action in which every orbit is dense); these diffeomorphisms have mapping tori which have minimal flows. They also show that if M admits a locally free C^∞ T^2-action, then M has a minimal C^∞ flow.

Problem 3.112 (S. Matsumoto) A minimal set X of a flow ϕ is called *isolated* if X admits a neighborhood U with the following property: if the whole orbit of a point x is contained in U, then x is a point of X.

Question: *Does a smooth flow on a 3-dimensional manifold admit an isolated minimal set which is not a closed orbit (homeomorphic to S^1 or a point)?*

Remarks: The answer is not known even for open manifolds (and for compact minimal sets).

Problem 3.113 (G. Kuperberg) *Is there a C^2, volume preserving flow on S^3 with no closed orbit? Can it be C^∞ or even real analytic?*

Remarks: G. Kuperberg has produced such a flow in the C^1 case [630, Kuperberg,1995] which is C^0 conjugate to Schweitzer's C^1 flow [976, 1974,Ann. of Math.]. K. Kuperberg has given an example which is analytic, but not volume preserving [632,1994,Ann. of Math.] (also [631,Kuperberg & Kuperberg,1995].

Chapter 4

4-Manifolds

- Problems 4.1–4.40 (1977), 4.41–4.68 (1982), 4.71–4.148 (new).
- Existence (given homotopy type), 4.1–4.10, 4.49, 4.69, 4.71–4.72, 4.92–4.98.
- Uniqueness (given homotopy type), 4.11–4.16, 4.42, 4.51, 4.80, 4.83–4.91.
- Surfaces in 4-manifolds, 4.21–4.31, 4.36, 4.41, 4.46, 4.58, 4.73, 4.81, 4.98, 4.105–4.113, 4.127, 4.128, 4.140, and 1.48–1.51, 1.93–1.97, 5.6.
- Complex surfaces, 4.36–4.40, 4.50, 4.78, 4.99–4.107, 4.124, 4.133–4.135.
- Exotic $\mathbb{R}^4$'s, S^4's and Casson handles, 4.41–4.43, 4.45, 4.76–4.80, 4.141.
- Smooth handlebodies, 4.16, 4.18, 4.21, 4.28, 4.43, 4.73, 4.88–4.91, 4.98, 4.115, 4.118.
- Structures on topological manifolds, 4.44, 4.71–4.75, 4.82, 4.83.
- Smooth bordisms and splitting 4-manifolds by 3-manifolds, 4.2–4.5, 4.16, 4.32, 4.33, 4.49, 4.74, 4.98, 4.114–4.119.
- Algebraic topology, 4.53, 4.97, 4.105, 4.126.

- Diffeomorphisms, 4.34, 4.35, 4.125–4.126.
- Group actions, 4.55, 4.56, 4.121, 4.123–4.125.
- Bundles, 4.57, 4.122, 4.148, and 2.17, 2.18.
- Fundamental groups, 4.9, 4.10, 4.17, 4.59, 4.60, 4.84, 4.109, 4.119–4.121.
- Imbedding 4-manifolds, 4.19, 4.20, 4.62, 4.63.
- Donaldson, Seiberg-Witten invariants, 4.128–4.133.
- Symplectic manifolds, 4.64, 4.113, 4.135–4.142.
- Differential geometry, 4.65–4.68, 4.143.
- Arrangements, 4.144–4.147.

Introduction

At the suggestion of Gompf, the adjectives *exotic* and *fake* are used as follows: N is called an *exotic* version of M when N and M are homeomorphic but not diffeomorphic, but N is called a *fake* version of M when N and M are only simple homotopy equivalent but not homeomorphic. Thus an exotic $\mathbb{R}^4$ is an exotic smooth structure on R^4, whereas a fake $\mathbb{CP}^2$ is a manifold homotopy equivalent but not homeomorphic to $\mathbb{CP}^2$.

It is also important to note that in dimension 4, *irreducible* means that every smoothly imbedded 3-sphere in a 4-manifold bounds a homotopy 4-ball, which is then homeomorphic to B^4 but is not known to be diffeomorphic to B^4.

Perhaps the best known, yet complicated, smooth 4-manifold is what is now called the $K3$ surface (it was called the Kummer surface in the old list, but that just refers to one of many complex structures on $K3$); this simply connected manifold with intersection form $2E_8 \oplus 3\left(\begin{smallmatrix}0&1\\1&0\end{smallmatrix}\right)$ has many descriptions, e.g. any nonsingular quartic in $\mathbb{CP}^3$, but the most useful one is that of an elliptic surface [444,Harer, Kas, & Kirby,1986]. The $K3$ surface is naturally a fiber connected sum of two copies of a smaller 4-manifold called $E(1)$ (of course $K3 = E(2)$, and $E(n)$ is formed by taking the fiber connected sums of n copies of $E(1)$). $E(1)$ is better known as the rational elliptic surface or $\mathbb{CP}^2\#9(-\mathbb{CP}^2)$ (and was called the half-Kummer surface in the old list).

Problem 4.1 Existence. *What integral, symmetric, unimodular, bilinear forms are the intersection forms of simply connected, closed 4-manifolds?*

Remarks: See [769, Milnor & Husemoller, 1973] for the algebraic background and further references. Odd, indefinite forms are represented by connected sums of copies of $\mathbb{CP}^2$ and $-\mathbb{CP}^2$, but little is known otherwise. In particular, is $E_8 \oplus \langle 1 \rangle$ (the odd, definite form of signature 9), or $E_8 \oplus E_8 \oplus n\left(\begin{smallmatrix} 0 & 1 \\ 1 & 0 \end{smallmatrix}\right)$, $n \leq 2$, or Γ_{16} (the other signature 16, even, definite form) represented by a manifold? By Rohlin's Theorem [334, Freedman & Kirby, 1978] any closed, *smooth*, almost parallelizable (implied by simply connected and even) M^4 satisfies $\sigma(M^4) \equiv 0 \bmod 16$; this rules out E_8 for example.

$$E_8 \oplus E_8 \oplus 3 \begin{pmatrix} 0 & 1 \\ 1 & 0 \end{pmatrix} \cong \Gamma_{16} \oplus 3 \begin{pmatrix} 0 & 1 \\ 1 & 0 \end{pmatrix}$$

is represented by the complex Kummer surface.

Any such form is represented by a simply connected, smooth, M^4 with ∂M^4 equal to a homology 3-sphere; hence, the next problems.

Update: An (even, odd) form is realized by exactly (one, two) simply connected, closed, topological 4-manifold [329, Freedman, 1982, J. Differential Geom.].

In the smooth case, no definite forms are realized other than $\pm \oplus n\langle 1 \rangle$ (which is realized by $\pm \# n\mathbb{CP}^2$), [247, Donaldson, 1983, J. Differential Geom.] (the same is true even if $\pi_1 \neq 0$, [250, Donaldson, 1987b, J. Differential Geom.]. All odd, indefinite forms are direct sums of $\langle 1 \rangle$ and $\langle -1 \rangle$ and hence realized. All even, indefinite forms are of the form $nE_8 \oplus m\left(\begin{smallmatrix} 0 & 1 \\ 1 & 0 \end{smallmatrix}\right)$, and by Rohlin's Theorem, the cases of odd n are not realized, so we can assume that $n = 2k$. Connected sums of the $K3$ surface and $S^2 \times S^2$ realized the cases when $m \geq 3k$. If $k \neq 0$ (and we only assume $H_1(M; \mathbb{Z})$ has no 2-torsion), then $m \geq 3$ by [248, Donaldson, 1986, J. Differential Geom.] and [250, Donaldson, 1987b, J. Differential Geom.] (the 2-torsion assumption is no longer necessary using Seiberg–Witten theory). Furuta [352, 1995] has claimed that $m \geq 2k$ for smooth manifolds with no restriction on the fundamental group. The remaining cases are still open (see Problem 4.103 and Problem 4.117).

Problem 4.2 *Which homology 3-spheres (with Rohlin invariant 0) bound contractible (or acyclic) 4-manifolds?*

Remarks: Essentially nothing is known, except for some families of homology 3-spheres which do bound contractible 4-manifolds, e.g. (Casson and Harer), all Brieskorn spheres $\Sigma(p,q,r)$ where (p,q,r) equals $(2,3,13)$, or $(2,3,25)$, or $(p,ps+1,ps+2)$ for p odd, or $(p,ps-1,ps-2)$ for p odd, or $(p,ps-1,ps+1)$ for p even and s odd (and always $p>0$, $s>0$); in fact, these bound a 4-manifold with one 0-handle, one 1-handle, and one 2-handle, except $(2,3,25)$ which uses two 1-handles and two 2-handles.

Akbulut's candidate for a homology 3-sphere which does not bound an acyclic 4-manifold is $\Sigma(2,3,11)$ which can be obtained by -1-surgery on the knot below (see Figure 4.2.1.

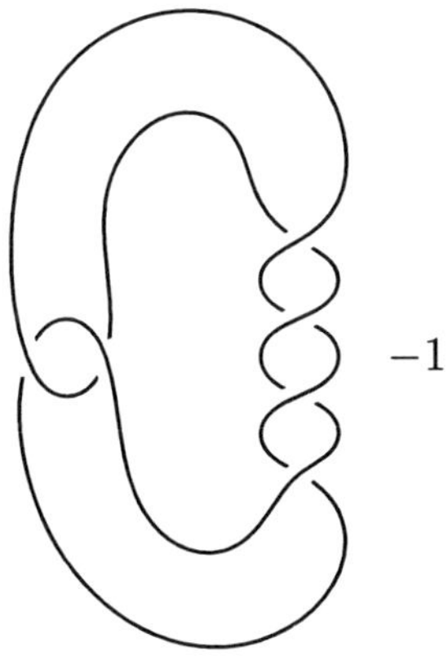

Figure 4.2.1.

The homology 3-sphere $\Sigma(2,7,13)$, which can also be obtained by $+1$-surgery on the $(2,7)$-torus knot, lies in the Kummer surface, bounding a manifold with form Γ_{16} on one side, and the even form $3\left(\begin{smallmatrix}0&1\\1&0\end{smallmatrix}\right)$ on the other. Does it bound an even 4-manifold of rank ≤ 4?

Update: All homology 3-spheres bound contractible, topological 4-manifolds, [329,Freedman,1982,J. Differential Geom.]. In the smooth case many don't, some do, and in most cases we don't know. For example, any homology 3-sphere, which is described by a framed link whose linking form is eliminated as a candidate for a closed, smooth 1-

connected 4-manifold (see Problem 4.1 Update), cannot bound a smooth contractible (or acyclic) 4-manifold. Akbulut's example, $\Sigma(2,3,11)$, does not bound a contractible 4-manifold since it bounds one with intersection form $2E_8 \oplus 2\binom{0\,1}{1\,0}$ [19,Akbulut,1991a,J. Differential Geom.]). And $\Sigma(2,7,13)$ which bounds the definite form Γ_{16} cannot bound a 1-connected, smooth, 4-manifold with $\beta_2 < 6$.

Y. Matsumoto [710,Matsumoto,1982b,J. Fac. Sci. Univ. Tokyo Sect. IA Math.], suggests that the *hyperbolic genus* of a homology 3-sphere M^3 be defined to be the minimal k such M^3 bounds a smooth 4-manifold W^4 with $H_1(W;\mathbb{Z}) = 0$ and intersection form equal to the direct sum of k copies of $\binom{0\,1}{1\,0}$and, if necessary, one E_8. Thus this Problem is a special case of:

Calculate the hyperbolic genus of any homology 3-sphere.

On the other hand here are some interesting examples which do bound contractible, smooth, 4-manifolds: $\Sigma(p, ps-1, ps+1)$ for p even and s odd, and $\Sigma(p, ps \pm 1, ps \pm 2)$ for p odd and any s [189,Casson & Harer,1981,Pacific J. Math.]; $\Sigma(2, 2s \pm 1, 4(2s \pm 1) + 2s \mp 1)$ for s odd, and $\Sigma(3, 3s \pm 1, 6(3s \pm 1) + 3s \pm 2)$, and $\Sigma(3, 3s \pm 2, 6(3s \pm 2) + 3s \pm 1)$ [1011,Stern,1978,Notices Amer. Math. Soc.], and also [27,Akbulut & Kirby,1979,Michigan Math. J.].

Perhaps the most interesting unknown case is the connected sum $M\# -M$ where M has Rohlin invariant one; $M\# -M$ bounds $(M - intB^4) \times I$, but no known example bounds a contractible smooth 4-manifold, (see Problems 4.49 and 4.114).

Problem 4.3 (S. Kaplan) *Does every homology* 3*-sphere bound an even, definite* 4*-manifold?*

Remarks: Many of the interesting examples are links of isolated singularities in complex surfaces, and hence bound negative definite forms [493,Hirzebruch,1966].

Update: Yes, topologically, since they all bound contractible 4-manifolds, [329,Freedman,1982,J. Differential Geom.]. No, smoothly, in the following sense: take a homology 3-sphere, such as $\Sigma(2,3,7)$ which smoothly imbeds in the $K3$ surface with $E_8 \oplus \binom{0\,1}{1\,0}$ on one side and $E_8 \oplus 2\binom{0\,1}{1\,0}$ on the other. If this homology 3-sphere bounds an even definite 4-manifold X^4 with $H_1(X;\mathbb{Z})$ having no 2-torsion, then adding one of the two sides (in $K3$) to X^4 (getting the orientations right) we would have a smooth 4-manifold contradicting [250,Donaldson,1987b, J. Differential Geom.]. The 2-torsion assumption is no longer necessary using Seiberg–Witten gauge theory.

Problem 4.4 *Find a homology 3-sphere H of Rohlin invariant one such that $H\#H$ bounds a PL acyclic 4-manifold.*

Remarks: If such an H exists, manifolds of dimension ≥ 6 are triangulable ([365,Galewski & Stern,1980,Ann. of Math.] or [718,Matumoto, 1976], and R. Edwards' triple suspension theorem). In our current state of ignorance the following conjecture is conceivable: H bounds an acyclic 4-manifold iff $H\#H$ does (H not necessarily of Rohlin invariant one).

Update: No progress, but note that in the Remarks dimension ≥ 6 should be replaced by ≥ 5 for closed manifolds [365,Galewski & Stern, 1980,Ann. of Math.] because of the double suspension theorem [273, Edwards,1975,Notices Amer. Math. Soc.], [176,Cannon,1979,Ann. of Math.] or [233,Daverman,1986].

Problem 4.5 (Casson) *Which rational homology 3-spheres bound rationally acyclic 4-manifolds?*

Remarks: If M^3 bounds a rationally acyclic W^4, then the linking form on $H_1(M;\mathbb{Z})$ must be null concordant (implying $|H_1(M;\mathbb{Z})| =$ square). If $H_1(W;\mathbb{Z})$ is cyclic, then the Casson–Gordon invariants [186,1978], [188,1986] must be ± 1 or 0. The question is already interesting for lens spaces. $L(m^2,q)$, m odd, bounds a W^4 if $q = km \pm 1$ with k, m coprime, or if $q = (m\pm 1)d$ with $d | 2m \mp 1$, or if $q = (m\pm 1)d$ or $(2m\mp 1)(m\pm 1)/d$ with $d | m \pm 1$ and d odd [ibid.].

Update: Casson & Gordon's results were reproved and strengthened using gauge theory in [311,Fintushel & Stern,1987,Topology], and further generalized in [707,Matic,1988,J. Differential Geom.] and [944, Ruberman,1988,Topology].

A Chern–Simons type obstruction can be found in [312,Fintushel & Stern,1990,J. London Math. Soc.].

Problem 4.6 (Freedman) (A) *Let $f : (M, \partial M) \to X, \partial X)$ be a degree 1 normal map from a smooth (or TOP) 4-manifold to a Poincaré space. Suppose $f|\partial$ is a $\mathbb{Z}[\pi_1(X)]$-equivalence. A surgery obstruction*

$$\sigma(f) \in \Gamma_4(\mathbb{Z}[\pi_1(X)] \to \mathbb{Z}[\pi_1(X)]) \cong L_4(\pi_1(X))$$

is defined. If $\sigma(f) = 0$, is f normally bordant to a homotopy equivalence rel ∂?

(B) *Can (A) be reduced to an equivalent question about link concordance?*

Remarks: An affirmative answer to (A) would yield *all* sought-after closed, smooth (or TOP) 4-manifolds (compare Problem 4.1).

There are Γ-group surgery problems with vanishing obstructions $\sigma(f) \in \Gamma_4(\mathbb{Z}[\mathbb{Z}] \to \mathbb{Z}[e])$, which are not normally bordant rel∂ to homotopy equivalences (this follows from [188,Casson & Gordon,1986], and [179,Cappell & Shaneson,1974,Ann. of Math.]). There are also problems with no solution and zero obstruction in $\Gamma_4(\mathbb{Z}[\{a, b : aba = bab\}] \xrightarrow{\gamma} \mathbb{Z}[\mathbb{Z}])$ where $\gamma(a) = \gamma(b) = 1$. Can a similar failure occur for a Wall group problem as in (A)?

Casson has shown ([423,Guillou & Marin,1986c; pages 201–244]) that if certain sequences of links contain a slice link (see Problem 1.39) then simply connected surgery will work in dimension 4. Since slicing these links is itself a nonsimply connected surgery problem, there is hope that a nonsimply connected generalization of Casson's work would yield a *universal surgery problem* (i.e., a problem set up to slice a certain link) whose solution would be equivalent to the solution of all 4-dimensional Wall group problems with zero obstructions.

Update: Call a fundamental group *good* (group theorists call them *elementary*) if it belongs to the smallest class of groups which contain $\mathbb{Z}$ and finite groups and is closed under direct limit, subgroups, extensions and quotients. Perhaps the most interesting finitely presented group which is not good is $\mathbb{Z} * \mathbb{Z}$. Also, there are good groups which have exponential growth, e.g. $G = \langle a, t \mid tat^{-1} = a^2 \rangle$ (note that G is an extension of two good groups, $0 \to Z\left[\frac{1}{2}\right] \to G \to Z \to 0$ with homomorphisms $\left(\frac{1}{2}\right)^n \to t^{-n}at^n$ and $t \to 1$, $a \to 0$; also, G is an example of a group of a knotted S^2 in S^4, but which is not the group of a knot in S^3 [932,Rolfsen,1976; page 345]). (In fact, all finitely generated good groups are virtually nilpotent or have exponential growth.)

Then the answer to **(A)** for TOP is yes if $\pi_1(X)$ is good. In fact, surgery *works* in the topological category for good groups, [336,Freedman & Quinn,1990]. Freedman & Teichner [339,1995] have recently shown that surgery works for a larger class of groups, namely the closure of groups of subexponential growth under the operations of extensions and direct limits (see Problem 5.9).

(A) fails in the smooth case, even if $\pi_1 = 0$. Many examples can be constructed along the lines of the following: Let K be a $K3$ surface (its intersection form is $2E_8 \oplus 3(\begin{smallmatrix}0&1\\1&0\end{smallmatrix}))$ and let M^4 be any submanifold with homology sphere boundary whose form is $2E_8$. Then there is a degree 1 normal map $f : K \to \widehat{M}$ with zero surgery obstruction, where $\widehat{M} = M \cup \mathrm{cone}(\partial M)$, but there is no smooth 4-manifold homotopy equivalent to $\widehat{M}$ by [247,Donaldson,1983,J. Differential Geom.].

(B) can be reduced in the topological case to a link slice problem called the *AB–Slice Problem*, see [330,Freedman,1984]

Problem 4.7 *Is there a TOP, closed, almost parallelizable* 4*-manifold of signature 8?*

Remarks: Yes, if one can prove TOP transversality in the missing case when a 4-dimensional preimage is expected (see [966,Scharlemann, 1976b,Invent. Math.]) (does Sullivan's proof that TOP manifolds are Lipschitz (dim $\neq 4, 5$) help here?), or if a Rohlin invariant one homology 3-sphere bounds an acyclic 4-manifold, perhaps by being TOP imbedded in S^4.

Update: Yes, and it is unique if simply connected. The Poincaré homology sphere bounds the E_8 plumbing and also a contractible topological 4-manifold, so the union is the desired 4-manifold [329,Freedman, 1982,J. Differential Geom.].

Topological transversality holds in all cases [913,Quinn,1988,Bull. Amer. Math. Soc.] and [336,Freedman & Quinn,1990; pages 151–160].

Problem 4.8 *Does there exist a manifold proper homotopy equivalent* *(or even homeomorphic) to $S^3 \times \mathbb{R}$ but not diffeomorphic?*

Remarks: Casson has shown that either such a manifold exists or another manifold, $Q^4 \simeq S^2 \times S^2 - pt$, exists having a exotic end (see Problem 1.39).

Update: Every 4-manifold proper homotopy equivalent to $S^3 \times \mathbb{R}$ is homeomorphic to it also [329, Freedman, 1982, J. Differential Geom.]. Freedman's original example [328, 1979, Ann. of Math.] is not diffeomorphic to $S^3 \times \mathbb{R}$ because it has a *transverse* smooth, imbedding of the Poincaré homology sphere but not S^3.

Other examples, diffeomorphically distinct, can be constructed with ends like exotic $\mathbb{R}^4$'s and/or transverse homology 3-spheres with Rohlin invariant 0 or 1. Since exotic $\mathbb{R}^4$'s are not classified yet up to diffeomorphism it is not surprising that smooth structures on $S^3 \times \mathbb{R} = \mathbb{R}^4 - 0$ are not classified either.

Problem 4.9 (M. Cohen) *Does there exist a 4-dimensional h-cobordism with nontrivial Whitehead torsion?*

Update: If the h-cobordism is W^4 with $\partial W^4 = M_1 \cup M_0$, then $G = \pi_1(W) = \pi_1(M_0)$ can be infinite or finite. In the infinite case, it is conjectured and known in many cases that $Wh(G) = 0$ (see Problem 3.32). Kwasik & Schultz [635, 1992b, Topology] show that any h-cobordism W has zero torsion if M_0 and M_1 are geometric 3-manifolds.

Problem 4.10 (Thurston) *If a closed, orientable 4-manifold M^4 is a $K(\pi, 1)$, must the Euler characteristic be ≥ 0?*

Remarks: If M^4 has nonpositive curvature then it is a $K(\pi, 1)$ and $\chi(M) \geq 0$ [198, Chern, 1955, Abh. Math. Sem. Univ. Hamburg]. This argument fails in higher dimensions [371, Geroch, 1976, Proc. Amer. Math. Soc.] The Hopf Conjecture is that sectional curvature ≤ 0 implies that $(-1)^k\chi(M^{2k}) \geq 0$.

Is it possible that $(-1)^k\chi(M^{2k}) \geq 0$ for all $M = K(\pi, 1)$?

Update: Still open, but see [235, Davis, 1984] and [192, Charney & Davis, 1995].

Problem 4.11 Uniqueness. *If two closed, simply connected 4-manifolds are homotopy equivalent, are they homeomorphic? If so, is the homotopy equivalence homotopic to the homeomorphism?*

Remarks: They are h-cobordant; thus they are diffeomorphic after connected sum with enough $S^2 \times S^2$'s [1094, Wall, 1964b, J. London Math. Soc.]. Without the simply connected assumption, simply homotopy equivalence does not even imply a smooth, normal bordism (Problems 4.13–4.15; also [178, Cappell & Shaneson, 1971, Comment. Math. Helv.]).

If the manifolds are not closed, but have homeomorphic boundaries and the homeomorphism extends to a homotopy equivalence, then we can ask if the homeomorphism extends to a homeomorphism of the 4-manifolds (unlikely; see Problem 4.16).

Kodaira has described [595, 1963, Ann. of Math.] some possible counterexamples, logarithmic transforms of elliptic surfaces. Here is a description for topologists (A. Kas): Let $S \xrightarrow{\pi} \mathbb{CP}^1 = S^2$ be an elliptic surface with an analytic projection of $\mathbb{CP}^1$ such that $\pi^{-1}(\text{point}) = \text{torus}$ T^2 except for a finite number of points. For example, we construct the Kummer surface by first taking the quotient of $T^4 = S^1 \times S^1 \times S^1 \times S^1$ by the involution which reflects each circle. The involution has 16 fixed points, so the quotient is a manifold except for 16 singularities equal to the cone on $\mathbb{RP}^3$. Replace the cone by the cotangent disk bundle of S^2, whose boundary is $\mathbb{RP}^3$ with the right orientation. This is the Kummer surface, and π is constructed from the projection of T^4 on any $S^1 \times S^1$. A similar construction using $S^2 \times T^2$, and $180°$ rotation on S^2, gives the *half-Kummer* surface, which is known to be diffeomorphic to $\mathbb{CP}^2 \# 9(-\mathbb{CP}^2)$.

The logarithmic transform is essentially S^1 cross the construction in Seifert fibered space theory in which a nonsingular S^1 fiber is replaced by a singular fiber of multiplicity m.

Let D be a 2-ball in $\mathbb{CP}^1$ so that $\pi^{-1}(D) = S^1 \times S^1 \times D$. The logarithmic transform $L_a(m)$ of S is $L_a(m) = S - \pi^{-1}(\text{int}D) \cup_h S^1 \times S^1 \times D$ where a is the center of D and $h : S^1 \times S^1 \times \partial D \to \partial(S - \pi^{-1}(\text{int}D)) =$

$S^1 \times S^1 \times \partial D$ is given by

$$h(\theta_1, \theta_2, \varphi) = \begin{pmatrix} 1 & 0 & 0 \\ 0 & 0 & -1 \\ 0 & 1 & m \end{pmatrix} \begin{pmatrix} \theta_1 \\ \theta_2 \\ \varphi \end{pmatrix}, \qquad \theta_i \text{ and } \varphi \in [0, 2\pi].$$

The map $\pi : S - \pi^{-1}(\text{int}D) \to \mathbb{CP}^1 - \text{int}D$ extends to $L_a(m)$ by defining π on $S^1 \times S^1 \times D$ by $\pi(\theta_1, \theta_2, r\varphi) = r(\theta_2 + m\varphi)$ where $r \in [0, 1]$.

The logarithmic transform of the half-Kummer surface is still a rational surface and hence is still diffeomorphic to the half-Kummer surface. However, two logarithmic transforms, $L_a(m)$ followed by $L_b(n)$, $(m, n) = 1$, is homotopy equivalent but not known to be diffeomorphic to the half-Kummer surface. Similarly one logarithmic transform of the Kummer surface is homotopy equivalent, but not known to be diffeomorphic.

Another case is the nonsingular quintic in $\mathbb{CP}^3$, which is known to be homotopy equivalent to $9\mathbb{CP}^2 \# 44(-\mathbb{CP}^2)$ but not necessarily homeomorphic.

It is not hard to construct examples of homotopy equivalent, simply connected, closed 4-manifolds, and presumably they are diffeomorphic. The point to these examples is that it is of independent interest to construct the expected diffeomorphism.

Update: If their intersection form is even or the form is odd and the Kirby–Siebenmann invariants agree, then the homotopy equivalence is homotopic to a homeomorphism; if the invariants disagree, then the manifolds are not homeomorphic [329, Freedman, 1982, J. Differential Geom.].

Regarding the examples in the Remarks, let $E(n)$ denote the fiber connected sum of n copies of the rational surface $\mathbb{CP}^2 \# 9(-\mathbb{CP}^2)$. Thus $E(2)$ is the $K3$ surface (called Kummer in the old list) and $E(1)$ was called the *half-Kummer* above.

First it was shown that for $E(2)$ all the log transforms with a single p were smoothly distinct [345, Friedman & Morgan, 1994]; then it was shown that all log transforms of $E(2)$ for distinct coprime pairs (p, q) were distinct [789, Morgan & O'Grady, 1994] and [71, Bauer, 1994, J. Reine Angew. Math.].

The fact that one or two log transforms were always smoothly distinct was shown for $n \geq 2$ by Morgan & Mrowka [788,1993,Internat. Math. Res. Notices, bound within Duke Math. J.] using [Morgan & O'Grady, ibid.], by Stipsicz & Szabó [1013,1994b,Duke Math. J.], and later by Lisca (half the cases [663,1994,Math. Ann.] and the other half [664,1995,Invent. Math.]); the case $n = 1$ was done by Friedman [344, 1995,J. Amer. Math. Soc.]. Later Fintushel & Stern [314,Fintushel & Stern,1994a] completely calculated the Donaldson polynomials for $E(n), n \geq 2$ without using algebraic geometry, and thereby gave another proof of the classifications.

The quintic does not even decompose smoothly as a connected sum into two pieces with b_2^+ positive, let alone into $9\mathbb{CP}^2\#44(-\mathbb{CP}^2)$ [251, Donaldson,1990,Topology].

Problem 4.12 (Mandelbaum & Moishezon) *Suppose M_1^4 and M_2^4 are closed, simply connected, and homotopy equivalent. Is $M_1\# \pm \mathbb{CP}^2$ homeomorphic to $M_2\# \pm \mathbb{CP}^2$?*

Remarks: The choice of $\mathbb{CP}^2$ or $-\mathbb{CP}^2$ may be crucial. The right choice gives an odd, indefinite intersection form so that we have a homotopy equivalence with a connected sum of copies of $\pm\mathbb{CP}^2$. Is there a homeomorphism? Mandelbaum & Moishezon [691,1976,Topology] construct a diffeomorphism, for the case $+\mathbb{CP}^2$, for a large class of complex surfaces.

Update: M_1 and M_2 are homeomorphic if their intersection form is even. When the form is odd, they are homeomorphic if they have the same Kirby–Siebenmann invariant, but otherwise $M_1\#N^4$ is not homeomorphic to $M_2\#N$ for any N [329,Freedman,1982,J. Differential Geom.].

In the smooth case $M_1\# \pm \mathbb{CP}^2$ may always be diffeomorphic to $M_2\# \pm \mathbb{CP}^2$, because the right choice of $\pm\mathbb{CP}^2$ destroys the Donaldson polynomial and the Seiberg–Witten invariants which are our only tools for distinguishing homotopy equivalent smooth 4-manifolds. See Problem 4.103 for more cases where a diffeomorphism exists.

Problem 4.13 (Cappell & Shaneson) *There are homotopy $\mathbb{RP}^4$'s which are not diffeomorphic to $\mathbb{RP}^4$ [181, Cappell & Shaneson, 1976,*

Ann. of Math.]. Which of these homotopy $\mathbb{RP}^4$*'s are homeomorphic (diffeomorphic), and which are homeomorphic to* $\mathbb{RP}^4$*?*

Remarks: One of these homotopy $\mathbb{RP}^4$'s has a double cover which is diffeomorphic to S^4 [392,Gompf,1991a,Topology]. Find an elegant way to describe this exotic involution on S^4.

Update: Up to homeomorphism there are exactly two homotopy $\mathbb{RP}^4$'s, namely $\mathbb{RP}^4$ and a non-smooth homotopy $\mathbb{RP}^4$ [434,Hambleton, Kreck, & Teichner,1994,Trans. Amer. Math. Soc.]; thus all the examples of Cappell & Shaneson are homeomorphic to $\mathbb{RP}^4$.

It is still not known which, if any, of these exotic $\mathbb{RP}^4$'s are diffeomorphic, but one of them is obtained from $\mathbb{RP}^4$ by a generalized Gluck twist as in Problem 4.51 (B), (for a different approach, see [1014,Stolz,1988, Invent. Math.]). The Cappell–Shaneson construction provides many possibly exotic S^4's, about which little is known beyond [Gompf, ibid.] and the comprehensive paper [12,Aitchison & Rubinstein,1984]. Fintushel & Stern [308,1981,Ann. of Math.] gave the first example of an involution on S^4 with an exotic $\mathbb{RP}^4$ as quotient. Also see Problem 4.85.

Problem 4.14 (Cappell & Shaneson) (A) *There is a homotopy equivalence* $h : S^2 \times \mathbb{RP}^2 \to S^2 \times \mathbb{RP}^2$ *constructed by*

$$S^2 \times \mathbb{RP}^2 \to (S^2 \times \mathbb{RP}^2) \vee S^4 \overset{id \vee \alpha}{\longrightarrow} S^2 \times \mathbb{RP}^2$$

where α *generates* $\pi_4(S^2)$*.* h *is not homotopic to a PL homeomorphism because it has a nontrivial normal invariant (the induced normal map* $h^{-1}(\mathbb{RP}^2) \to \mathbb{RP}^2$ *has a nontrivial Kervaire invariant). Is* h *homotopic to a homeomorphism?*

(B) *Construct a homotopy* $S^2 \times \mathbb{RP}^2$ *by removing a normal* B^3*-bundle of* $\mathbb{RP}^1$ *and sewing in a suitable* $(T^3 - B^3)$*-bundle over* S^1*. Is the manifold or the homotopy equivalenc exotic?*

Remarks: Note that the existence of the (PL) exotic homotopy equivalence in part (A) implies that the manifold above is s-cobordant to $S^2 \times \mathbb{RP}^2$.

Update: The answer to (A) is yes, for this is a topological surgery problem for a *good* group (see Problem 4.6 Update).

For (B), the homotopy equivalence is exotic [16,Akbulut,1984].

Problem 4.15 (Y. Matsumoto) *Is Scharlemann's* exotic $S^1 \times S^3 \# S^2 \times S^2$ *an exotic manifold or an exotic self-homotopy equivalence of* $S^1 \times S^3 \# S^2 \times S^2$ *[965,Scharlemann,1976a,Duke Math. J.] ?*

Remarks: After connected sum with $S^2 \times S^2$, Scharlemann's manifold is diffeomorphic to $S^1 \times S^3 \# 2(S^2 \times S^2)$ [307,Fintushel & Pao,1977,Proc. Amer. Math. Soc.].

Update: Scharlemann's example is constructed by surgering a circle α in $P^3 \times S^1$ where P^3 is the Poincaré homology sphere and $\langle\alpha\rangle$ normally generates $\pi_1(P)$. There are 118 such $\langle\alpha\rangle$ and some give 4-manifolds diffeomorphic to $S^1 \times S^3 \# S^2 \times S^2$, and hence exotic self homotopy equivalences . R. Lee (see [178,Cappell & Shaneson,1971,Comment. Math. Helv.; Prop. 3.2 on page 515] had already constructed an exotic self-homotopy equivalence of $S^3 \times S^1 \# S^2 \times S^2$. Note that all of these exotic self homotopy equivalences are homotopic to homeomorphisms [336, Freedman & Quinn,1990]. Exotic smooth structures on $S^1 \times S^3 \# S^2 \times S^2$ can also be constructed by surgery on null-homotopic, smoothly imbedded S^2's in $S^2 \times S^2$ which do not bound smoothly imbedded B^3's [964, Sato,1994,Proc. Amer. Math. Soc.].

Note that Scharlemann's example, and Akbulut's example of an exotic $S^1 \tilde{\times} S^3 \# S^2 \times S^2$ are both Gluck constructions on the standard manifolds [16,Akbulut,1984], [17,Akbulut,1985], [18,Akbulut,1988,Topology].

Problem 4.16 (Akbulut & Kirby) *Does every diffeomorphism of the boundary of a contractible 4-manifold* X^4 *extend over* X^4*?*

Remarks: If not, there is a counterexample to the relative h-cobordism theorem in dimension 5. Here is a candidate for a diffeomorphism which does not extend: In the symmetric link below, we can add 2-handles (to B^4) to both circles with framing 0. The boundary of this 4-manifold has an obvious involution obtained by switching circles. Let the contractible manifold X^4 be obtained by surgering one of the two obvious 2-spheres; X^4 is a well-known Mazur manifold.

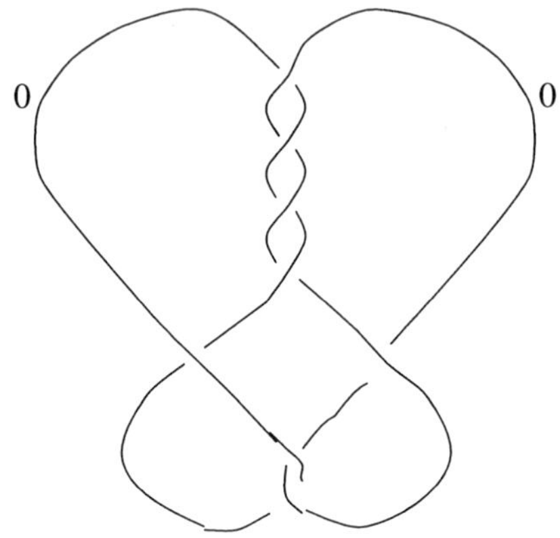

Figure 4.16.1.

Update: Every diffeomorphism extends to a homeomorphism by [329, Freedman,1982,J. Differential Geom.]. However Akbulut [19,1991a,J. Differential Geom.; Figure 1] has shown that the diffeomorphism constructed above (he uses the mirror reflection of the example above) does not extend to a diffeomorphism of the 4-manifold. From this he constructs two 1-connected, 4-manifolds with diffeomorphic boundaries (and rank $H_2 = 1$) which are not diffeomorphic, [21,Akbulut,1991c,J. Differential Geom.]. These are, at this time, the smallest such.

Problem 4.17 (Thurston) *Can a homology 4-sphere ever be a $K(\pi,1)$? Who knows an example of a rational homology 4-sphere which is a $K(\pi,1)$?*

Remarks: Many homology 3-spheres are $K(\pi,1)$'s, e.g., any Brieskorn $\Sigma(p,q,r)$ wih p,q,r pairwise coprime and infinite fundamental group.

Update: Still open, but see [236,Davis,1985,Proc. Amer. Math. Soc.] for a suggestion. Also note that if the fundamental group of a finite aspherical complex has an infinite amenable normal subgroup, then the Euler characteristic must be zero [194,Cheeger & Gromov,1986,Topology]; thus the fundamental group of an aspherical homology 4-sphere cannot have an infinite amenable normal subgroup.

Luo has given examples of rational homology spheres [685, Luo, 1988, Proc. Amer. Math. Soc.].

Note that there exist finitely presented groups G_n such that

$$H_*(G_n) \equiv H_*(K(G_n, 1); \mathbb{Z}) \cong H_*(S^n; \mathbb{Z})$$

for all n [75, Baumslag, Dyer, & Miller, III, 1983, Topology], [558, Kan & Thurston, 1976, Topology].

Problem 4.18 *Does every simply connected, closed* 4*-manifold have a handlebody decomposition without* 1*-handles? Without* 1*- and* 3*-handles?*

Remarks: Because there are nontrivial groups G which cannot be trivialized by adding the same number of generators and relations [373, Gerstenhaber & Rothaus, 1962, Proc. Nat. Acad. Sci. U.S.A.], there are contractible 4-manifolds V^4, with $\pi_1(\partial V^4) = G$, that require 1-handles (Casson). On the other hand, nonsingular complex hypersurfaces in $\mathbb{CP}^3$ need no 1-handles [954, Rudolph, 1976, Topology], or even 3-handles [444, Harer, Kas, & Kirby, 1986], [439, Harer, 1978, Math. Ann.], [28, Akbulut & Kirby, 1980, Math. Ann.]). This is true (Mandelbaum) for complete intersections (the intersection of n hypersurfaces in $\mathbb{CP}^{n+2}$ which are in general position). If 1-handles are unnecessary, then there is a geometric proof [692, Mandelbaum & Moishezon, 1983] of some of Rohlin's inequalities [931, 1971, Functional Anal. Appl.] using Tristram's work [1062, 1969, Math. Proc. Cambridge Philos. Soc.].

Update: No progress, but there are two relevant papers by Trace [1060, 1980, Proc. Amer. Math. Soc.] and [1061, 1982, Pacific J. Math.]. A good example to work on is the *Dolgachev surface* in [444, Harer, Kas, & Kirby, 1986].

Problem 4.19 (A) *Does any integral, unimodular, symmetric, bilinear form contain a characteristic element* α *(i.e.,* $\alpha \cdot x = x \cdot x \pmod 2$ *for all* x*) such that* $\alpha \cdot \alpha = \sigma$.

Remarks: Yes for indefinite forms and definite forms of rank $\leq$ 16.

(B) *Does every orientable, closed, smooth* M^4 *smoothly imbed in* $\mathbb{R}^7$*?*

Remarks:

1. M^4 smoothly imbeds in $\mathbb{R}^7 \Leftrightarrow$ there exists $\alpha \in H_2(M;\mathbb{Z})$ such that $\alpha \cdot x = x \cdot x \pmod 2$ for all $x \in H_2(M;\mathbb{Z})$, and $\alpha \cdot \alpha = \sigma(M^4)$ [110,Boéchat & Haefliger,1970]. The simply connected case is already interesting, where characteristic elements (integral duals to the second Stiefel–Whitney class) are known to exist, with $\alpha \cdot \alpha \equiv \sigma(M^4) \pmod 8$.
2. M^4 PL imbeds in $\mathbb{R}^7$ [491,Hirsch,1965,Math. Proc. Cambridge Philos. Soc.].
3. M^4 PL imbeds in $\mathbb{R}^6 \Leftrightarrow M^4$ is a spin manifold (in which case the imbedding can be chosen to have a single nonlocally flat point). M^4 smoothly imbeds in $\mathbb{R}^6 \Leftrightarrow M^4$ is spin and $\sigma(M^4) = 0$ (Cappell & Shaneson).

Update:

(A) In 1981 Odlysko and Sloane [852,1981,J. Reine Angew. Math.] verified (A) for even definite forms of rank ≤ 72.

(B) Yes, because Remark 1 implies that we need only consider 4-manifolds with definite intersection forms (otherwise there is a characteristic element satisfying the Boéchat & Haefliger condition). But Donaldson [247,1983,J. Differential Geom.], [250,1987b, J. Differential Geom.] has shown that all such manifolds have forms $\pm \oplus k\langle 1\rangle$, and these have a characteristic element α, the sum of the generators, for which $\alpha \cdot \alpha = \sigma$.

Note that this was the last case of the *hard* Whitney imbedding theorems which state that M^m smoothly imbeds in $\mathbb{R}^{2m}$ [1112, Whitney,1944,Ann. of Math.; page 236], and if M is orientable, then M imbeds in $\mathbb{R}^{2m-1}$; the latter is easy for $m = 2$, is due to Hirsch [490,Hirsch,1961,Ann. of Math.] for $m = 3$, follows from [426,Haefliger & Hirsch,1963,Topology], [702,Massey,1960, Amer. J. Math.], [703,Massey,1962,Proc. Amer. Math. Soc.], [704,Massey & Peterson,1963,Bol. Soc. Mat. Mexicana], [1123, Wu,1963,Sci. Sinica] for $m \geq 5$, and was open for $m = 4$.

A reference for Remark (3) is [943,Ruberman,1982,Math. Proc. Cambridge Philos. Soc.].

The topological case remains open, but here is a related theorem: If M is a closed, simply connected topological 4-manifold with odd intersection pairing, then M admits a locally linear involution iff the Kirby–Siebenmann stable triangulation obstruction vanishes and, as above, there is an indivisible characteristic element $\alpha \in H_2(M;\mathbb{Z})$ such that $\alpha \cdot \alpha = \sigma(M)$ [270, Edmonds, 1988, Topology Appl.].

For codimension one imbeddings, see Problem 4.63.

Problem 4.20 (R. Fenn) *Does there exist an M^4 which immerses in $\mathbb{R}^6$ with just one triple point (like Boy's surface in $\mathbb{R}^3$)?*

Update: No, in the orientable case. Herbert [479, 1981] and White [1109, 1975] proved that the algebraic number of triple points of a generic immersion is $-p_1(M)/3 = -\sigma(M)$. But $p_1(T_M \oplus \nu_M) = p_1(M) + \chi^3(\nu_M) = 0$ so $-p_1(M) = \chi^2(\nu_M)$. But the double point set is an integral dual ξ to $\omega_2(M)$, so $\xi \cdot \xi \equiv \sigma(M) \pmod 8$, and $\chi^2(\nu_M) = \xi \cdot \xi$. Thus $-3\sigma(M) \equiv \sigma(M) \pmod 8$, so the signature is even, as is the number of triple points.

But yes in the non-orientable case [517, Hughes, 1983, Quart. J. Math. Oxford Ser. (2)]. Immerse $\mathbb{RP}^2 \times \mathbb{RP}^2$ as Boy's surface cross itself, which (by the last paragraph) has an odd number of triple points. Remove these in pairs by ambient surgery to get an immersion of $\mathbb{RP}^2 \times \mathbb{RP}^2 \# n(S^1 \times S^3)$ with a single triple point.

Problem 4.21 (T. Price) *Let M^4 be a 4-manifold with boundary. Characterize the following isotopy classes of imbedded circles, J, in ∂M^4.*

(A) *J bounds an imbedded PL (not necessarily locally flat) 2-ball in M^4. (This is* Dehn's lemma *in dimension 4.)*

Remarks: A classical, unsolved case is Zeeman's example: Add a handle to $S^1 \times B^3$ along the curve C drawn below; does $J = S^1 \times *$, $* \in \partial B^3$, bound a PL 2-ball? J does bound a 2-ball which is wild along the boundary J [378, Giffen, 1977].

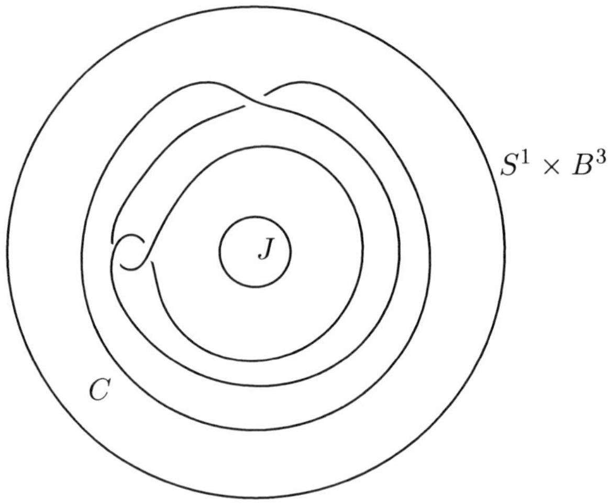

(B) *J bounds a smooth immersed 2-ball with null-homologous image.*

(C) *J bounds an imbedded, orientable, smooth surface F^2 in M^4 with $H_1(F;\mathbb{Z}) \to H_1(M;\mathbb{Z})$ being the zero map.*

Remarks: (A) $\Rightarrow$ (B) $\Rightarrow$ (C). There are other possible definitions of null-homologous singularities, e.g., define the *self-intersection number* of an immersion $f : B^2 \to M^4$ with double points $p_1, \ldots, p_k$ to be $\sum_{i=1}^k \varepsilon_i(g_i + g_i^{-1}) \in Z[H_1(M)]$ where $\varepsilon_i = \pm 1$ is the sign of p_i, and g_i is represented by the image of an arc in B^2 joining the two points in $f^{-1}(p_i)$; then

(D) J bounds a smooth immersed 2-ball with self-intersection number zero.

Note that (B) $\Rightarrow$ (D) $\Rightarrow$ (C). One can replace homology by homotopy throughout.

Update: Akbulut [20, 1991b, Topology] has shown that in the example above, J does not bound a PL 2-ball. This fact follows quickly from the example in Problem 4.16 which is very deep and uses gauge theory. No other progress on this problem is known when M^4 is contractible. For other M^4 the questions are still wide open. Note that (C) is related to Problem 4.112.

Problem 4.22 (Gordon) Let Σ^3 be a homology 3-sphere which bounds an acyclic 4-manifold V^4 such that $\pi_1(\Sigma^3) \to \pi_1(V^4)$ is surjective. Let K be a knot in Σ^3. Define K to be homotopically ribbon in V if there is a smoothly imbedded B^2 in V, $\partial B^2 = K$, such that $\pi_1(\Sigma^3 - K) \to \pi_1(V - B^2)$ is surjective.

(A) *Does* K slice in V *imply* K homotopically ribbon in V*?*

(B) *Does (A) hold at least for contractible* V*?*

Remarks: The classical *slice implies ribbon* conjecture (Problem 1.33) splits into two parts, *slice implies homotopically ribbon (in* B^4*)* and *homotopically ribbon implies ribbon.*

Update: Casson & Gordon [187,1983,Invent. Math.] show that either the answer to (B) is no, or there exists a curve in the boundary of the contractible 4-manifold which does not bound a PL disk. However Akbulut (see previous Problem) has shown that the latter can happen.

Problem 4.23 Let $f : S^2 \to \mathbb{CP}^2$ be a smooth imbedding which represents the generator of $H_2(\mathbb{CP}^2; \mathbb{Z})$.

Conjecture: $(\mathbb{CP}^2, f(S^2))$ *is pairwise diffeomorphic to* $(\mathbb{CP}^2, \mathbb{CP}^1)$. *Perhaps,* $f(S^2)$ *is even isotopic to* $\mathbb{CP}^1$.

Remarks: The conjecture may be easier than the 4-dimensional Poincaré conjecture which implies it.

Update: The conjecture is true in the topological case, using Freedman's [329,1982,J. Differential Geom.] topological 4-dimensional Poincaré conjecture (after *blowing down* $f(S^2)$), even if $f(S^2)$ is only locally flat (see Problem 4.31 Update).

If f is symplectic for some symplectic structure, then Gromov [414, 1985,Invent. Math.] shows that the symplectic structure is standard and f is isotopic to the inclusion $\mathbb{CP}^1 \hookrightarrow \mathbb{CP}^2$.

Problem 4.24 (Gluck) Let $K^2 \hookrightarrow S^4$ be a knotted 2-sphere. If we cut out a tubular neighborhood of K^2 and sew it back in with a *twist* (i.e., the nontrivial element of $\pi_1(SO(3)) = \mathbb{Z}/2\mathbb{Z}$ gives the *twist* $S^1 \times S^2 \to S^1 \times S^2$), then the resulting manifold is a homotopy 4-sphere Σ^4.

Question: *Must it be S^4?*

Remarks: (P. Melvin) This is equivalent to asking whether $(\mathbb{CP}^2, \mathbb{CP}^1)\#(S^4, K^2)$ is pairwise diffeomorphic to $(\mathbb{CP}^2, \mathbb{CP}^1)$ (see the previous Problem 4.23).

Update: Yes topologically, but the question is still wide open in the smooth case. One smoothly gets S^4 if K is 0-null-bordant (see Problem 1.105) [745,Melvin,1977] (for spun knots and the original treatment, see [387,Gluck,1962,Trans. Amer. Math. Soc.]). Note that $\Sigma^4 \# \pm \mathbb{CP}^2$ is diffeomorphic to $\pm\mathbb{CP}^2$.

Problem 4.25 (Y. Matsumoto) *Does there exist a smooth, compact W^4, homotopy equivalent to S^2, which is spineless, i.e., contains no PL imbedded S^2 representing the generator of $H_2(W)$?*

Remarks: There is such an example for T^2 instead of S^2 [708,Matsumoto,1975,Bull. Amer. Math. Soc.].

Update: No progress. However there exists a non-compact example [717,Matsumoto & Venema,1979,Invent. Math.], as well as many compact examples in genus 1 or more [566,Kawauchi,1980,Trans. Amer. Math. Soc.]. Compare Problem 4.73.

Problem 4.26 (L. Taylor) Construct a fake Hopf bundle by realizing $\Gamma = 3\gamma_0 + \gamma_1 + \ldots + \gamma_8$ in $\mathbb{CP}^2\#8(-\mathbb{CP}^2)$ by a *PL* imbedded sphere and taking a regular neighborhood (γ_i is the generator of $H_2(\pm\mathbb{CP}^2)$; the *Hopf bundle* is $B^4 \cup$ 2-handle attached to the trefoil knot with $+1$ framing). Twice the core of this *Hopf bundle* can be represented by a smoothly imbedded double torus.

Can it be represented by a torus? A sphere?

Remarks: $r\Gamma$ cannot be represented by a smoothly imbedded sphere if $r = 1$ [572,Kervaire & Milnor,1961,Proc. Nat. Acad. Sci. U.S.A.] or if $r \geq 3$ [1062,Tristram,1969,Math. Proc. Cambridge Philos. Soc.];

[513,Hsiang & Szczarba,1971]. The double branched cover of this *Hopf bundle* along the imbedded surface can be used to construct a spin manifold of signature 16 and betti number 22 (double torus), 20 (torus), or 18 (sphere).

Update: The answer is no because of the last sentence of the Remarks and [248,Donaldson,1986,J. Differential Geom.]; note that there is a 2-torsion assumption that can be satisfied [614,Kotschick & Matic,1995, Math. Proc. Cambridge Philos. Soc.] or can be avoided using Seiberg–Witten invariants.

Problem 4.27 (A) (Weintraub) *Does every simply connected, closed 4-manifold have a basis for H_2 consisting of PL imbedded 2-spheres? Smooth?*

Remarks: Yes, in the *PL* case, if there is a 2-dimensional spine, or if there are no 1-handles (see Problem 4.18).

(B) (Fenn) *Which even elements of $\pi_2(\mathbb{RP}^2 \times T^2) \cong \mathbb{Z}$ can be represented by PL imbedded spheres?*

Remarks: Odd elements cannot be represented [299,Fenn,1973, Math. Proc. Cambridge Philos. Soc.] .

Update: (A) No in the smooth case using gauge theory.

(B) The odd elements cannot be represented by locally flat imbeddings because Wall's self-intersection obstruction at the group element of order 2 is nontrivial; this argument also works for PL spheres.

Problem 4.28 (Y. Matsumoto) Construct M^4 by adding two 2-handles with 0-framing to B^4 along the link below.

(A) *Can the generators of $H_2(M^4)$ be represented by a smoothly imbedded wedge of two 2-spheres?*

(B) *Can one of the generators be represented by a smoothly imbedded S^2?*

(C) *In $M^4 \# kS^2 \times S^2$, k arbitrary, can a* half-basis *of $H_2(M^4)$ be represented by $k+1$ smoothly imbedded, disjoint 2-spheres?*

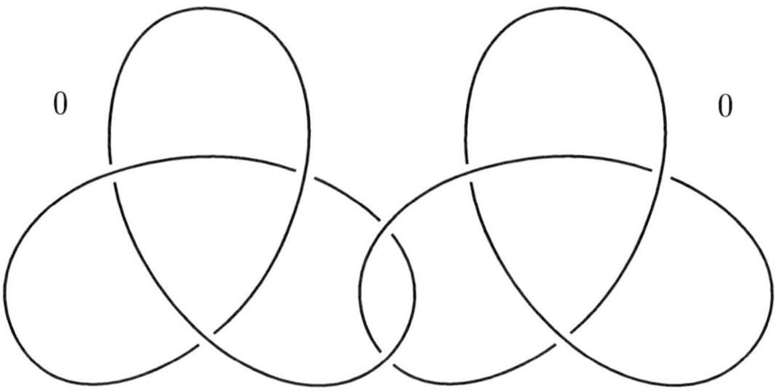

Remarks: (A) $\Rightarrow$ (B) $\Rightarrow$ (C). The generators of $H_2(M^4)$ are represented by two Casson flexible handles, i.e., by an open subset proper homotopy equivalent to $S^2 \times S^2-$ point (see Problem 1.39). A no for (A) implies the Casson handles are exotic.

(A) implies that ∂M^4 bounds a contractible 4-manifold, and (B) implies that ∂M^4 bounds an acyclic 4-manifold (in both cases by surgering an S^2). (C) holds iff ∂M^4 bounds an acyclic smooth 4-manifold.

Update: The answer to (A) is no [23, Akbulut, 1995a], by an application of Donaldson's Theorem C in [248, 1986, J. Differential Geom.], in fact, Akbulut shows that ∂M^4 cannot bound a contractible 4-manifold, nor a homology 4-ball W such that $\pi_1(\partial M) \to \pi_1(W)$ is onto.

Note that if one framing is changed for 0 to 6, then the resulting homology 3-sphere bounds a contractible 4-manifold, [701, Maruyama, 1984, J. Tsuda College; page 11] and [710, Matsumoto, 1982b, J. Fac. Sci. Univ. Tokyo Sect. IA Math.; page 294].

Problem 4.29 (M. Kato) *If F is a smooth, orientable, closed surface in S^4, is $H_2(\pi_1(S^4 - F); \mathbb{Z}) = 0$?*

Remarks: If so, then the class of fundamental groups of surface complements in S^4 is exactly Kervaire's class of finitely presented groups of weight one with $H_i(G;\mathbb{Z}) = H_i(\mathbb{Z};\mathbb{Z})$, $i = 1, 2$, which occur as knot groups in higher dimensions. That Kervaire's groups are realized by surfaces follows from arguments of T. Yajima [1126, 1969, Osaka J. Math.], [1127, 1970, Proc. Japan Acad. Ser. A Math. Sci.].

Update: Examples of surfaces $F \to S^4$ with $H_2(\pi_1(S^4-F);\mathbb{Z}) \neq 0$ have been given by [689, Maeda & Yajima, 1976, Math. Sem. Notes, Kwansei Gakuin Univ.], [688, Maeda, 1977, Math. Sem. Notes Kobe Univ.], [154, Brunner, Mayland, Jr., & Simon, 1982, Pacific J. Math.], [402, Gordon, 1981a, Math. Proc. Cambridge Philos. Soc.], and [666, Litherland, 1981, Quart. J. Math. Oxford Ser. (2)]. In particular, Litherland shows that if A is an abelian group with $2g$ generators, then there is a closed surface of genus g, F_g, and a smooth imbedding $F_g \to S^4$ such that $H_2(\pi_1(S^4 - F_g);\mathbb{Z}) = A$.

Problem 4.30 *Given an imbedding of T^2 in $\mathbb{R}^4$ with 4 critical points (with respect to projection to a coordinate axis), is it standard (up to isotopy)?*

Remarks: Observe the example

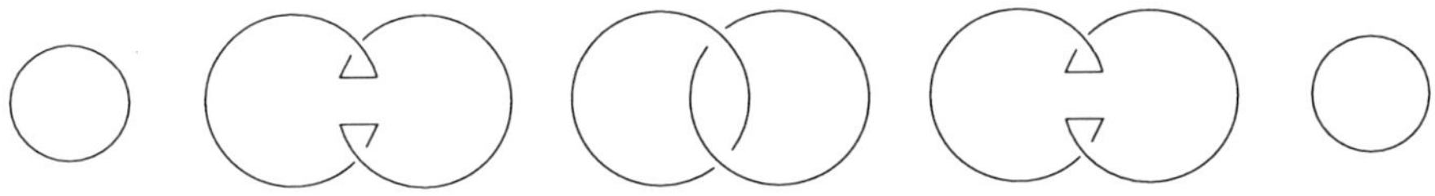

which *is* standard.

Update: Kreck & Teichner [621, 1995] have shown that any locally flat topological imbedding of an oriented surface in S^4, whose complement has abelian fundamental group, is standard. A smooth torus with 4 critical points has such a group, and hence is standard.

Also note that a smooth imbedding of $\mathbb{RP}^2$ with 3 critical points is standard [109, Bleiler & Scharlemann, 1988, Topology].

Problem 4.31 Conjecture: *Any locally flat surface in a 4-manifold has a normal bundle.*

Remarks: This is true for codimension two imbeddings in other dimensions [584, Kirby & Siebenmann, 1974].

Update: The conjecture is true [336, Freedman & Quinn, 1990; page 137]. (Note that the proof in [577, Kirby, 1970] fails in *Step 3* on page 419, and it hasn't been repaired.)

Problem 4.32 (Schoenflies) Conjecture: *If S^3 is PL imbedded in S^4, then its closed complements are PL 4-balls.*

Remarks: Note that they are TOP 4-balls since the S^3 is (PL) locally flat.

Update: The conjecture has been proved in the special case where there is a smooth function $f : S^4 \to \mathbb{R}$ whose restriction to the S^3 is Morse with k 0-handles and $\leq k+1$ 1-handles (so that the middle level has genus ≤ 2) [967,Scharlemann,1984,Topology].

Problem 4.33 (Fenn) *If S^3 is locally flatly imbedded in $S^2 \times S^2$, does it bound a TOP 4-ball? One may wish to assume a smooth imbedding.*

Remarks: Suppose the Casson continuum [188,Casson & Gordon, 1986] is cellular in $S^2 \times S^2$; then its end is $S^3 \times \mathbb{R}$. For the converse we need an affirmative answer to this problem.

Update: The S^3 must bound a contractible topological 4-manifold which is then homeomorphic to B^4 by [329,Freedman,1982,J. Differential Geom.].

Problem 4.34 *What is $\pi_0(Diff(S^4))$? Indeed, is $Diff(S^4) \simeq O(5)$?*

Remarks: The involution of S^4 giving the Cappell–Shaneson exotic $\mathbb{RP}^4$ may not be isotopic to the antipodal map (see Problem 4.13).

Update: No progress, but see Problem 4.126.

Problem 4.35 (Hatcher) On the torus T^n, $n \geq 5$, there are many homeomorphisms concordant but not isotopic to the identity [462,Hatcher, 1978].

Are there such examples on T^4?

Update: Yes, Kwasik gave examples of diffeomorphisms of T^4 which are smoothly pseudo-isotopic (i.e. concordant) to the identity, but not topologically isotopic [633,Kwasik,1987,Math. Proc. Cambridge Philos. Soc.]. In the simply connected case, any topological pseudo-isotopy can be isotoped to the identity [873,Perron,1986,Topology], [911,Quinn, 1986,J. Differential Geom.], and Kwasik in the above paper.

There are also examples on connected sums of metacyclic prism 3-manifolds of diffeomorphisms which are pseudo-isotopic but not isotopic [636,Kwasik & Schultz,1995].

Problem 4.36 (A) Thom Conjecture: *The minimal genus of a smooth imbedded surface in* $\mathbb{CP}^2$ *representing* $n \in H_2(\mathbb{CP}^2;\mathbb{Z}) = \mathbb{Z}$ *is* $(n-1)(n-2)/2$.

Remarks: Any nonsingular algebraic curve in $\mathbb{CP}^2$ of degree n has genus $(n-1)(n-2)/2$. The minimal genus is at least $n^2/4-1$ if n is even, and at least $(n^2(h^2-1)/4h)-1$ if $n \equiv 0 \pmod h$ and h is a power of an odd prime [931,Rochlin,1971,Functional Anal. Appl.], and [513,Hsiang & Szczarba,1971] .

(B) Let S be a simply connected complex, algebraic surface and let $\alpha \in H_2(S;\mathbb{Z})$. Suppose α is represented by an algebraic curve of genus g (given by the formula $2g-2 = -c_1 \cdot \alpha + \alpha \cdot \alpha$ where $c_1 =$ first Chern class).

Conjecture: *g is the minimal genus of any smoothly imbedded surface representing α.*

Remarks: This conjecture is risky. A counterexample could come from a simply connected surface S which is diffeomorphic to $M^4 \# \mathbb{CP}^2$ with odd indefinite intersection form on M, for then all primitive, ordinary classes in $S\# - \mathbb{CP}^2$ are represented by 2-spheres [1093,Wall,1964a,J. London Math. Soc.]. Apparently, no such surface is known (see next problem).

Update:

(A) True as proved by Kronheimer & Mrowka [624,1994a,Bull. Amer. Math. Soc.] and by Morgan, Szabó & Taubes [793,Morgan, Szabó, & Taubes,1995] using Seiberg–Witten gauge theory. The conjecture fails if *smooth imbedding* is replace by *locally flat imbedding*. Rudolph gave the first counterexamples and also the smallest degree counterexample for d=5 [957,Rudolph,1984,Comment. Math. Helv.]. Lee & Wilczynski have counterexamples whose complements have fundamental group $\mathbb{Z}/d\mathbb{Z}$ [646,1995]; their counterexamples sometimes show that the Rohlin and Hsiang & Szczarba bounds (mentioned above) are sharp.

(B) True if $\alpha \cdot \alpha \geq 0$, by arguments which are easier if $b_2^+ \geq 3$, using Seiberg–Witten gauge theory as above (this was done first by Kronheimer & Mrowka, and independently by Morgan, Szabó & Taubes).

There is a version for immersed spheres with no condition on $\alpha \cdot \alpha$; it estimates the number of positive double points ([316,Fintushel & Stern,1995a] and, using Seiberg–Witten invariants, [317,Fintushel & Stern,1995b,Turkish J. Math.]).

For $\alpha \cdot \alpha < 0$ there are estimates on the minimal genus which in general fall short of the generalized conjecture [614,Kotschick & Matic,1995,Math. Proc. Cambridge Philos. Soc.].

For non Kähler surfaces, the conjecture is trivially false because, for example, there is a holomorphic $S^1 \times S^1$ in the Hopf surface $S^3 \times S^1$.

Note that when this problem was written in 1977, Thom's name was not generally associated with the conjecture; however the association developed later although it is not clear whether Thom ever made the conjecture. In the early 70's, the handful of workers in the smooth topology of closed, simply connected 4-manifolds were well aware that all their examples came from complex surfaces, and that they didn't know how to show that smooth constructions couldn't be complex. The obvious questions in those days included the conjectures in this problem, as well as what is now called the 11/8–Conjecture (Problem 4.92) and the 3/2–Conjecture (Problem 4.93).

Problem 4.37 A simply connected, complex analytic surface S may contain exceptional curves (complex $\mathbb{CP}^1$'s with self-intersection -1) in which case S is diffeomorphic to $M^4 \# r(-\mathbb{CP}^2)$.

Can S be decomposed as a connected sum in any other way?

Remarks: (Yau) If $\beta_2(S) = 1$, then $S = \mathbb{CP}^2$; otherwise S has an indefinite intersection form and S decomposes homotopically.

Update: A complex, analytic surface is Kähler iff b_1 is even, and in this case the surface can even be deformed to an algebraic surface (the deformation is in [596,Kodaira,1964,Amer. J. Math.] and *even b_1 implies Kähler* follows from the Kodaira classification plus work of [773, Miyaoka,1974,Proc. Japan Acad. Ser. A Math. Sci.], [1057,Todorov, 1980,Invent. Math.] and [1002,Siu,1983,Invent. Math.]). Since S is algebraic (and simply connected), it follows that if S decomposes, one of the summands, N, must have negative definite intersection form [251, Donaldson,1990,Topology] which must be diagonal over $\mathbb{Z}$ [247,Donaldson,1983,J. Differential Geom.]. However this summand N is not yet known to be diffeomorphic to a connected sum of $-\mathbb{CP}^2$'s.

If S is not simply connected, but b_1 is even, then the above conclusions are all true and $\pi_1(N)$ has no non-trivial finite quotient [609, Kotschick,1993,Internat. Math. Res. Notices, bound within Duke Math. J.]. Gromov [416,1989,C.R. Acad. Sci. Paris Sér. I Math.] proves that one of the summands must be simply connected.

There exists a minimal Class VII_0 surface with $b_1 = 1$, $b_2^+ = 0$, which smoothly splits off a $-\mathbb{CP}^2$ [562,Kato,1978].

Recently, Kotschick [611,1995a] proved that a minimal, symplectic, four-manifold with $b_2^+ > 1$, and with residually finite fundamental group, is irreducible (every imbedded 3-sphere bounds a contractible manifold). This implies that complex surfaces with even b_1 which satisfy the assumptions of the theorem have no connected sum decompositions at all (unless one of the summands is a homotopy sphere).

Problem 4.38 (A. Kas) Let F^2 be a closed orientable 2-manifold. Let $\gamma_1, \ldots, \gamma_n$ be imbedded circles in F^2. let $\tau_i : F^2 \to F^2$ $(i = 1, \ldots, n)$ be a left-handed Dehn (Lickorish) twist about the circle γ_i. Assume that $\tau_n \circ \ldots \circ \tau_2 \circ \tau_1 : F^2 \to F^2$ is isotopic to the identity.

Conjecture: *Let $E \subset H_1(F^2; \mathbb{Q})$ be the subspace spanned by the γ_i $(i = 1, \ldots, n)$. Then the alternating intersection form is nondegenerate on E.*

Remarks: This conjecture gives a topological proof of the hard Lefschetz theorem for complex algebraic surfaces. Let S be an algebraic surface and let $F \subset S$ be a generic hyperplane section. The states that intersection with F defines an isomorphism from $H_3(S; \mathbb{Q})$ to $H_1(S; \mathbb{Q})$

(see Mumford's appendix to Chapter VI, in [1137, Zariski, 1935; pages 151–152]).

Update: No known progress. However, the hard Lefschetz theorem fails for symplectic 4-manifolds; if these have Lefschetz pencils, then a proof along the lines of this problem (using pencils) is unlikely.

Problem 4.39 (T. Matumoto) *Find new examples of complex surfaces with positive signature.*

Remarks: The only known simply connected example is $\mathbb{CP}^2$. There are non-simply connected examples due to Hirzebruch [494, 1978], also [135, Borel, 1963, Topology] and [597, Kodaira, 1967, J. Analyse Math.].

Notice that if S is a complex surface with $c_1 \equiv 0 \pmod 2$, then there exists a holomorphic line bundle $\frac{1}{2}K$ (now called $K^{1/2}$) such that

$$\sigma(S) = -8(2 \dim H^0(S; \mathcal{O}(\frac{1}{2}K)) - \dim H^1(S; \mathcal{O}(\frac{1}{2}K))).$$

It follows that $H^1(S; \mathcal{O}(\frac{1}{2}K)) = 0$ implies that $\sigma(S) < 0$.

Update: The basic numerical invariants of a compact, complex surface S are its Chern numbers

$$c_2(S) = \chi(S)$$

and

$$c_1^2(S) = 3\sigma(S) + 2\chi(S) = 3\sigma(S) + 2c_2(S).$$

(Since $p_1 = c_1^2 - 2c_2$, the latter equality is familiar to topologists as the Hirzebruch signature theorem: $p_1 = 3\sigma$.) In 1976 Miyaoka [774, 1977, Invent. Math.] and Yau [1133, 1978, Comm. Pure Appl. Math.] proved that $c_1^2 \leq 3c_2$, the best possible inequality since compact quotients of the complex ball satisfy $c_1^2 = 3c_2$, and conversely. Since, at that time, all known surfaces with non-negative signature (equivalently $c_1^2 \geq 2c_2$) had infinite fundamental groups, this became conjectured for all such surfaces (the Bogomolov Watershed Conjecture).

This was disproved in 1985 by Moishezon & Teicher [778, 1987, Invent. Math.] who constructed a simply connected surface with zero signature and infinitely many with positive signature.

This raised a *geographical question*: which pairs of integers (x, y) occur as the Chern numbers (c_1^2, c_2) of a simply connected, complex surface (the above inequality $c_1^2 \leq 3c_2$, the Noether inequality, and the congruence $c_1^2 + c_2 \equiv 0 \pmod{12}$ must be satisfied by $(x, y) = (c_1^2, c_2)$)?

This question was nearly answered by Z. Chen [197,1987,Math. Ann.] who used a different construction of surfaces with $\sigma > 0$ due to Xiao Gang (unpublished). Chen showed that asymptotically, for large x, all lattice points (x, y) with $x \leq 2.7y$ can be realized by simply connected surfaces with $(x, y) = (c_1^2, c_2)$.

Kotschick has investigated topological properties of these surfaces [608,1992c,Topology] and [607,1992b,Math. Ann.] and in the latter shows that the Moishezon–Teicher surfaces are spin whereas the Chen surfaces are not.

Problem 4.40 (A) (Moishezon) Let C be an algebraic curve in $\mathbb{CP}^2$ which has only ordinary quadratic singularities.

Conjecture: *$\pi_1(\mathbb{CP}^2 - C)$ is commutative.*

Remarks: A point p is an ordinary quadratic singularity (or ordinary double point) if there are local coordinates near $p = (0, 0)$ such that C is given locally by $x^2 + y^2 = 0$. It would follow that if C is irreducible of degree n, then $\pi_1(\mathbb{CP}^2 - C) = \mathbb{Z}/n\mathbb{Z}$. The conjecture is known for small n and large n.

Using a theorem of Severi, now considered unproven, Zariski published a proof of the conjecture [1136,Zariski,1929,Amer. J. Math.]; also [1137,Zariski,1935; page 210]. Thus the conjecture can be proved by proving a topological analogue of Severi's theorem. Let F_g be a compact, Riemann surface of genus g. Call a smooth immersion $g : F \to \mathbb{CP}^2$ semi-algebraic if there exists a point $q \in \mathbb{CP}^2 - g(F)$ such that the projection (along lines through q) $\pi : g(F) \to \mathbb{CP}^1$ satisfies, (1) $\pi g : F \to \mathbb{CP}^1$ is holomorphic of, say, degree n, (2) in local coordinates, $\pi g(z) = z$ or z^2, (3) g and πg are in general position, and (4) if d is the number of double points of $g(F)$, then $g = \frac{1}{2}(n-1)(n-2) - d$.

(B) *Let $g_1, g_2 : F \to \mathbb{CP}^2$ be two semialgebraic immersions of degree n. Find an ambient isotopy of $\mathbb{CP}^2$ carrying $g_1(F)$ to $g_2(F)$. (True for $n \leq 3$.)*

Update:

(A) The conjecture is true; an algebraic proof is given in [350,Fulton, 1980,Ann. of Math.], and a geometric version in [240,Deligne, 1981].

(B) No progress.

Problem 4.41 *There exists a smooth, proper imbedding of the Poincaré homology sphere P minus a point in $\mathbb{R}^4$ with a possibly exotic smooth structure (Freedman). Exhibit this smooth imbedding, or (easier) ignore the differentiability and construct a locally flat imbedding into $\mathbb{R}^4$. Is there a smooth proper imbedding of P–pt. into $\mathbb{R}^4$?*

Remarks: If yes, that would give an example of a smooth S^2 in S^4 which is topologically unknotted, but smoothly knotted since it would have the punctured Poincaré homology sphere as *Seifert surface.*

Update: No one has yet constructed an imbedding of the punctured Poincaré homology sphere P. But Gompf has pointed out that there cannot be a smooth proper imbedding in $\mathbb{R}^4$. For if so, following the Remark, then surgery on the smooth S^2 linking $\infty = S^4 - \mathbb{R}^4$ would give a smooth, homotopy $S^3 \times S^1$ containing P as a *slice* which contradicts Taubes' end theorem [1036,Taubes,1987,J. Differential Geom.].

Problem 4.42 Let $r\dot{B} = \{x \in \mathbb{R}^4 \mid |x| < r\}$ and give $r\dot{B}$ the smooth structure inherited from R^4_θ, one of the exotic $\mathbb{R}^4$'s.

(A) *What is the largest value of r, say ρ, for which $r\dot{B}$ is diffeomorphic to $\mathbb{R}^4$, and what happens at ρS^3? (This depends on fixing an atlas representing θ.)*

(B) *Is $r\dot{B}$ diffeomorphic to R^4_θ or to $s\dot{B}$ for any $\rho < r < s$?*

Remarks: If so, then a furling argument gives an exotic structure on $S^3 \times S^1$. If not, then the reals inject into the moduli space of smooth structures on $\mathbb{R}^4$.

(C) *Does every smoothly imbedded S^3 in R^4_θ bound a smooth B^4? Or, avoiding the smooth 4-dimensional Schoenflies conjecture, can it be engulfed in a standard $\mathbb{R}^4$ in R^4_θ.*

Update:

(A) No progress.

(B) For some R^4_θ's, these exotic $\mathbb{R}^4$'s are never diffeomorphic, [1036, Taubes, 1987, J. Differential Geom.] providing at least a 2-parameter family of exotic $\mathbb{R}^4$'s, [389, Gompf, 1985, J. Differential Geom.].

(C) No progress.

Problem 4.43 (A) *Can any exotic $\mathbb{R}^4$ be covered by a finite number of coordinate charts? In particular, can an exotic $\mathbb{R}^4$ be the union of two copies of $\mathbb{R}^4$?*

(B) *Find a handlebody decomposition of an exotic $\mathbb{R}^4$.*

(C) *Describe in some usable way a complete Riemannian metric on an exotic $\mathbb{R}^4$. What can be said about the topology of the cut locus for this metric?*

(D) *Does there exist an exotic $\mathbb{R}^4$ which cannot be split by a smooth proper $\mathbb{R}^3$ into two exotic pieces?*

Update:

(A) Freedman produced examples of exotic $\mathbb{R}^4$'s with only finitely many coordinate charts when he constructed exotic $\mathbb{R}^4$'s which imbed in S^4 (see the exposition in [578, Kirby, 1989; pages 98–101]); since the interior of a Casson handle is diffeomorphic to $\mathbb{R}^4$, then one gets examples by adding finitely many Casson handles to a compact smooth 4-manifold (and deleting the remaining boundary). A simple example with 3 charts is obtained by adding the simplest Casson handle (one positive kink at each stage) to the complement of a smooth ribbon-slice for the $(-3, -3, 3)$ pretzel knot [101, Bižaca & Gompf, 1995], (one chart is the 0-handle tubed to the Casson handle, the second chart is a neighborhood of the 1-handles tubed together, and the third chart is derived similarly from the 2-handles).

Of course an exotic $\mathbb{R}^4$ can be covered with 5 charts in the same sense that any smoothly–triangulated m-manifold can be covered by $(m+1)$ smooth charts (cover the vertices with small disjoint charts which are then tubed together to form one chart; next cover the barycenters of the 1-simplices and tube to get the second chart; and so on).

(B) Bižaca was the first to find an explicit description of an (extremely large) handlebody decomposition of an exotic $\mathbb{R}^4$ [99, Bižaca, 1994, J. Differential Geom.], but the construction was quickly simplified to that in the next to the last paragraph. Note that a Casson handle which has a branch with all positive kinks is exotic, and any of these can be used to make examples, as above.

(C) No progress.

(D) No progress.

Problem 4.44 (A) *Can every homeomorphism of $\mathbb{R}^4$ be approximated by a Lipschitz homeomorphism?*

(B) *Does Donaldson's theorem hold in the Lipschitz category?*

Remarks: (Sullivan) If the answer to (A) is yes, then every topological 4-manifold has a Lipschitz structure, (see [1017, Sullivan, 1979]), negating (B). Recall that in higher dimensions the answer to (A) is yes; in fact Lipschitz can be replaced by PL or DIFF [221, Connell, 1963, Ann. of Math.] in (A), and furthermore, TOP = Lipschitz in dimensions $\neq 4$ [Sullivan, ibid.].

Update:

(A) No.

(B) Yes.

Both follow from [253, Donaldson & Sullivan, 1989, Acta Math.]. They extend enough of gauge theory to quasiconformal 4-manifolds (or Lipschitz) so as to exhibit both non-existence and non-uniqueness of quasiconformal structures on topological 4-manifolds; this gives a negative answer to (A) (page 183 [ibid.]).

Note that we still do not know that every quasiconformal 4-manifold has a smooth structure, or that two smooth 4-manifolds are diffeomorphic if they are quasiconformally equivalent (see Problem 4.75).

Problem 4.45 *Does there exist an exotic differentiable structure on S^4? On $S^3 \times S^1$? On any other closed orientable, smooth 4-manifold?*

Remarks: There are plenty of candidates. For example, the Gluck construction on any knotted S^2 in S^4 gives a homotopy 4-sphere (for a specific example without 3-handles, see [29, Akbulut & Kirby, 1985, Topology]), or any presentation of the trivial group which cannot be trivialized by Andrews–Curtis moves gives a smooth homotopy 5-ball whose boundary may be exotic. For possibly exotic $S^3 \times S^1$'s, see Problem 4.42.

The existence of many exotic smooth structures on non-compact 4-manifolds makes an affirmative answer seem likely.

Exotic smooth structures on non-orientable 4-manifolds abound; several topologists have found isolated examples [181, Cappell & Shaneson, 1976, Ann. of Math.], [308, Fintushel & Stern, 1981, Ann. of Math.], [18, Akbulut, 1988, Topology], [309, Fintushel & Stern, 1984], and [618, Kreck, 1984] has found large classes.

Update: The problem is still open for S^4 and $S^3 \times S^1$, but there are plenty of examples on *larger* closed, orientable 4-manifolds. The first example, the rational surface $\mathbb{CP}^2 \# 9(-\mathbb{CP}^2)$ and its logarithmic transforms, was given by Donaldson in 1985 [249, 1987a, J. Differential Geom.]. The smallest (smallest rank H_2) closed example is the Barlow surface (homeomorphic to $\mathbb{CP}^2 \# 8(-\mathbb{CP}^2)$) [605, Kotschick, 1989, Invent. Math.]. For manifolds with boundary, the smallest example is Akbulut's [21, 1991c, J. Differential Geom.] (see Problem 4.16). All examples rely on gauge theory, in particular Donaldson polynomials [252, Donaldson & Kronheimer, 1990]. (Also, see Problem 4.85.)

Problem 4.46 (Freedman) *Is a positive untwisted double of the Borromean rings topologically slice?*

Remarks: This is a simple case of the kind of slicing problem one runs into with some approaches to the topological s-cobordism conjecture. The answer is yes if either non-simply connected surgery or the proper s-cobordism theorem holds.

Update: No progress. This still seems to be a crucial test case for extending Freedman's work to all fundamental groups (see Chapter 12 in [336,Freedman & Quinn,1990], and also [335,Freedman & Lin,1989, Topology]).

Problem 4.47 (Freedman) *Let X be the cone on the unlink of n components in S^3. Suppose X is imbedded properly in B^4 and is locally flat except at the cone point $*$. Suppose the local homotopy at $*$ is free. Does this imply that the imbedding is* flat, *i.e. has a neighborhood homeomorphic to a neighborhood of the standard imbedding of X in B^4?*

Remarks: If yes, then topological non-simply connected surgery works and we *almost* get the s-cobordism theorem; conversely, the s-cobordism theorem for all π_1 would imply *yes*. Note that each disk in X is flat by itself.

Update: No progress, but note that the imprecise condition *local homotopy at $*$ is free* should be that the local homotopy type of the complement of X is that of a wedge of circles.

Problem 4.48 (Freedman) *Find a homotopy theoretic criterion for when $M^3/Y \subset \mathbb{R}^4$ has a one-sided mapping cylinder neighborhood, where Y is an acyclic set in the 3-manifold M.*

Remarks: (Quinn) has such a criterion when Y is a CE set. A *reasonable* criterion would give the topological s-cobordism theorem. An interesting acyclic set is obtained by starting with a genus two handlebody Y_0; get Y_1 by reimbedding Y_0 in itself according to any two distinct words in the commutators of the two generators of $\pi_1(Y_0)$; get Y_2 by reimbedding Y_0 in Y_1 according to the same two words, or any other such pair. Continue, and let $Y = \cap_{k=0}^{\infty} Y_k$.

Update: No progress. This problem had the same goal as Problems 4.46 and 4.47, but this approach, as Freedman puts it, should die a graceful death.

Problem 4.49 *If M^3 is a homology 3-sphere, does $M\#(-M)$ bound a smooth contractible 4-manifold?*

Remarks: It bounds a topological, contractible 4-manifold (Freedman) and it smoothly bounds $M^3 \times I$.

Update: Not always; Akbulut observed [1036,Taubes,1987,J. Differential Geom.] that if M bounds a smooth, simply connected W^4 having a non-diagonalizable, definite intersection form, e.g. the Poincaré homology sphere , then $M\#(-M)$ does not bound a smooth contractible 4-manifold, nor, for that matter, a smooth, simply connected, definite manifold.

Problem 4.50 *Is each simply connected, smooth, closed 4-manifold (other than S^4) realized as a connected sum of complex surfaces (with or without their preferred orientations)?*

Remarks: Probably the answer is no, but Donaldson's work makes *yes* a bit more likely. Furthermore, yes is indicated by analogy with dimension 2 where every orientable closed 2-manifold is a complex curve.

Update: The answer is no. Gompf & Mrowka [397,1993,Ann. of Math.] have constructed an infinite family of smooth 4-manifolds, homeomorphic to the $K3$ surface, but not diffeomorphic to a connected sum of complex surfaces.

However it is still possible that each smooth, simply connected, closed 4-manifold is homotopy equivalent to a connected sum of complex surfaces (see the 11/8–Conjecture, Problem 4.103). Equivalently, are more simply connected homotopy types realized by smooth manifolds than by connected sums of complex surfaces? Without the assumption of simple connectivity, the latter is certainly true, because all finitely presented groups are fundamental groups of closed, smooth, 4-manifolds, but relatively few are fundamental groups of complex surfaces.

Problem 4.51 (Akbulut) (A) *If M_1^4 and M_2^4 are simple homotopy equivalent, closed, smooth 4-manifolds, can we pass from M_1 to M_2 by a series of Gluck twists on imbedded 2-spheres?*

Remarks: No for certain lens spaces cross S^1 (S. Weinberger).

(B) *Same question for a generalized Gluck twist, which is defined as follows:*

Split M^4 along a smooth submanifold N^3 with closed complements W_1 and W_2. In W_1, find a properly imbedded, smooth 2-ball D_1.

Twist D_1 by removing $D_1 \times B^2$ and sewing back by spinning D_1 k-times while traversing ∂B^2. Then find a ∂D_2 with $\partial D_2 = \partial D_1$ and twist back by $-k$. Thus N remains unchanged and we can reglue along N.

In this way, a Cappell–Shaneson exotic $\mathbb{RP}^4$ can be changed to $\mathbb{RP}^4$ by splitting $\mathbb{RP}^4 = S^1 \tilde{\times} B^3 \cup \mathbb{RP}^2 \tilde{\times} B^2$ and twisting $\mathbb{RP}^2 \tilde{\times} B^2$ along $* \times B^2$ to $\mathbb{RP}^2 \times B^2$ $(k = 1)$ and then twisting back by a strange B^2 in $\mathbb{RP}^2 \times B^2$ [16, Akbulut, 1984].

Update: There are now more cases where the answer to (A) is no. If one can pass from M_1 to M_2 by a Gluck twist on a smooth imbedded S^2, then it follows that $M_1 \# (-\mathbb{CP}^2)$ is diffeomorphic to $M_2 \# (-\mathbb{CP}^2)$; but Donaldson invariants are preserved under blowing up, so we need only begin with M_1 and M_2 having different invariants. The only known manifolds for which the answer is yes are for $S^1 \tilde{\times} S^3 \# S^2 \times S^2$ and its exotic version [18, Akbulut, 1988, Topology].

(B) may still be true. Note that the genus 1 version of the Gluck twist is called the logarithmic transform (see Problem 4.11), but this does not generalize to higher genus (see Problem 3.49).

Problem 4.52 *Given M^m and N^n imbedded in Q^4, is there an isotopy making M^m topologically transverse to N^n when $m = 3$, $n = 2$ or $m = 3$, $n = 3$?*

Remarks: The answer is yes for other m and n [909, Quinn, 1982, J. Differential Geom.]. When Q is higher dimensional, see [696, Marin, 1977, Ann. of Math.].

Update: The answer is yes, topological transversality holds in all cases [913, Quinn, 1988, Bull. Amer. Math. Soc.] and [336, Freedman & Quinn, 1990].

Problem 4.53 (Mandelbaum) *What (minimal) knowledge of homotopy groups, intersection pairings, etc. determines the homotopy type of a closed, compact 4-manifold?*

Remarks: For $\pi_1(M^4) = 0$, the intersection form determines. For $\pi_1(M^4) = \mathbb{Z}/p\mathbb{Z}$, p prime, then π_1 and the intersection pairing $\pi_2(M) \otimes$

$\pi_2(M) \to Z[\pi_1]$ determine, (C. T. C. Wall). Is this theorem true for a larger class of fundamental groups? Give an example where π_1 and the intersection form do not suffice.

Let a generalized Lefschetz torus fibration $M^4 \xrightarrow{f} F_g$ be a map which is a torus bundle off a finite number of points in F_g (= surface of genus g) and over those points $f^{-1}(p)$ is an immersed 2-sphere with one transverse double point. Examples of these are complex elliptic surfaces with no multiple fibers, and (Y. Matsumoto) simply connected, smooth 4-manifolds without one and 3-handles. Mandelbaum & Harper have shown that a homotopy type of a generalized Lefschetz torus fibration is determined by the genus g and the intersection pairing $H_2(M;\mathbb{Z}) \otimes H_2(M;\mathbb{Z}) \to \mathbb{Z}$.

Update: In the oriented case, the following data $\{\pi_1$, intersection pairing $\pi_2 \otimes \pi_2 \to \mathbb{Z}[\pi_1]$, and k-invariant $k \in H^3(\pi_1;\pi_2)\}$ determine the homotopy type in the case that π_1 is finite and its 2-Sylow subgroup is cyclic or quaternion [432,Hambleton & Kreck,1988,Math. Ann.], [70, Bauer,1988]. This is not true more generally because the above data does not distinguish $T^2 \times S^2$ and the twisted S^2-bundle $T^2 \tilde{\times} S^2$. If one adds to the data either w_2 or the intersection form on H_2, then apparently there are no counterexamples.

Less is known in the non-orientable case, but see [434,Hambleton, Kreck, & Teichner,1994,Trans. Amer. Math. Soc.], [573,Kim, Kojima, & Raymond,1992,Trans. Amer. Math. Soc.] and [486,Hillman,1994; Chap. IX]

Hillman [ibid.] determines the homotopy type from simple data in certain geometric cases. For example, a closed orientable 4-manifold M is simple homotopy equivalent to an F-bundle over B (where F and B are orientable surfaces and F is not S^2) iff $\chi(M) = \chi(B)\chi(F)$ and $\pi_1(M)$ is an extension of $\pi_1(B)$ by a normal subgroup isomorphic to $\pi_1(F)$.

Also, [72,Baues,1991] in principle classifies homotopy types of 4-complexes and hence 4-manifolds, but computations are difficult.

Problem 4.54 *Find a geometric proof that* $\Omega^4_{spin} = \mathbb{Z}$.

Remarks: There exists such a proof that $\Omega^4 = \mathbb{Z}$ ([747,Melvin,1984]), but it is not clear how to modify it to get the spin case.

Update: A proof is given in [578,Kirby,1989]; a spin 4-manifold with signature zero is immersed in $\mathbb{R}^6$, and then spin borded to an imbedding which bounds a spin *Seifert surface.*

Given the current interest in $Spin^{\mathbb{C}}$, it is worth remarking that $\Omega^4_{Spin^{\mathbb{C}}} \cong \mathbb{Z} \oplus \mathbb{Z}$. The isomorphism takes a pair (X^4, L) (where L is a complex line bundle with $c_1(L) \equiv w_2(X) \pmod 2$) to $(\sigma(X), (c_1^2(L) - \sigma(X))/8)$. Surely a proof exists in [1015,Stong,1968] (doing the calculation may be easier), or can be found along the lines of Lemma 1, in [578,Kirby,1989; page 65].

Problem 4.55 *Describe the Fintushel–Stern involution on S^4 in* equations. *(See [308,Fintushel & Stern, 1981, Ann. of Math.].)*

Update: No progress.

Problem 4.56 (Melvin) Let M^4 be a smooth closed orientable 4-manifold which supports an effective action of a compact connected Lie group G. Suppose that $\pi_1 M$ is a free group.

Question: *Is M diffeomorphic to a connected sum of copies of $S^1 \times S^3$, $S^2 \times S^2$ and $S^2 \tilde{\times} S^2$?*

Remarks: The answer to both questions is yes for $G \neq S^1$ or T^2; also for $G = T^2$ provided the orbit space of the action (a compact orientable surface) is not a disc with ≥ 2 holes [746,Melvin,1981,Math. Ann.].

Update: Still open.

Problem 4.57 *Classify closed 4-manifolds which fiber*

(A) *over a circle with fiber an S^1-manifold,*

(B) *over a surface.*

Remarks: If (in (A)) the monodromy is periodic and equivariant, then M supports a nonsingular T^2-action and is generally classified by $\pi_1 M$ [858,Orlik & Raymond,1974,Topology]. Exceptions arise when $\pi_1 M/\text{center}$ is finite, e.g. for $S^1 \times L$, L a lens space.

Update: The homotopy classification of 4-dimensional bundle spaces is considered in detail in Chapters III and IV of [486,Hillman,1994].

Mapping tori may be characterized in terms of fundamental group, Euler characteristic, Stiefel–Whitney classes and first k-invariant. (See Theorems III.2–4 of [ibid.] and their Corollaries, and also [486,Hillman, 1994]. In particular, a closed 4-manifold M is simple homotopy equivalent to the mapping torus of a self homeomorphism of an aspherical Seifert fibered 3-manifold if and only if $\chi(M) = 0$ and $\pi_1(M)$ satisfies some clearly necessary conditions, and there are only finitely many topological s-cobordism classes of 4-manifolds in each such homotopy type. (See Theorem V.15 of [ibid.].)

Surface bundles are determined by π and χ in the aspherical cases; when the base or fiber is S^2 or $\mathbb{RP}^2$ other characteristic classes are needed, and the picture is not yet complete for non-orientable surface bundles over $\mathbb{RP}^2$. (See Chapter IV of [ibid.]). When the base is an aspherical surface, $Wh(\pi) = 0$, and in most cases there are only finitely many topological s-cobordism classes of such manifolds with the same π and χ as a given bundle space. In particular, S^2-bundles over the torus or Klein bottle are determined up to homeomorphism by π_1, χ and w. (See Chapter V of [ibid.].)

The situation is less satisfactory for S^1-bundles over 3-manifolds, but partial results are given in Theorems III.8 and V.16 of [ibid.].

Problem 4.58 (Melvin) Let $P \subset S^4$ be the standardly imbedded $\mathbb{RP}^2$ (e.g. $P = q(\mathbb{RP}^2)$, where $q : \mathbb{CP}^2 \to S^4$ is the quotient map by complex conjugation) and $K \subset S^4$ be an odd twist spun knot. Denote by $(S^4, P\#K)$ the pairwise connected sum $(S^4, P)\#(S^4, K)$.

Is $(S^4, P\#K)$ pairwise diffeomorphic to (S^4, P)?

Remarks:

- They have the same 2-fold branched covers, namely $(\mathbb{CP}^2, \mathbb{RP}^2)$ (Melvin), so a negative answer yields an exotic involution on $\mathbb{CP}^2$ with fixed point set $\mathbb{RP}^2$ and quotient S^4.

- $\pi_1(S^4 - P\#K) = \mathbb{Z}/2\mathbb{Z}$, so $S^4 - N(P\#K)$ is s-cobordant *rel* boundary to $S^4 - N(P)$ [640,Lawson,1984,Math. Ann.], where $N(\ )$ denotes an open tubular neighborhood.

Update: It follows from Freedman's topological s-cobordism theorem for $\pi_1 = \mathbb{Z}/2\mathbb{Z}$ that $(S^4, P\#K)$ is pairwise homeomorphic to (S^4, P), so the real question is whether $(S^4, P\#K)$ is an exotic smooth imbedding of $\mathbb{RP}^2$ in S^4.

Finashin, Kreck & Viro ([305,1987,Bull. Amer. Math. Soc.] and details in [306,1988]) later produced exotic smooth imbeddings of a connected sum of 10 copies of $\mathbb{RP}^2$ in S^4 using gauge theory; the imbeddings are distinguished by their 2-fold branched covers which are log transforms of $E(1)$ (see the Update to Problem 4.11).

Problem 4.59 (Hillman) *Minimize the Euler characteristic over all closed 4-manifolds M with $\pi_1(M^4) = G$ given.*

Remarks: Hopf's Theorem gives $H_2(\widetilde{M}) \to H_2(M) \to H_2(\pi_1(M)) \to 0$ which puts a lower bound on the rank of $H_2(M)$, given $\pi_1(M)$. But this minimum is not always achieved (Hillman), e.g. let $G = \mathbb{Z} \oplus \mathbb{Z}$ so that $H_2(\mathbb{Z}\oplus\mathbb{Z}) = \mathbb{Z}$, but $\chi(M) = -1$ is not possible by an Euler characteristic argument on the equivariant homology of the universal covering space. Note that this problem generalizes the problem of which groups are the fundamental group of a homology 4-sphere.

Update: Multiplicative properties of $q(\pi) = \min\{\chi(M) \mid M \textit{ is an orientable closed } 4\textit{-manifold with } \pi_1(M) \cong \pi\}$ were used by Hausmann & Weinberger [468,1985,Comment. Math. Helv.] to obtain new restrictions on the groups of homology 4-spheres. If π has a finite 2-complex or closed orientable 4-manifold as a $K(\pi,1)$-space then $q(\pi) = \chi(K(\pi,1))$. (See Theorem II.12 of [486,Hillman,1994]). Hence orientable bundle spaces usually realize the minimal value of q for their fundamental group. (For aspherical bundle spaces this is clear; S^2-bundles over aspherical surfaces have groups of cohomological dimension 2 and for mapping tori see Theorem III.1 of [ibid.]).

Kotschick and F. E. A. Johnson have considered this invariant in connection with a similar invariant $p(\pi) \leq q(\pi)$ involving the signature, and show that if π is the fundamental group of an aspherical closed orientable 3-manifold then $q(\pi) = 2$. (See [610,Kotschick,1994], [549, Johnson & Kotschick,1993,Math. Proc. Cambridge Philos. Soc.], and [549,Johnson & Kotschick,1993,Math. Proc. Cambridge Philos. Soc.]. See also [681,Lück,1994,Topology].)

Problem 4.60 (Hass) *Let M^4 be closed, smooth and satisfy $\pi_2(M) = 0$ but $\pi_3(M) \neq 0$, i.e. $L(p,q) \times S^1$. Is there a smooth imbedded 3-manifold L^3, with finite cover S^3, representing a non-zero element of $\pi_3(M)$?*

Update: Ruberman [945,1990,Pacific J. Math.] gives a topological counterexample; that is, there is a closed topological 4-manifold which is simple homotopy equivalent to $S^1 \times L(3,1)$, but which contains no quotient of S^3 by a finite group acting linearly which represents a non-trivial element of π_3.

Relaxing the condition that $\pi_2 = 0$, Ruberman [ibid.] also shows that $S^2 \times T^2$ contains no imbedded smooth 3-manifold with finite π_1 representing a non-trivial element of $\pi_3(S^2 \times T^2)$.

Problem 4.61 (Hughes) (A) *Find representatives for each regular homotopy class of immersions of S^n in $\mathbb{R}^{n+k}$.*

Remarks: This is trivial for $k > n$ and solved by Whitney–Graustein for S^1 in $\mathbb{R}^2$. For $n = k$, Smale's solution is to add double points to get $\mathbb{Z}$ for n even or one, and $\mathbb{Z}/2\mathbb{Z}$ otherwise. For S^2 in $\mathbb{R}^3$, Smale's famous theorem (that $\mathrm{Imm}(S^n, \mathbb{R}^{n+k}) = \pi_n(V_{n+k,n})$) shows there is just one class. For S^3 in $\mathbb{R}^4$, Hughes [516,Hughes,1982] gives two generators gotten by capping off the track of an eversion of S^2 in $\mathbb{R}^3$, and capping off twice an eversion. The inclusion of the first of these solves the case S^3 in $\mathbb{R}^5$. The next interesting case is S^4 in $\mathbb{R}^5$.

(B) *Find representatives for all bordism classes of immersions of n-manifolds in $\mathbb{R}^{n+k}$.*

Remarks: This group is $\pi^s_{n+k}(MSO(k))$ (assuming orientability) [1105,Wells,1966,Topology]. This has been solved for $n = 1$, and all k, and 2-manifolds in $\mathbb{R}^3$ [457,Hass & Hughes,1985, Topology]. Several bordism invariants have been developed ([185,

Carter, 1986, Trans. Amer. Math. Soc.] gives a good summary of n-tuple point invariants).

(C) *For a surface in $\mathbb{R}^3$, a neighborhood of a double curve is an immersed $B^1 \vee B^1$-bundle over S^1. In general a k-tuple set will have an immersed $B^n \vee \overset{k}{\ldots} \vee B^n$-bundle. Does the multiple point set with this structure determine the bordism class of the immersion?*

Remarks: Yes for 2-manifolds in $\mathbb{R}^3$. The codimension one case is investigated in [265, Eccles, 1980, Math. Proc. Cambridge Philos. Soc.] and [J. S. Carter, ibid.].

(D) *Can one find explicit coordinates for Boy's surface, i.e. find a smooth function from S^2 to $\mathbb{R}^3$ taking S^2 onto Boy's surface as a 2-1 cover.*

Remarks: Morin & Francis [805, Morton, 1978, Topology] have a complicated function whose image is not the standard Boy's surface.

(E) *The number of quadruple points of an immersed S^3 in $\mathbb{R}^4$ is a $\mathbb{Z}/2\mathbb{Z}$ invariant under regular homotopy. Is it a $\mathbb{Z}/24\mathbb{Z}$ $(= \pi_3^s)$ or even a $\mathbb{Z}$ invariant?*

Update: **(A)** The reflection $r_0 : S^n \to \mathbb{R}^{n+1}$ is trivial up to regular homotopy iff $n = 0, 2, 6$. In particular, the non-trivial element in $Imm(S^4, \mathbb{R}^5) = \pi_4(SO(5)) = \mathbb{Z}/2\mathbb{Z}$ is realized by r_0 [556, Kaiser, 1988, Archiv. Math. (Basel)].

The regular homotopy class of r_0 and more generally, the actions of pre- and post-composition with reflections on $Imm(S^n, \mathbb{R}^{n+k})$ can be explicitly computed. For $k > 1$ it is given by the action of the James automorphisms [540, James, 1976] on homotopy groups of Stiefel manifolds (see [555, Kaiser, 1985]). For $k = 1$, the result can be stated [556, Kaiser, 1988, Archiv. Math. (Basel)] in terms of natural generators of $\pi_n(SO(n+1))$. In particular, the subsets of $Imm(S^n, \mathbb{R}^{n+1})$ which are generated by r_0 under connected sum and reflection are groups when $n \neq 1, 3, 7$. All elements are realized this way when $n \not\equiv 2, 4, 5, 6 \pmod 8$. For $n \equiv 1, 3, 7 \pmod 8$ and $n \geq 8$, the reflection generates an infinite cyclic direct summand, and for $n \equiv 0 \pmod 8$, a direct summand of order 2.

There are at most two elements in $Imm(S^n, \mathbb{R}^{n+1})$ which are realized by imbeddings [Kaiser, ibid.].

The image of $\pi_n(SO(n))$ in $\pi_n(SO(n+1)) \cong Imm(S^n, \mathbb{R}^{n+1})$ can be realized by using Hughes' track construction on regular homotopies of the standard imbedding $S^{n-1} \to \mathbb{R}^n$. For $n \neq 1, 3, 7$, the group of regular homotopy classes is the direct sum of the subgroup generated by the reflection and track constructions [555,Kaiser,1985].

The action of composition with reflection for immersions $M^n \to \mathbb{R}^{n+1}$ can be completely determined [556,Kaiser,1988,Archiv. Math. (Basel)] (for some cases, see [649,Li & Peterson,1985,Kexue Tongbao (Chinese)]). The action is discussed for $M^n \to \mathbb{R}^{2n-2}$ in [557,Kaiser & Li,1991,Acta Math. Sci. (English Ed.)].

Hughes & Melvin [518,1985,Comment. Math. Helv.] determine which classes in $\mathrm{Imm}(S^n, \mathbb{R}^{n+2})$ contain representatives which are imbeddings.

(B) If the bordism class contains an algebraic representative (an immersion given by polynomial components), then any immersion can be ϵ-isotoped to an algebraic immersion [26,Akbulut & King,1992,Topology].

(C) No information.

(D) Apéry [42,1986,Adv. Math.] has given a parametrization of Boy's surface in $\mathbb{R}^3$ by three degree four polynomials, and also Boy's surface as the zero set of a polynomial of degree six in three variables (both degrees are minimal). Color pictures can be found in [43,Apéry, 1987].

(E) No information.

Problem 4.62 (A) *Do the cyclic branched covers of 2-spheres in S^4 imbed in S^5?*

(B) *Does every mapping torus of a 3-manifold imbed in S^5?*

Remarks: If a knot K is doubly null concordant (the slice of an unknotted S^3 in S^5) then all of its cyclic branched covers imbed in S^5, so (A) concerns an obstruction to K being doubly null-concordant.

Update: No progress.

Problem 4.63 (A) *Find a smooth, closed, spin, signature zero 4-manifold X^4 which does not imbed punctured in S^5.*

(B) *Find an X^4 such that $X \# kS^2 \times S^2$ imbeds smoothly in S^5 but X does not.*

Remarks: X smoothly imbeds in S^5 if its fundamental group is simple enough, e.g. $H_1(X;\mathbb{Z})$ is the direct sum of no more than two cyclic groups [209, Cochran, 1984b, Topology]. However there are examples with $\pi_1 = \mathbb{Z}/p\mathbb{Z} \oplus \mathbb{Z}/p\mathbb{Z} \oplus \mathbb{Z}/p\mathbb{Z}$, p odd where X does not smoothly imbed in S^5, does not imbed stably ($\# k(S^2 \times S^2)$), and sometimes is known to imbed punctured [208, Cochran, 1984a, Invent. Math.].

Update: No progress.

Problem 4.64 (A) *What 4-manifolds have a symplectic structure?*

Remarks: A symplectic structure is given by a 2-form Ω with $d\Omega = 0$ for which $\Omega \wedge \Omega$ is a volume form. Thus it is necessary that $H^2(M^4;\mathbb{R})$ contain an element Ω with $\Omega \wedge \Omega \neq 0$.

(B) *Does every contact structure on a 3-manifold M^3 extend to symplectic structure on a bounding 4-manifold?*

Remarks: A contact structure is a 1-form α such that $\alpha \wedge d\alpha$ is nowhere zero. We would require that $\alpha(v) = \Omega(v,n)$ for n an outward pointing normal to $M^3 = \partial W^4$ and $v \in T_M$.

Update:

(A) All finitely presented groups are represented as fundamental groups of symplectic 4-manifolds [396, Gompf, 1995]. These and other examples of Gompf provide many examples of 4-manifolds heretofore not known to have symplectic structures.

(B) A contact structure on M is a field of 2-planes in T_M which is nowhere integrable; it forms the kernel of many 1-forms α which differ by multiplication by smooth functions $f : M \to \mathbb{R}$. So (B) should ask whether a given contact structure has an associated 1-form α such that $\alpha(v) = \Omega(v, n)$. This question is still open (see Problem 4.142).

Problem 4.65 (A) *Find a differential geometric invariant which distinguishes the ends of smooth non-compact 4-manifolds. For example, if X^4 is a simply connected topological manifold with a definite intersection form, then X^4-point is smooth but its end is not standard; can this be detected in a direct differential geometric way?*

(B) *Find a differential geometric proof of Rohlin's theorem.*

Remarks: There is such a proof using the $\hat{A}$ genus, and Taubes has found a nice proof by getting the quaternions to act on the sequence $\Omega^0(\mathfrak{g}) \xrightarrow{d_A} \Omega^1(\mathfrak{g}) \xrightarrow{p-d_A} \Omega^2_-(\mathfrak{g}_-)$ from Donaldson's work, but maybe there is a proof more in the spirit of (A).

Update:

(A) No progress.

(B) Nothing new.

Problem 4.66 *How do metrics (e.g. Riemannian, Lorentz, constant curvature) behave under standard topological constructions such as connected sum, plumbing, handle addition? Same question for η-invariants, moduli spaces, etc.*

Update: Much has probably been done on this open ended problem, and the editor has not attempted to update it.

Problem 4.67 (Hopf) *Does there exist a metric of strictly positive sectional curvature on $S^2 \times S^2$?*

Update: The conjecture was probably attributed to Hopf for the first time in writing by Chern in [199,1971; pages 44–45] and the only known examples of 4-manifolds with strictly positive sectional curvature are the standard ones: S^4, $\mathbb{RP}^4$ and $\mathbb{CP}^2$.

In 1970, Weinstein [1104,1970,J. Differential Geom.] showed that no such metrics could be induced by an immersion into $\mathbb{R}^6$. Most other results since then have been obtained in either one of two areas: perturbation of the nonnegative sectional curvature product metric or restriction on the existence of α-pinched manifolds (sectional curvature K_σ satisfying $0 < \alpha \leq K_\sigma \leq 1$, possibly after renormalization).

- Berger [81,1966,C.R. Acad. Sci. Paris Sér. I Math.] showed that no perturbation of the product metric could have increasing curvature at all planes; that is, if
$$\frac{dK_\sigma}{dt}|_{t=0} \geq 0 \quad \text{then} \quad \frac{dK_\sigma}{dt}|_{t=0} = 0 \quad \text{for all planes } \sigma .$$
- Bourguignon, Deschamps & Sentenac [139,Bourguignon,1975], [142,Bourguignon, Deschamps, & Sentenac,1972,Ann. Sci. École Norm. Sup. (4)], [143,Bourguignon, Deschamps, & Sentenac, 1973,Ann. Sci. École Norm. Sup. (4)] showed that if a product metric (including products of non-standard metrics) has no infinitesimal isometries then there are no one-parameter analytic perturbations of this metric with strictly positive sectional curvature. Observe that in the case of the product of standard metrics in S^2 (which has infinitesimal isometries) it is possible to obtain a variation of the metric with increasing sectional curvature on all mixed planes [143,Bourguignon, Deschamps, & Sentenac,1973, Ann. Sci. École Norm. Sup. (4)], but with no control over the curvature of non-mixed planes.
- Hsiang & Kleiner [514,1989,J. Differential Geom.] showed that there are no positive sectional curvature metrics with infinitesimal isometries and Seaman [984,1988,Michigan Math. J.] showed that no such metric could exist with a harmonic 2-form of constant length.

The Sphere Theorem obtained from compound efforts of Rauch, Berger & Klingenberg (for a modern exposition see [245,do Carmo,1992; Chap 13], [193,Cheeger & Ebin,1975; Chap 6] and [587,Klingenberg,1982; Sec. 2.8]) says that if M is complete, simply connected and pinched by $\frac{1}{4}$ then it is homeomorphic to the n-sphere.

- Berger [80,1963,C.R. Acad. Sci. Paris Sér. I Math.] showed that if M^4 has an α-pinched metric with $\alpha > \frac{4}{17}$ then M^4 has a definite intersection form, ruling out such metrics for $S^2 \times S^2$. Later Bourguignon [140,Bourguignon,1981a] rewrote the proof bringing it down to $\frac{4}{19}$.
- Ville [1076,1989,Ann. Inst. Fourier (Grenoble)] showed that under the above assumptions of $\frac{4}{19}$ pinched, M is homeomorphic to either S^4 or $\mathbb{CP}^2$, and Seaman showed that both Bourguignon and Ville's result were valid under the assumption of $\alpha = 0.1883$ [983, Seaman,1987,Ann. Global Anal. Geom.] and [985,Seaman,1989, Geom. Dedicata].

Problem 4.68 (A) *There exists an anti-self-dual Einstein metric on the Kummer surface. Describe it explicitly.*

Remarks: Its existence follows from Yau's proof of the Calabi conjecture [1132,Yau,1977,Proc. Nat. Acad. Sci. U.S.A.] .

(B) *If M^4 is compact, closed and has an Einstein metric, then $\chi(M) \geq \frac{3}{2}|\sigma(M)|$ [497,Hitchin,1974,J. Differential Geom.]. Are there any other topological restrictions?*

(C) *Does $\#p\mathbb{CP}^2\#q(-\mathbb{CP}^2)$ have an Einstein metric?*

Remarks: If $p > 3$ and $q = 0$, then no. If $p = 1$ then the manifold is complex and a Kähler–Einstein metric exists only when $q \in \{0, 3, 4, 5, 6, 7, 8\}$.

An Einstein metric has the property that sectional curvatures are equal on orthogonal 2-planes. A good reference is [141,Bourguignon,1981b,Invent. Math.].

Update:

(A) A good attempt can be found in [554,Joyce,1995].

(B) There are more restrictions, related to the fundamental group via Gromov's notion of simplicial volume (see [82,Besse,1987]).

(C) If $p = q = 1$, then there is an Einstein metric (which is not Kähler) [865,Page,1979,Phys. Lett. B]. This example is also described in [82,Besse,1987], which also contains information relevant to (A) and (B).

NEW PROBLEMS

Problem 4.69 (Weinberger) (A) *Is every 3-dimensional, ANR, homology manifold stably resolvable?*

(B) *Construct 4-dimensional, nonresolvable, ANR homology manifolds.*

Remarks: X is called a homology manifold if it is a finite dimensional, locally compact metric space satisfying $H_*(X, X-p) \cong H_*(\mathbb{R}^n, \mathbb{R}^n - 0)$ for all points $p \in X$. An m-dimensional, ANR, homology manifold X is resolvable if there exists an m-manifold M and a cell like map $f: M \to X$ (cell like means that for every open set U in X, $f: f^{-1}(U) \to U$ is a proper homotopy equivalence). Stably resolvable means resolvable after crossing with $\mathbb{R}^k$ for some k.

For $m \geq 5$, Quinn ([910,1983,Invent. Math.] [912,1987,Michigan Math. J.]) showed that resolutions are unique, and that they exist if an invariant $I(X) \in H_0(X; \mathbb{Z})$ vanishes.

Edwards [274,1980] characterized topological m-manifolds, $m \geq 5$, as resolvable, ANR, homology m-manifolds which satisfy the Disjoint Disk Property: for any $\epsilon > 0$ and any maps $f, g : D^2 \to X$ there are maps $f', g' : D^2 \to X$ with $d(f, f') < \epsilon, d(g, g') < \epsilon$ and $f'(D^2) \cap g'(D^2) = \emptyset$.

Bryant, Ferry, Mio & Weinberger [155,1993,Bull. Amer. Math. Soc.][156,1995] have shown that there exist nonresolvable, m-dimensional, ANR, homology manifolds X with arbitrary index $I(X) \equiv 1 \pmod 8$, for $m \geq 5$. Some of these X's are not even homotopy equivalent to a closed, topological, m-manifold. It is expected that this phenomena can be pushed down one dimension (B) but not two dimensions (A).

(C) *Are these X's homogeneous, or are all homogeneous, ANR, homology manifolds actually manifolds?*

Remarks: X is *homogeneous* if for any two points in X, there is a homeomorphism of X taking one point to the other.

Problem 4.70 *Let $f : X \to Y$ be a cellular map between polyhedra and suppose X or Y is a 4-manifold. Can f be approximated by a homeomorphism?*

Remarks: Yes in dimensions $\neq 4$ [478, Henderson, 1981, Topology Appl.].

A subset A of X is said to be cellular in X iff there exists a pseudoisotopy of X shrinking exactly A to a point (that is, a homotopy $g_t : X \to Y$ which is a homeomorphism for each $t < 1$ and for which g_1 has only one non-trivial point inverse, namely A). Then f is cellular iff all point inverses are cellular.

Problem 4.71 (Freedman) Conjecture: *All compact 4-manifolds are Hölder continuous.*

Remarks: Non-compact 4-manifolds are smoothable. A function $f : U \to V$ is α *Hölder continuous* on U if for all $x, y \in U$,

$$\sup_{x \neq y} \frac{|f(x) - f(y)|}{|x - y|^{\alpha}} < \infty.$$

Hölder continuous means that f is α Hölder continuous for all $0 < \alpha < 1$. Then a 4-manifold is Hölder continuous if its coordinate transformations are Hölder continuous everywhere. Note that Hölder continuous is not quite Lipschitz ($\alpha = 1$ Hölder continuous) and not all 4-manifolds are Lipschitz (see Problem 4.44 Update).

The more general conjecture which would prove existence and uniqueness is: any homeomorphism between open sets in $\mathbb{R}^4$ which is already Hölder continuous on a neighborhood of a closed set C can be isotoped relative to C to a homeomorphism which is Hölder continuous on a neighborhood V of a given closed set D, where the isotopy is constant outside V.

Problem 4.72 *Does there exist a closed, non-smoothable 4-manifold M^4 having a triangulation (necessarily non-combinatorial)?*

Remarks: Not if M has non-trivial Kirby–Siebenmann invariant (Casson) or an intersection form which cannot be realized by a closed, smooth 4-manifold (Donaldson). If M is triangulated, then all links are spheres except perhaps the links of vertices which are only known to be homotopy spheres. After removing open cones around the vertices, we have a PL, hence smooth 4-manifold with boundary a disjoint union of homotopy spheres. If the Kirby–Siebenmann invariant is one, then some boundary component must have Rohlin invariant one, contradicting Casson's Theorem [30,Akbulut & McCarthy,1990]. If M has a non-realizable form, then the form also cannot be realized by a smooth 4-manifold with homotopy sphere boundary components [1036,Taubes, 1987,J. Differential Geom.].

In dimensions ≥ 5, all manifolds are triangulable if there exists an oriented homology 3-sphere H of Rohlin invariant one such that $H\#H$ bounds an acylic 4-manifold (see Problem 4.4).

Problem 4.73 (Ancel) Definition: X is a *pseudo-spine* of a compact manifold M if X is a compact subset of $intM$ and $M - X$ is homeomorphic to $\partial M \times [0,1)$. In contrast, a *topological spine* X is a compact subset of $intM$ such that M is homeomorphic to the mapping cylinder of a map from ∂M to X, (see Problem 4.25).

(A) *Does the Mazur 4-manifold [724,1961,Ann. of Math.] have disjoint pseudo-spines? More generally, does every compact contractible n-manifold have disjoint pseudo-spines?*

Remarks: This question is motivated by the following question: does the interior of a compact contractible manifold which is not the ball ever cover a compact manifold? (If so, then there is a covering translation which moves a pseudo-spine off itself.) D. Wright [1119,1992,Topology] showed the answer to the latter question is *no* without answering (A).

Note that if a compact, contractible, PL manifold M collapses to a subpolyhedron X in $intM$, then X is a pseudo-spine of M. If $dimM \geq 5$ and $dimX \leq dimM - 3$, then M is a ball (∂M is simply connected by general position, so this follows from the known Poincaré Conjectures and the Schoenflies Theorem).

Guilbault [421,1995,Topology] has constructed compact, contractible n-manifolds containing disjoint spines, for $n \geq 9$, which are not n-balls.

Ancel & Guilbault [34,1995a,Pacific J. Math.] have proved that for $n \geq 5$, every compact contractible n-manifold has a wild arc pseudo-spine. Also the Mazur 4-manifold has a wild arc pseudo-spine.

(B) *Does every compact, contractible 4-manifold have an arc pseudo-spine?*

The Mazur 4-manifold has a handlebody decomposition consisting of a single 0-handle, 1-handle and 2-handle. This suggests splitting question (B) into the following two questions:

(B.1) *Does every PL compact contractible 4-manifold have a handlebody decomposition with no 3- or 4-handles? (compare with Problem 4.18).*

(B.2) *Does every compact contractible 4-manifold that has a handlebody decomposition with no 3- or 4- handles have an arc pseudo-spine?*

(C) *If M^4 is a compact 4-manifold which is homotopy equivalent to a compact surface F^2, does M have a pseudo-spine homeomorphic to F^2?*

Remarks: There are interesting examples which support an affirmative answer:

1. the Giffen disk in the Mazur 4-manifold [233,Daverman,1986; pages 103–106];
2. Y. Matsumoto's example [708,1975,Bull. Amer. Math. Soc.] of a compact 4-manifold, homotopy equivalent to T^2, with a wild T^2 pseudo-spine but no locally flat or PL torus spine.
3. Kawauchi's example [566,1980,Trans. Amer. Math. Soc.] of a locally knotted PL 2-sphere spine of a compact 4-manifold with no locally flat 2-sphere spine.

Let $X(n_1, \dots n_k)$ denote the space obtained by attaching k disks to a circle, the ith disk being attached by a degree n_i covering map. Ancel & Guilbault [35,1995b] have generalized the Giffen disk construction by proving that if a compact 4-manifold M^4 is obtained by attaching k disjoint 2-handles to $S^1 \times B^3$, the ith 2-handle being attached along a curve that wraps n_i times homotopically around the S^1 factor ($n_i \neq 0$), then M^4 has a pseudo-spine that is homeomorphic to $X(n_1 \dots n_k)$.

(D) *If a compact 4-manifold M^4 is homotopically equivalent to $X(n_1 \dots n_k)$, $n_i \neq 0$, then does M^4 have a pseudo-spine that is homeomorphic to $X(n_1 \dots n_k)$?*

Problem 4.74 *Is every closed 4-manifold M^4 the union of a smooth 4-manifold Y and an acyclic topological manifold Z, joined along their common homology sphere boundary? Can acyclic be improved to contractible?*

Remarks: If the $\pi_1(M) = 0$, then the answer is yes: pick any framed link in S^1 whose linking form is the intersection form on $H_2(M;\mathbb{Z})$ and add 2-handles to B^4 along the framed link to get Y^4; then ∂Y is a homology 3-sphere which bounds a topological contractible 4-manifold, Z^4.

Without the assumption that M^4 is closed, the answer is no (independently R. Stong, L. Taylor, and others?). Let K be any Alexander polynomial one knot in S^3 which is not smoothly slice in any acyclic smooth 4-manifold; there are many such K, e.g. the untwisted double of the trefoil knot (Akbulut) (see also [215,Cochran & Gompf,1988, Topology] and [960,Rudolph,1993,Bull. Amer. Math. Soc.]). Since K is topologically slice [330,Freedman,1984] and the slice has a normal bundle [336,Freedman & Quinn,1990; 9.3], it follows that the complement in B^4 is our topological manifold M^4 with smooth boundary equal to 0-surgery on K. For the complement of M^4 in S^4 can be smoothed (it is just B^4 union a 2-handle attached to K), and if this smoothing extends over all of M but an acyclic piece, Z, then the acylic piece Z can be replaced by $S^4 - Z$, producing an acylic smooth manifold in which K is smoothly slice.

$*\mathbb{RP}^4$ is a good test case (see Problem 4.82).

It is interesting to note that M is not triangulable; if it is, then it is smooth on the complement of the vertices in the interior of M, and these can be *amalgamated* by shrinking arcs so that M would be smooth in the complement of a contractible set.

Problem 4.75 (A) (Existence) *Does every quasiconformal 4-manifold have a smooth structure?*

(B) (Uniqueness) *Given smooth 4-manifolds M and N, a homeomorphism $h_0 : M \to N$, closed sets $C \subset D \subset M$ and open sets $U \subset V \subset M$ such that $C \subset U$ and $D \subset V$. Assume h_0 is quasiconformal on U. Does there exist an isotopy $h_t : M \to N$ such* *that h_1 is quasiconformal in a neighborhood of D, and $h_t = h_0$ near C and on $N - V$?*

Remarks: For $C = D = U = V = M$, (B) asks whether compatible smooth structures are unique on a quasiconformal 4-manifold. The more general relative version in (B) implies (A). Also, see the Update on Problem 4.44.

Problem 4.76 *Show that all Casson handles with exactly one kink at each stage are exotic (in the sense that the attaching circle does not bound a smooth 2-ball in the Casson handle).*

Remarks: This is known if all the kinks are positive, or if all are negative [100,Bižaca,1995,Proc. Amer. Math. Soc.], but for the continuum of Casson handles in between, it is not. A positive answer would show that all Casson handles are exotic, because every Casson handle imbeds in one of these.

Problem 4.77 An exotic smooth structure on $\mathbb{R}^4$ crossed with $\mathbb{R}^1$ is diffeomorphic to $\mathbb{R}^5$.

(A) *How can we usefully* see *the exotic $\mathbb{R}^4$ in $\mathbb{R}^5$?*

The diffeomorphism induces a smooth, non-zero vector field (the image of the tangents to $\mathbb{R}^1$) on $\mathbb{R}^5$, whose orbit space is diffeomorphic to the original exotic $\mathbb{R}^4$. So an exotic $\mathbb{R}^4$ can be described by 5 differentiable functions of 5 variables (the components of the vector field).

(B) **(Arnold)** *Can the vector field be* improved *beyond being C^∞? Can its components be analytic? Trigonometric? Polynomial? Can it be written explicitly?*

Problem 4.78 *Which smooth structures on $\mathbb{R}^4$ have compatible Stein structures?*

Remarks: A Stein manifold of complex dimension n is a proper, non-singular, analytic subvariety of $\mathbb{C}^N$ for some N, e.g. an affine, algebraic manifold over $\mathbb{C}$. Any Stein manifold has a plurisubharmonic function, which will then be a Morse function with critical points of index $\leq n$, and generic level sets are all strictly pseudo-convex.

In dimensions $2n > 4$, an open, smooth, almost complex manifold admits a Stein manifold structure iff it admits a proper Morse function with critical points of index $\leq n$ [278,Eliashberg,1990b,Internat. J. Math.], but in dimension 4 this condition is not sufficient. On the other hand, an affine, algebraic surface which is contractible and 1-connected at infinity must have the standard $\mathbb{C}^\infty$-structure [915,Ramanujam,1971, Ann. of Math.].

All exotic $\mathbb{R}^4$'s admit complex structures (and even Kähler metrics) since any contractible 4-manifold immerses smoothly in $\mathbb{R}^4 = \mathbb{C}^2$. Gompf has shown that an uncountable family of exotic $\mathbb{R}^4$'s have Stein structures, (note that these all are smooth subsets of S^4). Moreover, Gompf has shown that any 4-manifold with a handlebody decomposition with only *1*- and *2*-handles has interior homeomorphic to a Stein manifold.

Problem 4.79 (Gompf) **(A)** *Is there a closed 4-manifold covered by an exotic $\mathbb{R}^4$?*

Remarks: Most exotic $\mathbb{R}^4$'s do not cover simply because there are uncountably many exotic $\mathbb{R}^4$'s and only countably many compact 4-manifolds.

(B) *Is there a smooth involution on the standard $\mathbb{R}^4$, topologically equivalent to Id. $\times$ $-$Id. on $\mathbb{R}^2 \times \mathbb{R}^2$, whose quotient is an exotic $\mathbb{R}^4$?*

Remarks: The answer is yes if the adjectives (standard, exotic) are replaced by (standard, standard), or (exotic, standard), or (exotic, exotic); the first case is obvious, the second is due to Freedman (but see [395,Gompf,1993,J. Differential Geom.] for this and other examples involving exotic $\mathbb{R}^4$'s which do or do not imbed in S^4), and the third could arise from double branch covering an exotic $\mathbb{R}^4$ along a smooth $\mathbb{R}^2$ imbedded in a neighborhood of a smooth arc running to ∞.

Problem 4.80 (Gompf) Freedman & Taylor [338,1986,J. Differential Geom.] constructed a universal half space H, that is, a smoothing on

$\frac{1}{2}\mathbb{R}^4 = \{(x_1, x_2, x_3, x_4) \mid x_4 \geq 0\}$ such that any other smooth 4-manifold homeomorphic to $\frac{1}{2}\mathbb{R}^4$ smoothly imbeds in H. They prove H is unique up to diffeomorphism, and that int$H = U$ is a universal smoothing of $\mathbb{R}^4$.

(A) *Is U unique (as a smoothing of $\mathbb{R}^4$ into which all other smoothings of $\mathbb{R}^4$ smoothly imbed) up to diffeomorphism?*

Remarks: U has the property that its end-connected-sum with any other exotic $\mathbb{R}^4$ is U, and it is unique in that sense (since any other U' with that property would satisfy $U' = U'$ end-sum $U = U$).

There is a relation $\leq$ on exotic $\mathbb{R}^4$'s defined by $R \leq R'$ iff any compact submanifold of R smoothly imbeds in R'. R is said to be *compact equivalent* to R' if $R \leq R' \leq R$, and $\leq$ is a partial order on compact equivalence classes.

(B) If $R \leq R'$, does it follow that all of R must smoothly imbed in R'?

The class of U is the unique maximal element in this partial ordering, and, assuming that (B) is true, (A) is equivalent to *the class of U contains only U*. The unique minimal element is the class of $\mathbb{R}^4$ and it is one of uncountably many classes each of which has uncountably many elements [395, Gompf, 1993, J. Differential Geom.].

(C) *Do all compact equivalence classes have uncountably many elements?*

Remarks: Obviously (A), (B) and (C) are not all true.

Problem 4.81 (Teichner) *Let $f : S^2 \to \mathbb{RP}^2 \to \mathbb{RP}^2 \times B^2$ be the composition of the covering map and the inclusion. Is $2f$ homotopic to a locally flat topological imbedding? A smooth imbedding?*

Remarks: f is homotopic to an immersion with one double point which represents the generator in the group $\mathbb{Z}/2\mathbb{Z}$, so f is not homotopic to an imbedding.

Problem 4.82 (Teichner) *Does $(*\mathbb{RP}^4)\#(*\mathbb{CP}^2)$ have a smooth structure?*

Remarks: $*M$ refers to a manifold of the homotopy type of M but opposite Kirby–Siebenmann invariant; it does not always exist (for spin 4-manifolds, the Kirby–Siebenmann invariant is determined by the signature, a homotopy invariant, and in the non-spin case, Teichner gives examples in [1042,Teichner,1995]), and it is also not necessarily unique if $\pi_1 \neq 0$ [ibid.]. $*\mathbb{CP}^2$ was defined by Freedman to be a 0-handle, union a 2-handle added to the trefoil knot with framing one, union a contractible 4-manifold. $*\mathbb{RP}^4$ is constructed [434,Hambleton, Kreck, & Teichner,1994,Trans. Amer. Math. Soc.] from $\mathbb{RP}^4\#E_8$ by surgering away the extra homology. $(*\mathbb{RP}^4)\#(*\mathbb{CP}^2)$ is a *fake* manifold, that is, it is homotopy equivalent to $\mathbb{RP}^4\#\mathbb{CP}^2$ but not homeomorphic to it; however after connected summing with $S^2 \times S^2$, it is homeomorphic to the standard manifold [ibid.].

Thus $(*\mathbb{RP}^4)\#(*\mathbb{CP}^2)$ has zero Kirby–Siebenmann invariant and may have a smooth structure. All other non-orientable 4-manifolds with fundamental group equal to $\mathbb{Z}/2\mathbb{Z}$ have a smooth structure iff their Kirby–Siebenmann invariants are zero, [ibid.].

A positive answer to the imbedding question in Problem 4.81 implies that $*\mathbb{RP}^4\# * \mathbb{CP}^2$ is smooth. $*\mathbb{RP}^4$ is a good test case for Problem 4.74 since it is not known whether it is a smooth 4-manifold union a contractible TOP manifold, as in the case of $*\mathbb{CP}^2$.

Problem 4.83 (Weinberger) *If X^4 is a closed 4-manifold and is homotopy equivalent to a hyperbolic 4-manifold, does X have a smooth structure? Is X then a hyperbolic manifold?*

Remarks: In dimension ≥ 5 the answer is yes to both questions, the first in [290,Farrell & Hsiang,1981a,Ann. of Math.] (also see [302,Ferry & Weinberger,1991,Invent. Math.]), and the second in [294,Farrell & Jones,1989,J. Amer. Math. Soc.].

The same questions can be asked assuming only that X^4 is a $K(\pi_1, 1)$ (see Problem 5.29).

In dimension 3, Gabai has shown that any closed, orientable 3-manifold M, homotopy equivalent to a hyperbolic 3-manifold N, is itself hyperbolic if in addition N contains a closed geodesic in a sufficiently thick hyperbolic tube [360,Gabai,1994b,Bull. Amer. Math. Soc.].

This problem is related to:

Question (Borel Rigidity Conjecture): If X^n is closed and aspherical, then any homotopy equivalence $h : X^n \to Y^n$ is homotopic to a homeomorphism (X and Y are manifolds) (see Problem 5.29).

Problem 4.84 (Teichner) Let M^4 and N^4 be homotopy equivalent, closed 4-manifolds with the same Kirby–Siebenmann invariant. We can ask whether they are equivalent in various senses.

(A) Are M and N homeomorphic? No, and the first examples have fundamental group Q_{16} ($= \Delta^*(2,2,4) = \langle x, y \mid x^4 = y^2 = (xy)^2 \rangle =$ binary triangle group) [1041, Teichner, 1992]. In fact, M and N are not even stably homeomorphic (homeomorphic after connected sums with copies of $S^2 \times S^2$). These examples are not spin, so

Question: *If M and N are also spin, are they stably homeomorphic, or even homeomorphic?*

Remarks: In the spin case with Kirby–Siebenmann invariant zero, J. Davis [234, 1994] proves that M and N are stably diffeomorphic (M and N become smooth after stabilizing) for all fundamental groups for which the Novikov conjecture holds. Kreck [619, 1995] has shown that in the orientable case with zero Kirby–Siebenmann invariant, then stably homeomorphic implies stably diffeomorphic.

(B) Suppose in addition that M and N are rational homology spheres. An interesting set of examples are those formed from $S \times S^1$ where S is a spherical 3-manifold (a quotient of S^3 by a free linear action of a finite subgroup π of $SO(4)$); let M be the result of surgering $* \times S^1$ with the product normal framing, and N the result using the other framing of the normal bundle (note that N can be obtained from M by a Gluck twist on the surgery 2-sphere). Then Teichner, [ibid.], shows that $M \simeq N$ iff the 2-Sylow subgroup of π is cyclic iff M is stably diffeomorphic to N; furthermore M is diffeomorphic to N if π is cyclic [879, Plotnick, 1986, Trans. Amer. Math. Soc.].

Question: *Is M diffeomorphic to N in the case $\pi = \Delta^*(2,2,2n+1) = \langle x, y \mid x^{2n+1} = y^2 = (xy)^2 \rangle$ (π is not cyclic, but has cyclic 2-Sylow subgroup)?*

Remarks: They may be diffeomorphic, but if not, gauge theory is not likely to help because $M\#(-\mathbb{CP}^2)$ is diffeomorphic to $N\#(-\mathbb{CP}^2)$ since one is a Gluck twist on the other.

Problem 4.85 (Gompf & Taylor) *Are there any exotic smooth structures on closed, orientable 4-manifolds that can be distinguished by classical (non gauge theoretic) invariants?*

Remarks: Given a homeomorphism $h : M' \to M$, exotic could mean that h is not isotopic to a diffeomorphism, or it could mean that h is not homotopic to a diffeomorphism, but in this case it should at least mean that M' is not diffeomorphic to M.

In this last sense of exoticity, the answer is *yes* in the non-orientable case: there exist exotic $\mathbb{RP}^4$'s [181,Cappell & Shaneson,1976,Ann. of Math.], [308,Fintushel & Stern,1981,Ann. of Math.], as well as exotic $S^1 \tilde{\times} S^3 \# S^2 \times S^2$'s [16,Akbulut,1984], [309,Fintushel & Stern,1984]; and in the orientable case with boundary [390,Gompf,1986,Math. Ann.].

Scharlemann [965,1976a,Duke Math. J.] gives orientable examples on $S^1 \times S^3 \# S^2 \times S^2$ which are exotic in the sense that h is not homotopic to a diffeomorphism. Lashof & Taylor [639,1984] use Scharlemann's and Akbulut's examples to construct smoothings on any $X^4 \# S^2 \times S^2$ with $H^3(X; \mathbb{Z}/2\mathbb{Z}) \neq 0$ which are exotic in the sense that h is not isotopic to a diffeomorphism.

For open 4-manifolds, Freedman gave an exotic $S^3 \times \mathbb{R}$ [328,1979, Ann. of Math.]; more generally, since smoothing theory works for open 4-manifolds, any open X^4 with $H^3(X^4; \mathbb{Z}/2\mathbb{Z}) \neq 0$ has exotic smoothings in the isotopy sense.

All of these examples are predicted by surgery theory and involve Rohlin's Theorem. The prototypical examples to find would be exotic $S^3 \times S^1$'s and $S^3 \tilde{\times} S^1$'s.

However it would be most interesting to find (not using gauge theory) an exotic M' which is not predicted by surgery theory.

Problem 4.86 *Do all closed, smooth 4-manifolds have more than one smooth structure? Easier is whether all algebraic surfaces have more than one smooth structure.*

Remarks: The first question contains the smooth Poincaré conjecture. A list of algebraic surfaces with more than one smooth structure is given in [433,Hambleton & Kreck,1990,Invent. Math.], but it is now far from complete.

Problem 4.87 *Does every non-compact, smooth 4-manifold have an uncountable number of smoothings?*

Remarks: Gompf [395,1993,J. Differential Geom.] has shown that for any 4-manifold M, $M - pt$ has uncountably many smoothings (also see [242,Ding,1995a], [243,Ding,1995b]). Of course compact 4-manifolds can have only countably many smoothings because there exist only countably many smooth, compact 4-manifolds (see [195,Cheeger & Kister, 1970,Topology]).

Problem 4.88 (Poénaru) Define a manifold to be *geometrically simply connected* if it has a handle body decomposition without 1-handles. Call it *geometrically simply connected at long distance* (GSCLD) if its interior has an exhaustion by compact geometrically simply connected manifolds.

Let Δ be a homotopy 3-ball.

Conjecture (Poénaru): *If $\Delta \times I$ is GSCLD, then $\Delta \times I$ is geometrically simply connected.*

Remarks: This is a special case of the conjecture that any $\Delta \times I$ is geometrically simply connected, which is related to Problem 4.18.

Poénaru had claimed to prove the 3-dimensional Poincaré Conjecture if the above conjecture is true; it is a crucial step in Poénaru's program to prove the Stable PC: $\Sigma \times I$ is geometrically simply connected where Σ is a homotopy 3-sphere (see [362,Gabai,1995b] and its references to many preprints and papers by Poénaru) (also see Problem 3.1).

Note that if B is a Schoenflies 4-ball (that is, a component of the complement of a smoothly imbedded S^3 in S^4), then *int* B is diffeomorphic to $\mathbb{R}^4$ [723,Mazur,1959,Bull. Amer. Math. Soc.], [153,Brown, 1960,Bull. Amer. Math. Soc.], so B is GSCLD. Proving the Conjecture for B, instead of $\Delta \times I$, would thus establish that B has a handle decomposition without 1-handles, i.e. the *1/2-Shoenflies Theorem.*

Problem 4.89 Smooth Poincaré Conjecture: *A smooth homotopy 4-sphere Σ is diffeomorphic to S^4.*

Remarks: Σ is homeomorphic to S^4, so the conjecture is that S^4 has only one smooth structure. This problem is a special case of Problem 4.11 which asks whether homotopy equivalent, simply connected, closed 4-manifolds are homeomorphic, or, implicitly, diffeomorphic. Homeomorphism was answered by Freedman, and Donaldson theory has given many counterexamples to diffeomorphism for sufficiently high second Betti number (see Problem 4.45). But the Conjecture above remains untouched.

Problem 4.24 concerns one possible source, the Gluck construction, of potential counterexamples to this Conjecture. Another source, potential counterexamples to the Andrews–Curtis Conjecture (Problem 5.2, D_0) (i.e. add 1 and 2-handles to B^4 according to any finite presentation of the trivial group and then double), suggests breaking this Problem into two parts:

(A) *Can every homotopy 4-sphere be described without using 3-handles?*

(B) *Is the double of a homotopy 4-sphere without 3-handles diffeomorphic to S^4?*

Remarks: (A) is a special case of Problem 4.18. An affirmative answer to both (A) and (B) together with the Schoenflies Conjecture (Problem 4.32) would prove the Conjecture. (B) may be easier to prove than the Andrews–Curtis Conjecture which implies it. For, given a presentation P of the trivial group (see Problem 5.2 for examples), one can build a homotopy 4-ball B with one and two-handles corresponding to the generators and relations.

(C) *Can B be constructed so that $\partial B = S^3$? So that B is diffeomorphic to B^4?*

Remarks: $B \times I$ is a homotopy 5-ball which is standard if the Andrews–Curtis Conjecture is true for the original presentation P.

The Gluck construction and presentations P give *bushel baskets* of potential counterexamples to the smooth 4-dimensional Poincaré Conjecture.

A corollary of the theorem in the next Problem 4.90 is that an arbitrary homotopy 4-sphere can be constructed from two smooth imbeddings of some homology 3-sphere Σ into $\mathbb{R}^4$, by taking the union, along Σ, of the bounded complements. (Note that in dimension 3, such a property is equivalent to the Poincaré conjecture, see Problem 3.1C.)

Problem 4.90 Form a 5-dimensional h-cobordism, W^5, between two contractible 4-manifolds, which is trivial over the boundary, as follows: choose a link in $S^3 = \partial B^4$ consisting of two components, K_0, K_1, each being the unknot, which link algebraically once. Add two 2-handles to B^4 along each unknot with framing zero and call the result X^4. X^4 has two obvious smoothly imbedded 2-spheres, namely the cores of the 2-handles union the obvious disks, D_0, D_1, in B^4 which the unknots bound. Starting with $X^4 \times I$, add a 3-handle to one 2-sphere in $X^4 \times 0$ and another 3-handle to the other 2-sphere in $X^4 \times 1$. The result is W^5. The portion of ∂W^5 which is $(\partial X^4) \times I$ is the (obviously trivial) side of the h-cobordism W; the bottom and top of the h-cobordism consist of the two contractible 4-manifolds, $\partial_0 W$ and $\partial_1 W$, made by adding a 2-handle to B^4 along K_1 (resp. K_0) and excising a thickened D_0 (resp. D_1) from B^4 (or equivalently putting a *dot* on K_0 (resp. K_1) in the language of the framed link calculus [578,Kirby,1989]).

(A) *Under what conditions on $K_0 \cup K_1$ is W a product h-cobordism relative to the given product structure on the boundary?*

Remarks: The construction above is a modest generalization of the h-cobordism W, using the link drawn in Figure 4.16.1, which was shown to be nontrivial by Akbulut [21,1991c,J. Differential Geom.]. (A suitable name for such W (and its generalizations) would be *Akbulut's corks*, for all non-trivial h-cobordisms are formed by pulling out one of Akbulut's corks and putting it back in with a twist, as remarked in the Theorem below.) A possible answer to (A) is that W is trivial iff $K_0 \cup K_1$ is concordant to the Hopf link (two unknots which link geometrically once).

The above construction has an obvious generalization: let L_0 and

L_1 be two unlinks of n components each, and suppose that the linking number of the ith component of L_0 with the jth component of L_1 is δ_{ij}. Form an h-cobordism W^5 as above, by adding 2-handles to each component with framing zero, obtaining X^4, and then adding 3-handles to n 2-spheres in $X^4 \times 0$ and 3-handles to the other n 2-spheres in $S^4 \times 1$.

(B) *Same question: under what conditions on* $L_0 \cup L_1$ *does the product structure on the* sides *of* W^5 *extend to a product structure on* W^5*?*

Remarks: The h-cobordisms W^5 are special cases of the h-cobordisms which appear in the following theorem of Curtis & Hsiang:

Theorem: Let V^5 be a smooth h-cobordism between two simply connected closed 4-manifolds, M_0 and M_1. Then there is a contractible *sub*$-h$-cobordism W^5 whose complement $V - intW$ is smoothly a product $(M_0 - int\partial_0 W) \times I$.

A greatly simplified proof was found by Freedman & Stong (and appears in [232, Curtis, Freedman, Hsiang, & Stong, 1996, Invent. Math.]) and other versions were independently found by Bižaca and Matveyev. There are also three addenda: W can be chosen so that,

(1) $V - W$ can be taken to be simply connected, and, in fact, to be made with no 3- and 4-handles [Curtis, Freedman, Hsiang & Stong, ibid.];

(2) W^5 is diffeomorphic to B^5 (Bižaca, Kirby);

(3) $\partial_0 W$ is diffeomorphic to $\partial_1 W$ (but not rel boundary) [721, Matveyev, 1995].

From (3), we see that M_1 is obtained from M_0 by cutting out a contractible manifold, $\partial_0 W$, and sewing it back in by a non-trivial diffeomorphism.

The following sketch (see [579, Kirby, 1996]) of a proof of the theorem will make clear the connection between the h-cobordisms W^5 in parts (A) and (B) and the Theorem.

Sketch: V has a handlebody decomposition (starting with M_0) with no 0-, 1- 4- or 5-handles, and we will assume for simplicity that there is only one 2- and one 3-handle. The cocore and core of these handles will intersect a middle level, say M, in a pair of 2-spheres, S_0 and S_1, where $S_0 \cap S_1 = p_1 \cdots p_m$ and the algebraic sum of the points is 1. A regular neighborhood in M of the 2-spheres can be built with a 0-, some 1-, and two 2-handles (called h_0, h_1). Extend this handlebody structure to all of M. Since M is 1-connected, it is possible to slide 2-handles (after adding cancelling 2-3-handles pairs to avoid Andrews–Curtis problems) while avoiding h_0, h_1, so as to get a contractible 4-manifold U containing the 0-handle and all the 1-handles, and just enough 2-handles (to kill π_1) but not h_0, h_1. Now add h_0 and h_1 to U, and then add the original 3-handle along S_1 on one side of U (thickened), and the original 2-handle, now a 3-handle, along S_0 to the other side of U; this forms W. It is clear that the complement of W is a product bordism for it has no critical points; furthermore, W is contractible because h_0 and h_1 are geometrically cancelled by the two 3-handles (in fact, since $U \times I = B^5$ (no Andrews–Curtis problems), W is also B^5).

Thus, an arbitrary h-cobordism produces a W which is only more general than the W in (B) in that 2- and 3-handles are added to U rather than B^4.

(C) *Can U always be chosen to be B^4 so that (B) is the general case? In other words, in the construction in the sketch, can 2-handles be slid so that all the 1-handles are geometrically cancelled?*

Problem 4.91 (Shkolnikov) *Suppose that M^4 is closed, has no* 1 *or* 3 *– handles and its signature is zero. Does it follow that M is diffeomorphic to $\#k(S^2 \times S^2)\#l(S^2 \tilde{\times} S^2)$?*

Remarks: The conclusion does not follow from just assuming that M is simply connected and $\sigma(M) = 0$. The Moishezon–Teicher examples ([778,Moishezon & Teicher,1987,Invent. Math.] and see Problem 4.39) have $\sigma = 0$ and are spin (Kotschick), so are homeomorphic to the connected sum above, but are not diffeomorphic because they have non-zero Donaldson invariants [251,1990,Topology]. These examples have huge second betti number and their handle structure is completely unknown. The non-spin case is similar and examples are older and easier [197, Chen,1987,Math. Ann.].

Problem 4.92 (11/8–Conjecture) *Given a smooth, closed, spin 4-manifold, X^4, with $H_1(X^4;\mathbb{Z}) = 0$, then $\beta_2/|\sigma| \geq 11/8$.*

Remarks: The intersection form is $2kE_8 \oplus m\binom{0\,1}{1\,0}$, so this conjecture can be rephrased as $m \geq 3k$. $11/8$ is realized by the $K3$ surface because rank $H_2(K3;\mathbb{Z}) = 22$ and $\sigma(K3) = 16$. It is known that for signature 16, β_2 cannot be less than 22 [248,Donaldson,1986,J. Differential Geom.].

Connected sums of copies of $K3$ give examples with signature equal to a multiple of 16, and the question is whether β_2 can be lower than in these examples. All complex surfaces satisfy the conjecture. This is all that remains of the question of what forms are realized by closed, simply connected, smooth 4-manifolds (see Update, Problem 4.1).

Recently, Furuta [352,1995] has claimed that $\beta_2/|\sigma| > 10/8$.

Problem 4.93 (3/2–Conjecture) *Given an irreducible, simply connected, closed, smooth, spin 4-manifold X^4, then $\chi(X)/|\sigma(X)| \geq 3/2$.*

Remarks: The form is $2kE_8 \oplus m\binom{0\,1}{1\,0}$, so the conjecture can be rephrased as $m \geq 4k-1$. $3/2$ is realized by the $K3$ surface. Irreducibility is necessary because $\chi(K3\#K3) = 46$ and $|\sigma(K3\#K3)| = 32$. Perhaps one can relax the simply connected condition to just $H_1(X;\mathbb{Z}) = 0$. This conjecture (for $H_1 = 0$) implies the 11/8–Conjecture (for $H_1 = 0$). Possible counterexamples of Akbulut and Gompf can be found in the Remarks to Problem 4.97.

Problem 4.94 (Mikhalkin) *Is it possible to find closed, smooth, (simply connected would be nice) 4-manifolds X, not of the form $M^4 \# \pm k\mathbb{CP}^2$, so that the fraction $b_2(X)/|\sigma(X)|$ is arbitrarily close to 1?*

Remarks: This is easy to do in the topological case by choosing X to be simply connected and to represent arbitrarily large, non-trivial, definite, intersection forms. Exotic smooth structures on $\mathbb{CP}^2\#9(-\mathbb{CP}^2)$ (see Problem 4.11) give 5/4, but note that Furuta has shown that 5/4 cannot be achieved by a spin manifold (see Problem 4.92).

The general geography question is: which pairs $(b_2(X), \sigma(X))$ are realized by irreducible X?

Problem 4.95 (Akbulut) *Does Rohlin's Theorem still hold for smooth, spin 4-manifolds with simply connected periodic ends? More specifically, suppose X^4 is spin, $\sigma(X^4) \equiv 8 \bmod 16$, and ∂X^4 is a homology 3-sphere. Can there be a simply connected, homology bordism W with $\partial W = \partial X \cup -\partial X$, so that X^4 union {countably many copies of W} would contradict an extended Rohlin's Theorem for periodic ends?*

Remarks: Taubes [1036,1987,J. Differential Geom.] shows that Rohlin's Theorem holds if the 4-manifold has a definite intersection form. If the extended Rohlin's Theorem does hold, then an exotic smooth homotopy $S^1 \times S^3$ cannot have a smooth, imbedded cross-sectional, homology 3-sphere M^3 with Rohlin invariant 1 (Taubes [ibid.] shows that such an M^3 cannot bound a definite smooth 4-manifold).

Problem 4.96 We will say that an intersection form is of type $(1, n)$ if it is equivalent (over $\mathbb{R}$) to $\langle 1 \rangle \oplus n\langle -1 \rangle$.

Are there any smooth, closed, irreducible, simply connected 4-manifolds of type $(1, n)$ other that $\mathbb{CP}^2$, $S^2 \times S^2$, the Barlow surface of type $(1, 8)$, the Dolgachev surfaces of type $(1, 9)$, or blowups of any of these?

If so, are any of them algebraic surfaces?

Remarks: There are no other algebraic surfaces which are diffeomorphic to any of the above list [346,Friedman & Qin,1995,Invent. Math.]. The Barlow surface is not diffeomorphic to $\mathbb{CP}^2 \# -8\mathbb{CP}^2$ [605,Kotschick, 1989,Invent. Math.]; furthermore, no blowup of the Barlow surface is diffeomorphic to a blowup of $\mathbb{CP}^2$ [612,Kotschick,1995b,C.R. Acad. Sci. Paris Sér. I Math.] There are infinitely many Dolgachev surfaces (see Update to Problem 4.11).

Problem 4.97 (A) Conjecture: *Every irreducible, smooth, closed, simply connected 4-manifold X^4 except S^4 has an almost complex structure.*

Remarks: An oriented X^4 has an almost complex structure iff $w_2(X)$ lifts to an integral class $c_1 \in H^2(X; \mathbb{Z})$ satisfying $c_1^2 = 2\chi(X) + 3\sigma(X)$. An indefinite form (or trivial definite form) has such a c_1 iff b_2^+ is odd (easy exercise using mod 8 arithmetic). Thus the conjecture is equivalent to:

(B) Conjecture: *A smooth, closed, simply connected 4-manifold with even b_2^+ and even b_2^- decomposes as a connected sum (or is S^4).*

Remarks: Here are two potential counterexamples which appear to be irreducible:

(1) **(Akbulut)** $\Sigma(2,3,13)$ bounds a smooth, contractible 4-manifold [189, Casson & Harer, 1981, Pacific J. Math.] whose union with the Milnor fiber has intersection form $2E_8 \oplus 4\left(\begin{smallmatrix}0&1\\1&0\end{smallmatrix}\right)$ and thus $b_2^+ = 4$.

(2) **(Gompf)** Remove an open tubular neighborhood of a smoothly imbedded S^2 in the $K3$ surface, with $S^2 \cdot S^2 = -2$; the boundary is $\mathbb{RP}^3$, so glue together two copies by an orientation reversing diffeomorphism of $\mathbb{RP}^3$. The result is not spin and has the intersection form $\oplus 6\langle 1\rangle \oplus 38\langle -1\rangle$, and if irreducible, would also violate the 3/2–Conjecture (Problem 4.93).

Problem 4.98 (Gompf) *Does every simply connected, smooth 4-manifold with $b_2^+ \geq 3$ have a Gompf nucleus?*

Remarks: A Gompf nucleus is the 4-manifold described by the framed link in Figure 4.98.1 which should always be homologically essential in the ambient 4-manifold [393, Gompf, 1991b, Topology].

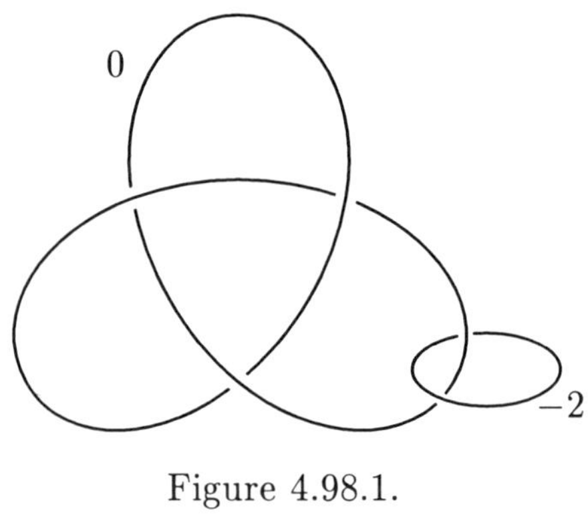

Figure 4.98.1.

This question is also interesting for simpler 4-manifolds, each of which is contained in a Gompf nucleus:

- a neighborhood of a cusp fiber, or
- a neighborhood of a fishtail fiber, or
- a neighborhood of a torus with self intersection zero.

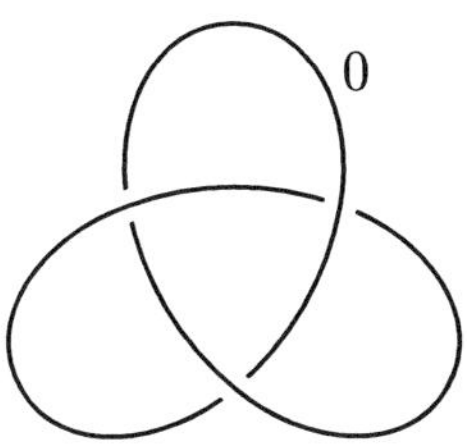

cusp fiber

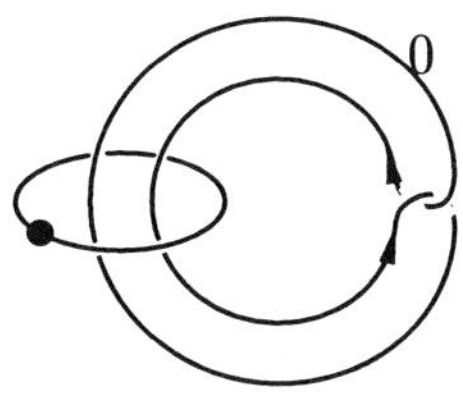

fishtail fiber

A detailed description of singular fibers and their neighborhoods in an elliptic surface can be found in [444,Harer, Kas, & Kirby,1986].

Problem 4.99 (Yau) *Classify smooth, oriented, 4-manifolds* X^4 *with boundary which have a projection to* B^2 *which is a surface bundle over* $B^2 - 0$ *and whose fiber over* 0 *consists of smooth imbedded surfaces in* X^4 *with normal crossings.*

Remarks: It may be easier to first solve the classification problem if blowing up (only with $-\mathbb{CP}^2$) is allowed.

The next step would be to understand how to glue the pieces together to get a closed 4-manifold with a singular fibration over a closed surface. Hopefully, the Seiberg–Witten invariants and the Miyaoka–Yau inequality will come into the gluing argument (and perhaps there are local invariants in the above local setting also). Jost & Yau [553,1993, Amer. J. Math.] have shown that if $X^4 \xrightarrow{\pi} F^2$ is a smooth map with finitely many singular fibers and X^4 is Kähler, then π is homotopic to a holomorphic projection with some conformal structure over F^2.

The classification in terms of the topological monodromy around the singular fiber has been carried out by Matsumoto & Montesinos

[715,1994,Bull. Amer. Math. Soc.] [716,1995] under the assumption that all the normal crossings are positive and each crossing has local coordinates (z_1, z_2) such that the projection is $(z_1, z_2) \to z_1^m z_2^n$, where m and n are the multiplicities of the irreducible components which meet at the crossing.

Problem 4.100 (Y. Matsumoto) A C^∞-Lefschetz fibration of fiber genus g over a surface F is determined up to C^∞-isomorphism by the conjugacy class of the monodromy representation

$$\rho : \pi_1(F - \{p_1, p_2, \cdots, p_s\}) \to \Gamma_g,$$

where $\{p_1, \cdots, p_s\}$ is the set of singular loci, and Γ_g is the mapping class group of genus g. In what follows, F will be S^2 and in this case, the monodromy representation is described by the s-tuple of elements of Γ_g, $\rho(p_1), \rho(p_2), \cdots, \rho(p_s)$, each of which is a right handed Dehn twist about a simple closed curve on the general fiber. This satisfies $\rho(p_1)\rho(p_2)\cdots\rho(p_s) = 1$. Conversely if the latter relation is satisfied, then we obtain a C^∞-Lefschetz fibration of fiber genus g over S^2. (Here we use the same notation p_i for the small loop around the point p_i.)

The mapping class group of genus 2, Γ_2, is generated by the 5 generators $\zeta_1, \zeta_2, \cdots, \zeta_5$, which are represented by full right-handed Dehn twists about canonical curves, (see [90,Birman,1974; page 184]). We have the following relations

$$(\zeta_1\zeta_2\zeta_3\zeta_4{\zeta_5}^2\zeta_4\zeta_3\zeta_2\zeta_1)^4 = 1$$

$$(\zeta_1\zeta_2\zeta_3\zeta_4)^{10} = 1.$$

The first relation is an immediate consequence of the relations in Birman's book, and the second one is algebraically derived from them. These two relations correspond to C^∞-Lefschetz fibrations $M_1 \to S^2$ and $M_2 \to S^2$ of fiber genus two, respectively. The total spaces M_1 and M_2 are simply connected 4-manifolds. They have the same signature, -24, and Euler number, 36, and therefore [329,Freedman,1982,J. Differential Geom.] are homeomorphic to $5\mathbb{CP}^2 \# 29(-\mathbb{CP}^2)$. However, they are not C^∞-isomorphic as Lefschetz fibrations, because the monodromy representation of $M_1 \to S^2$ is onto while that of $M_2 \to S^2$ is not.

Question: *Are M_1, M_2, and $5\mathbb{CP}^2 \# 29(-\mathbb{CP}^2)$ mutually diffeomorphic? (Probably not.)*

Remarks: If M_1 and M_2 have complex structures, then they are not diffeomorphic to $5\mathbb{CP}^2\#29(-\mathbb{CP}^2)$ by [248,Donaldson,1986,J. Differential Geom.].

The Lefschetz fibrations corresponding to the familiar relations

$$(\zeta_1\zeta_2\zeta_3\zeta_4{\zeta_5}^2\zeta_4\zeta_3\zeta_2\zeta_1)^2 = 1$$

$$(\zeta_1\zeta_2\zeta_3\zeta_4\zeta_5)^6 = 1$$

have total spaces $\mathbb{CP}^2\#13(-\mathbb{CP}^2)$, and $K3\#2(-\mathbb{CP}^2)$, respectively.

Problem 4.101 (Stern) For fixed odd $r > 1$, the following pairs $H(r)$ and $H'(r)$ of complex algebraic surfaces of general type are homotopy equivalent but deformation inequivalent. They cannot be distinguished by their Seiberg–Witten invariants; D. Gomprecht, Fintushel & Stern and Morgan & Szabó have shown they also have the same Donaldson polynomials.

(A) Question: *Are $H(r)$ and $H'(r)$ diffeomorphic?*

Remarks: These are known as the Horikawa surfaces [503,Horikawa, 1976a,Ann. of Math.], [504,Horikawa,1976b,Invent. Math.], [505, Horikawa,1978,Invent. Math.], [506,Horikawa,1979,Invent. Math.], and they are double branched covers of $S^2 \times S^2$ which is identified as $\mathbf{F}_{2r}$, the simply connected ruled surface with sections $s_\pm$ with self-intersection $\pm 2r$ and with fiber f. Let $H(r)$ be the double cover of $\mathbf{F}_0$ branched over a smoothing of $6s_+ + 4rf$ and let $H'(r)$ be the double cover of $\mathbf{F}_{2r}$ branched over a disconnected branch locus which is a smoothing of $5s_+ + s_-$. These are surfaces with $c_1^2 = 8r - 8$ and $c_2 = 40r - 4$ and Horikawa has shown [ibid.] that they are deformation inequivalent. If r is even these surfaces can be distinguished by the type of their intersection form. However, if r is odd, both these surfaces have odd intersection form and hence are homotopy equivalent.

Note that a yes answer to (A) provides a counterexample to the Conjecture in Problem 4.134 which states that diffeomorphic complex surfaces are deformation equivalent.

(B) Question: *Are there restrictions on self-diffeomorphisms f of a minimal Kähler surface X (of nonnegative Kodaira dimension) with canonical class K_X beyond the conditions $f^*K_X = \pm K_X$?*

Remarks: For example, consider the surface X which is the double cover of $S^2 \times S^2$ along a branch locus which represents $|2af_1 + 2bf_2|$, where the f_i are the fibers of the two different projections and a, b are positive integers with $a \neq b$. Is it true that for general a and b every orientation-preserving self-diffeomorphism of X preserves the pullbacks to X of f_1 and f_2 up to sign and not just K_X (which is a positive combination of the pullbacks)? Even for simply connected elliptic surfaces there is a gap of finite index between those isometries of $H^2(X;\mathbb{Z})$ known to arise from self-diffeomorphisms and the restrictions placed on such isometries by Donaldson theory or Seiberg–Witten theory (namely, that they preserve the homology orientation of the manifold). So the problem is to construct more diffeomorphisms for these double branched covers or to construct new invariants which rule out the existence of such diffeomorphisms.

Problem 4.102 (Yau) Conjecture: *The minimal, complex surfaces which are not Kähler, are the manifolds of class VII_0 as listed by M. Inoue [519, 1974, Invent. Math.].*

Remarks: A compact complex surface is in the class VII_o if it is minimal and $b_1 = 1$.

The conjecture is true for $b_2 = 0$ [112, Bogomolov, 1976, Math. USSR-Izv.], [113, Bogomolov, 1983, Math. USSR-Izv.], [650, Li, Yau, & Zheng, 1990, Illinois J. Math.], [651, Li, Yau, & Zheng, 1994, Comm. Anal. Geom.]. For $b_2 > 0$, see the table in [828, Nakamura, 1984, Invent. Math.; page 439].

Problem 4.103 Suppose X^4 is a simply connected, algebraic surface.

(A) *Is $X^4 \# \mathbb{CP}^2$ diffeomorphic to a connected sum of copies of $\mathbb{CP}^2$ and $-\mathbb{CP}^2$?*

Remarks: Yes for X^4 elliptic or a complete intersection [691, Mandelbaum & Moishezon, 1976, Topology].

(B) *If X^4 is also spin, is $X^4 \# S^2 \times S^2$ diffeomorphic to a connected sum of copies of $K3$ and $S^2 \times S^2$?*

Remarks: Yes for X^4 elliptic [690, Mandelbaum, 1979], (see also [394, Gompf, 1991c, J. Differential Geom.]).

(C) *If X and Y are simply connected algebraic surfaces (or symplectic 4-manifolds) other than $\mathbb{CP}^2$, does $X \# -Y$ decompose as in (A) for the non-spin case, or (B) for the spin case?*

Remarks: Yes, for elliptic surfaces [Gompf, ibid.] and for arbitrary algebraic surfaces provided one is non-spin and neither is general type other than a complete intersection [391, Gompf, 1988, Invent. Math.].

(D) *How about $X \# Y$?*

Remarks: Nothing is known here for irrational surfaces.

Problem 4.104 (Mikhalkin) Let X^4 be a simply connected, closed, symplectic 4-manifold, and let $\tau : X \to X$ be an anti-symplectic ($\tau(\omega) = -\omega$) involution with a smooth, non-empty, imbedded surface as fixed point set.

Question: *Is X/τ completely decomposable, i.e. is*

$$X/\tau = \#r\mathbb{CP}^2 \# s(-\mathbb{CP}^2) \quad \text{or} \quad \#n(S^2 \times S^2)?$$

Remarks: This mimics the case of conjugation of complex algebraic surfaces, with $Fix(\tau)$ a real algebraic surface, in which X/τ is completely decomposable for many cases [22, Akbulut, 1994, J. Reine Angew. Math.], (Finashin).

Note that when τ is free, then X irreducible implies that X/τ is also irreducible; if in addition X is complex algebraic, then the Seiberg–Witten invariants of X/τ are zero [1097, Wang, 1995, Math. Res. Lett.]. If τ is not free, then the Seiberg–Witten invariants of X/τ still vanish if the fixed point set of τ contains an orientable component of genus > 1 (because of the adjunction formula) and in some other cases (Mikhalkin).

Problem 4.105 (A) *Given an arbitrary smooth 4-manifold, X^4, is the self-intersection of a smoothly imbedded S^2 bounded below by a constant which depends only on X^4?*

Remarks: There is an upper bound of -2 for those X^4 (with no $-\mathbb{CP}^2$ summands) which have non-zero Donaldson or Seiberg–Witten invariants because of the vanishing theorems for those invariants.

In the simply-connected topological case, Freedman [329,1982,J. Differential Geom.] proved that any primitive, non-characteristic class with an even dual is represented by a locally flat imbedded 2-sphere, so there is (assuming $b_2^- > 1$) no lower bound.

(B) *What is the smoothly imbedded 2-sphere with minimal (most negative) self intersection in the $-E_8$ plumbing?*

Remarks: One gets -30 by taking the connected sum of the eight cores of the plumbing, where the connected sum (with appropriate orientations) takes place at the intersection points.

(C) *Same question for the plumbing of only two 2-spheres, each with self intersection -2. Can one improve on -6?*

(D) *Same question for the $K3$ surface.*

Remarks: It has a smoothly imbedded S^2 with self-intersection -66 (Ruberman constructs it by starting with the configuration on page 185 of [68,Barth, Peters, & de Ven,1984], removing one curve, e.g. E_1, and taking the connected sum (with alternating signs) of the rest). Recently, Mikhalkin has constructed one with self intersection -82. In fact, it appears in [19,Akbulut,1991a,J. Differential Geom.; Figure 33] where one can see a tree of 21 2-spheres with self intersection -2 which can be connected summed together; the manifold in Figure 33 was only known to be a homotopy $K3$, but Gompf verified that it is diffeomorphic to $K3$.

Problem 4.106 *Does there exist a simply connected, algebraic surface having elements $\alpha \in H_2(X;\mathbb{Z})$ such that $\alpha \cdot \alpha = -2$, but no smoothly imbedded 2-sphere representing any such class?*

Remarks: Without the simply connected assumption, there are algebraic surfaces which are $K(\pi,1)$'s and have no essential 2-spheres whatsoever. For example, surfaces of general type with $c_1^2 = 3c_2$ are quotients of the complex hyperbolic ball [1133,Yau,1978,Comm. Pure Appl. Math.]. (The complex hyperbolic ball is $\{z \in \mathbb{C}^3 \mid \|z\| = -1\}/S^1$ where $\|z\|^2 = -|z_0|^2 + |z_1|^2 + |z_2|^2$.)

Problem 4.107 (Catanese) *Does there exist an algebraic surface for which $inf(C \cdot C) = -\infty$ where the infimum is taken over all irreducible, complex curves C?*

Problem 4.108 (Ruberman) *Does every smooth, closed, simply connected 4-manifold contain a configuration of 2-spheres such that the complement is a* $K(\pi, 1)$*?*

Remarks: A configuration is just a collection of smoothly imbedded 2-spheres which intersect each other transversally. Rational surfaces have such configurations, as do double coverings of, for example, $\mathbb{CP}^1 \times \mathbb{CP}^1$ branched along unions of generators ($\{x\} \times \mathbb{CP}^1$ and $\mathbb{CP}^1 \times \{x\}$) with singularities resolved to get non-singular surfaces (this construction gives configurations in $K3$ and the Enriques surface) (Kharlamov).

Problem 4.109 *Let* X *be a simply connected 4-manifold and* Σ *an imbedded surface. If* $\Sigma \cdot \Sigma$ *is non-zero and is square free (all prime divisors appear once), does it follow that* $\pi_1(X - \Sigma)$ *is trivial?*

Remarks: Under the assumption, $\pi_1(X - \Sigma)$ will have no non-trivial representations in $SO(3)$.

Problem 4.110 In $\mathbb{CP}^2$, let F^2 (smoothly imbedded) represent $\alpha \in H_2(\mathbb{CP}^2; \mathbb{Z}) = \mathbb{Z}$, suppose genus($F$) is minimal and suppose that $\pi_1(\mathbb{CP}^2 - F) = \mathbb{Z}/\alpha\mathbb{Z}$.

Question: *Is* F *smoothly isotopic to an algebraic curve?*

Remarks: For $\alpha = 1$, this is the old Problem 4.23, which is still open. The topological case is also open.

Problem 4.111 (Eliashberg) (A) *Is there a Lagrangian imbedding of the Klein bottle* K *in* $\mathbb{R}^4$*?*

(B) *Is there a knotted Lagrangian imbedding of the torus* T^2 *in* $\mathbb{R}^4$*?*

Remarks: The usual symplectic form on $\mathbb{R}^4$ is $\omega = 1/2(dx_1 \wedge dx_2 + dx_3 \wedge dx_4)$, and an imbedding of a surface F is Lagrangian if $\omega = 0$ on each tangent plane in T_F. It follows from [1111,Whitney,1941] that an orientable F cannot be imbedded as a Lagrangian unless genus(F) $= 1$. The Lagrangian imbedding of T^2 is provided by the product of two circles in the (x_1, x_2) and (x_3, x_4)-coordinate planes.

$\mathbb{RP}^2$ cannot be imbedded as a Lagrangian, but $K\#nT^2$ can be if $n > 0$ [887,Polterovich,1991,Geom. Funct. Anal.]. To get Lagrangian imbeddings of non-orientable surfaces, one can first deform the imbedding of the T^2 into a Lagrangian immersion with $2n$ intersection points and then surger away the intersection points; positive intersection points contribute orientable handles while negative intersection points contribute non-orientable ones.

One can also ask for the classification of Lagrangian imbeddings up to Lagrangian, or even Hamiltonian, isotopy. See [282,Eliashberg & Polterovich,1995] for further discussion of this problem.

Problem 4.112 (Taylor) *Let M^4 be an orientable 4-manifold.*

(A) *Is there an imbedded (locally flat) surface F, dual to $w_2(M)$, such that $H_1(F;\mathbb{Z}/2\mathbb{Z}) \to H_1(M;\mathbb{Z}/2\mathbb{Z})$ is the zero map?*

(B) *Given $x \in H_2(M;\mathbb{Z}/2\mathbb{Z})$, x dual to $w_2(M)$, then for which covers of M, $\pi : \widetilde{M} \to M$, does there exist $y \in H_2(\widetilde{M};\mathbb{Z}/2\mathbb{Z})$ such that $\pi_*(y) = x$? In particular*

(B') does y exist for the universal cover of M?

(B'') does y exist for the cover corresponding to the kernel of $\pi_1(M) \to H_1(M;\mathbb{Z}/2\mathbb{Z})$?

Remarks: The answer to (A) is yes iff the answer to (B'') is yes (Taylor). (B'), hence (A), is known for $\pi_1(M)$ abelian or free; only the 2-Sylow subgroup of $\pi_1(M)$ matters when $\pi_1(M)$ is finite. (B') is not true in general, but (B'') may be.

A motivation for (A) is that older definitions of quadratic enhancements on $H_1(F;\mathbb{Z}/2\mathbb{Z})$ depended on each element $\alpha \in H_1(F;\mathbb{Z}/2\mathbb{Z})$ being represented by a loop in F which bounded a surface in M^4, e.g. [712,Matsumoto,1986], but later definitions [586,Kirby & Taylor,1990] did not need this condition; thus, were the later definitions more general, or is the answer to (A) yes?

Problem 4.113 (A) (Montesinos) *Is every closed, smooth, orientable 4-manifold X^4 an irregular, simple, 4-fold cover of S^4, branched along a closed surface in S^4?*

Remarks: This is true for 4-manifolds made with 0-, 1-, and 2-handles [779, Montesinos, 1978, Trans. Amer. Math. Soc.], so one can split a closed 4-manifold along the boundary of the 3- and 4-handles, realize both sides as irregular covers, and then worry about classifying such irregular covers of $\#n(S^1 \times S^2)$, [780, Montesinos, 1985]. Recently, Piergallini [875, 1995, Topology] has classified such covers and has shown that every X^4 is such a cover, but over an immersed surface with only transverse double points.

(B) (Akbulut) *Is every simply connected 4-manifold X^4 a cyclic branched cover of S^4, or $\#r(\mathbb{CP}^2)\#s(-\mathbb{CP}^2)$, or $\#n(S^2 \times S^2)$, branched over a smooth imbedded surface?*

Remarks: $\mathbb{CP}^2$ is the branched cover of S^4 along $\mathbb{RP}^2$ [629, Kuiper, 1974, Math. Ann.], and many complex surfaces arise this way (see [28, Akbulut & Kirby, 1980, Math. Ann.] for constructions).

(C) (Akbulut) *Does every irreducible, simply connected, closed, smooth 4-manifold X^4 have a branched cover which is a symplectic manifold.*

Problem 4.114 Conjecture: *Let M^3 be a homology 3-sphere which does not bound a smooth, contractible 4-manifold; then for any homology 3-sphere N^3, $M\#N$ also does not bound a smooth, contractible 4-manifold.*

Remarks: Although $M\#(-M)$ bounds an acyclic 4-manifold, $M \times I$, it is sometimes known not to bound a smooth, contractible 4-manifold (see Update to Problem 4.49).

Problem 4.115 *Is there a theory for smooth 4-manifolds which is analogous to splitting 3-manifolds along 2-spheres (to get irreducible 3-manifolds) and characteristic tori (to get pieces which are atoroidal and, conjecturally, geometric)?*

Remarks: A 4-manifold X^4 is currently said to be irreducible if any smoothly imbedded S^3 bounds a contractible 4-manifold in X; this avoids difficulties with the Schoenflies Conjecture (Problem 4.32). But it is not clear what should replace tori in dimension 4. It is plausible

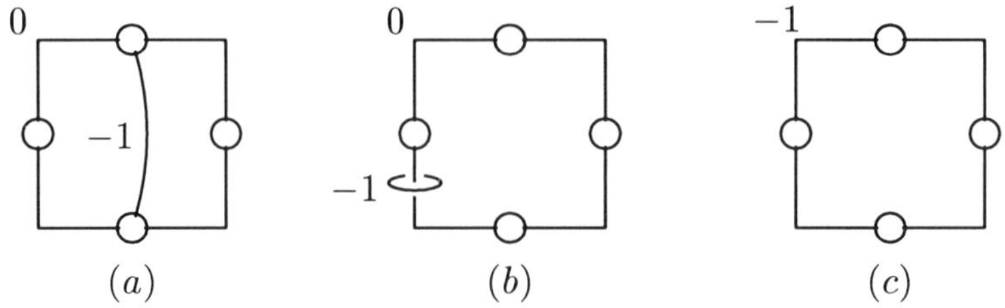

Figure 4.115.1.

to decompose along S^1 bundles over surfaces which do not bound B^2 bundles. But the following example of Fukaya and Y. Matsumoto shows the difficulties with that suggestion.

S^4 can be split into two copies of a neighborhood N of a fishtail fiber (a 2-sphere with one point of self intersection), where the splitting 3-manifold ∂N is the S^1 bundle over T^2 with Euler class 1. Describe N by the Whitehead link with one component denoting a 1-handle (put a dot on it) and the other denoting a 2-handle attached with framing zero. The double of N is not S^4, but if one glues instead by the diffeomorphism of ∂N which interchanges the components of the Whitehead link, then S^4 is obtained. To see that ∂N is the S^1 bundle over T^2, observe that N can be described by adding a 2-handle with framing -1 to $T^2 \times B^2$ (as in Figure 4.115.1 (a)), but ∂N could equally well be described by Figure 4.115.1 (b) where we have just added the 2-handle to a different factor of $S^1 \times S^1 \times S^1 = T^3 = \partial(T^2 \times B^2)$, and then blowing down the -1 circle gives the Euler class 1 bundle over T^2 (Figure 4.115.1 (c)).

This example can also be described as follows: suspend the Hopf map and compose with the Hopf map to get $S^4 \to S^3 \to S^2$; then all fibers are tori except those over the poles which are spheres with double points, and the inverse image of the equator is a torus bundle over S^1 with monodromy $\binom{0\,1}{1\,0}$which is known to be N. (See [709,Matsumoto, 1982a,Proc. Japan Acad. Ser. A Math. Sci.], [713,Matsumoto,1989, Sugaku Expositions], and more generally [711,Matsumoto,1984].)

Problem 4.116 (Ruberman) Given a smooth X^4, if $H_2(X;\mathbb{Z})$ has an orthogonal splitting then it can be realized by a splitting of X^4 by a homology 3-sphere Σ, that is, $X^4 = M_1 \bigcup_\Sigma M_2$ [337,Freedman & Taylor, 1977,Topology].

Find an X and a splitting of $H_2(X)$ where Σ cannot be chosen to be a Seifert fibered 3-manifold.

Remarks: Σ can always be chosen to be hyperbolic [945,Ruberman, 1990,Pacific J. Math.].

Problem 4.117 (Stern) *Which $\Sigma(p,q,r)$ imbed in the standard $K3$ surface?*

Conjecture: *Only finitely many.*

Problem 4.118 (Stern) The lens space $L(p^2, p-1)$, $p \geq 2$, bounds the smooth 4-manifold P obtained from plumbing according to the weighted linear graph

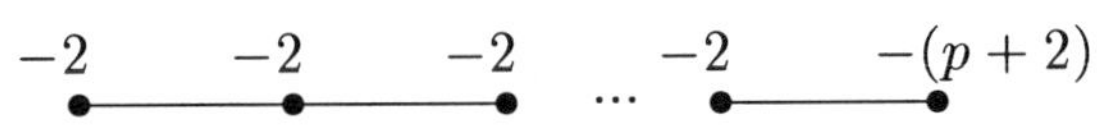

where there are $p-2$ nodes with weights -2 and the last node with weight $-(p+2)$. $L(p^2, p-1)$ also bounds the rational ball Q which is constructed with one 1-handle and a 2-handle going over the 1-handle p times according to the $(-1,p)$ torus knot.

Define the rational blow down in a smooth 4-manifold as the operation of replacing P by Q (assuming one can find a configuration of smoothly imbedded 2-spheres as in the above graph, which can often be done) and a rational blow up as the reverse operation.

Question *Can one obtain any simply connected, smooth, closed 4-manifold from $E(n)$ by a sequence of rational and ordinary blow ups and blow downs?*

Remarks: Fintushel & Stern [314, 1994a] have formulas for the change in the Donaldson series under rational blow ups and blow downs. Logarithmic transforms can be achieved by rational blow ups and downs [ibid.], (the case $p = 2$ is due to Gompf [396, 1995]). It is not known whether a Gluck twist on a smoothly imbedded S^2 can be achieved in this way.

Problem 4.119 *Find examples of aspherical 4-manifolds with π_1-injective, immersed, 3-manifolds with infinite fundamental group. Are these common?*

Problem 4.120 (Kotschick) For a finitely presentable group Γ, define $q^{CAT}(\Gamma)$ to be the minimum Euler characteristic of a closed oriented CAT 4-manifold X^4 with $\pi_1(X) = \Gamma$. (Compare Problem 4.59.)

(A) *Is $q^{TOP}(\Gamma) = q^{DIFF}(\Gamma)$?*

Let $p^{CAT}(\Gamma)$ be the minimal of $\chi(X^4) - \sigma(X^4)$ over CAT 4-manifolds X as above.

(B) *Is $p^{TOP}(\Gamma) = p^{DIFF}(\Gamma)$?*

Remarks: There are many examples (e.g. free groups or surface groups) of groups Γ for which equality holds [610, Kotschick, 1994].

(C) *Is there an integer $k = k(\Gamma_1, \Gamma_2)$ such that for all CAT manifolds X^4 as above with $\pi_1(X) = \Gamma_1 * \Gamma_2$, the connected sum $X \# k(S^2 \times S^2)$ CAT splits as a connected sum $X_1 \# X_2$ with $\pi_1(X_i) = \Gamma_i$?*

Remarks: There is always an l such that $X \# l(S^2 \times S^2)$ splits [487, Hillman, 1995]. The question is whether there is a universal l that works for all X with given fundamental group.

There are Γ_1, Γ_2 for which there exists an X which splits topologically but not smoothly. There are also examples which do not split topologically [620, Kreck, Lück, & Teichner, 1995].

(D) *What can be said about $q^{Symplectic}$ and p^{Sympl}?*

Problem 4.121 (Weinberger) *Let G be a group on Milnor's list [764, 1957a, Amer. J. Math.] of possible groups which can act on S^3. Show that unless G is a subgroup of $SO(4)$, $\mathbb{Z} \times G$ is not the fundamental group of any smooth 4-manifold with zero Euler characteristic.*

Remarks: Other groups, e.g. the metacyclic groups of order pq, p and q odd primes with q dividing $p-1$, come up in the topological case when $n > 4$ [435,Hambleton & Madsen,1986,Canad. J. Math].

Note that there are many topological homotopy $S^1 \times L$, L a lens space, other than $S^1 \times L$ itself. In some cases, they are not even of the form $S^1 \times$(homotopy lens space), which is relevant because, see the Update to Problem 3.37, it is not yet known whether cyclic groups can only act linearly on S^3. See [1102,Weinberger,1987,Israel J. Math.] for an example of the failure of Farrell's fibering theorem based on Casson's invariant.

The group $Q(8,7,29,1)$ is a good candidate for a counterexample (see the Update to Problem 3.37); $\mathbb{Z} \times Q(8,7,29,1)$ is the fundamental group of a *topological* 4-manifold with zero-Euler characteristic.

Problem 4.122 (Fukaya) *Is there a bundle over S^2, with a 4-manifold as fiber, which is topologically trivial but not smoothly trivial?*

Problem 4.123 (Fintushel & Stern) *Does there exist a pseudo-free, smooth, S^1 action on S^5 with more than three multiple orbits?*

Remarks: An S^1 action is pseudo-free if there are no fixed points and the multiple orbits (orbits of finite isotropy) are isolated. Seifert [986, Seifert,1933,Acta Math.] proved that all S^1 actions on S^3 are linear. Montgomery & Yang [782,1972] proved that every homotopy 7-sphere has pseudo-free S^1 actions with arbitrarily many multiple orbits, and this was extended to all S^{2k-1}, $k > 4$, by Petrie [874,1975].

If there exists a Seifert fibered, homology 3-sphere Σ with more than 3 multiple fibers which bounds an acyclic 4-manifold W^4 with an epimorphism $\pi_1(\Sigma) \to \pi_1(W^4)$, then the answer is yes; note that $\Sigma \times B^2 \cup W \times S^1 = S^5$ and S^1 acts diagonally on $\Sigma \times B^2$ and trivially on W.

Problem 4.124 (Edmonds) (A) *Does the fake $\mathbb{CP}^2$ (homotopy equivalent but not homeomorphic to $\mathbb{CP}^2$) admit a topological involution?*

(B) *Does the $K3$ surface admit a periodic diffeomorphism acting trivially on homology?*

Remarks: The answer to (B) is no for period 2 [946, Ruberman, 1995, Proc. Amer. Math. Soc.], [719, Matumoto, 1992]. The question is open for odd period.

Problem 4.125 (A) *Which closed, smooth, simply connected 4-manifolds X have big diffeomorphism groups with respect to some $\kappa \in H_2(X;\mathbb{Z})$?*

Remarks: Let $\mathrm{Diff}_\kappa(X)$ be the group of orientation preserving diffeomorphisms f of X such that $f_*(\kappa) = \kappa$, and let $\mathrm{Aut}_\kappa X$ be the group of automorphisms of $H_2(X;\mathbb{Z})$ which preserves the intersection form and κ. Then X has a *big diffeomorphism group* with respect to κ if the image of $\mathrm{Diff}_\kappa(X)$ in $\mathrm{Aut}_\kappa X$ has finite index.

Simply connected, minimal elliptic surfaces with $p_g \geq 1$ have big diffeomorphism groups with respect to their canonical class [345, Friedman & Morgan, 1994].

(B) Here is a possible partial answer to **(A)**. Let $\kappa_1, \ldots, \kappa_l$ be the basic classes on a smooth, closed $b_1 = 0$, $b_2^+ \geq 3$, 4-manifold X of simple type, and let H be the subspace of $H_2(X;\mathbb{Z})$ spanned by the basic classes.

Conjecture: *The image of $\mathrm{Diff}_H(X) = \mathrm{Diff}(X)$ in $\mathrm{Aut}_H(H_2(X;\mathbb{Z}))$ has finite index.*

Remarks: The converse is almost true, in that if $\mathrm{Diff}(X)$ has finite index in Aut_H then the basic classes belong to H. Do they span H? One can work over $\mathbb{C}$ and replace finite index by the condition that $\mathrm{Diff}(X)$ is Zariski dense in $\mathrm{Aut}_H(H_2(X;\mathbb{C}))$.

Problem 4.126 (D. Randall) Let $C^{PL}(M^m)$ denote the PL pseudo-isotopy (concordance) group for a closed, PL m-manifold M, and let $C^{TOP}(M^m)$ be the corresponding group in the topological category (this is the space of homeomorphisms of $M \times I$ which are the identity on $M \times 0$). It is known that C^{PL} and C^{TOP} have the same homotopy type for any closed, connected PL manifold of dimension $m \geq 5$ [164, Burghelea & Lashof, 1974, Trans. Amer. Math. Soc.].

(A) *Does $C^{PL}(M^4)$ have the same homotopy type as $C^{TOP}(M^4)$ for every closed, simply-connected, smooth M^4?*

Remarks: The map $C^{PL}(S^4) \longrightarrow C^{TOP}(S^4)$ is a homotopy equivalence iff $TOP(4)/PL(4)$ is homotopically equivalent to $K(\mathbb{Z}/2\mathbb{Z}, 3)$.

For 3-manifolds, the pseudo-isotopy groups need not have the same homotopy type; for example, $C^{PL}(S^3)$ is 3-connected while $\pi_2(C^{TOP}(S^3)) = \mathbb{Z}/2\mathbb{Z}$ [916,Randall & Schweitzer,1994] .

(B) Conjecture: *$C^{PL}(S^3)$ is contractible.*

Remarks: $C^{PL}(S^3)$ is contractible if $H_*(C^{PL}(S^3); \mathbb{Z})$ is a *finitely generated*, graded, abelian group [Randall & Schweitzer, ibid.].

(C) 4-dim Smale Conjecture: *The inclusion*

$$SO(5) \longrightarrow SDiff(S^4)$$

is a homotopy equivalence.

Remarks: Note that (B) is equivalent to (C) because $C^{PL}(S^3) \simeq \Omega(PL(4)/O(4))$ is contractible iff $\mathrm{Diff}(D^4, \partial D^4) \simeq \Omega^5(PL(4)/O(4))$ is contractible iff the Smale conjecture holds, since $\mathrm{SDiff}(S^4) \simeq SO(5) \times \mathrm{Diff}(D^4, \partial D^4)$.

The inclusion $SO(n+1) \hookrightarrow \mathrm{SDiff}(S^n)$ is a homotopy equivalence for $n = 1$ (trivial proof), $n = 2$ [1004,Smale,1959,Proc. Amer. Math. Soc.], $n = 3$ [464,Hatcher,1983,Ann. of Math.], and is not a homotopy equivalence for $n \geq 5$ [41,Antonelli, Burghelea, & Kahn,1972,Topology] and [164,Burghelea & Lashof,1974,Trans. Amer. Math. Soc.]. The inclusion $SO(n+1) \hookrightarrow STOP(S^n)$ is a homotopy equivalence iff $n \leq 3$, while $STOP(S^n)$ does not even admit the weak homotopy type of a *finite* complex for each $n > 3$ (the last statement follows from the fact [Randall & Schweitzer, ibid.] that $\pi_3(STOP(S^n))$ contains a copy of $\mathbb{Z}/2\mathbb{Z}$ for each $n > 3$ whereas π_3 of any finite, connected, H-space must be free abelian of finite rank ≥ 0).

(D) Conjecture: $C^{TOP}(S^3) \simeq K(\mathbb{Z}/2\mathbb{Z}, 2)$.

Remarks: Both $C^{PL}(S^3)$ is contractible and $C^{TOP}(S^3) \simeq K(\mathbb{Z}/2\mathbb{Z}, 2)$ iff both $PL(4)/O(4)$ is contractible and also $TOP(4)/PL(4) \simeq K(\mathbb{Z}/2\mathbb{Z}, 3)$.

Problem 4.127 (Kronheimer) Let X^4 be an oriented, closed 4-manifold with non-zero Donaldson invariants . Define $\|\alpha\|$ to be

$$\|\alpha\| = 2g - 2 - F_g \cdot F_g$$

where $\alpha \in H_2(X; \mathbb{Z})$ and F_g is an imbedded surface of minimal genus g which represents α.

Question: *Is* $\| \ \|$ *a seminorm on the positive cone in* H_2*?*

Remarks: A seminorm, unlike a norm, can be zero on nonzero elements. The main issue in the question is the triangle inequality: is $\|\alpha + \beta\| \leq \|\alpha\| + \|\beta\|$? If one does not restrict to the positive cone, the triangle inequality can easily fail. For example, in the $K3$ surface there are classes α and β satisfying $\alpha \cdot \alpha = \beta \cdot \beta > 0$ and $\|\alpha\| = \|\beta\| = 0$ but $\|\alpha + (-\beta)\| > 0$ because $-(\alpha \cdot \beta)$ is too large.

Problem 4.128 (A) *Is the Donaldson series for a smooth, closed, simply connected 4-manifold* X^4 *determined by the following data: the intersection form on* $H_2(X; \mathbb{Z})$*, and the minimal genera of smoothly imbedded surfaces representing an appropriately chosen finite set* $\mathcal{S}$ *of elements in* $H_2(X; \mathbb{Z})$*?*

Remarks: When X^4 is simple (see Problem 4.131), then there are finitely many classes $K_1, \ldots, K_p \in H^2(X; \mathbb{Z})$ and non-zero rationals $a_1, \ldots, a_p$ such that the Donaldson series is given by

$$D = exp(\frac{\mathcal{Q}}{2}) \sum_{s=1}^{p} a_s e^{K_s}$$

where $\mathcal{Q}$ is the intersection form and the *basic* classes K_s are all integral lifts of $w_2(X)$ [624, Kronheimer & Mrowka, 1994a, Bull. Amer. Math. Soc.], [626, Kronheimer & Mrowka, 1995a, J. Differential Geom.], and [316, Fintushel & Stern, 1995a]. If K is a basic class, so is $-K$.

The following inequality is also satisfied [Kronheimer & Mrowka, ibid.] where Σ is any smooth, imbedded, essential surface of genus g in X with $\Sigma \cdot \Sigma \geq 0$:

$$2g - 2 \geq \Sigma \cdot \Sigma + |K_s \cdot \Sigma|.$$

Are the basic classes characterized as those elements of $H^2(X;\mathbb{Z})$ which are lifts of w_2 and satisfy the inequality for all classes Σ with $\Sigma \cdot \Sigma \geq 0$? Is it enough to just use a finite set of the Σ which includes a basis for $H_2(X)$ (we can assume the intersection form is not negative definite)?

(Note that there is another version of the inequality for immersed 2-spheres with p double points of positive sign which holds with no restriction on $\Sigma \cdot \Sigma$ [Fintushel & Stern, ibid.],

$$2p - 2 \geq \Sigma \cdot \Sigma + |K_s \cdot \Sigma|.)$$

(B) Conjecture: *Any basic class K satisfies the equality*

$$K \cdot K = 2\chi + 3\sigma$$

Remarks: This equality holds in the known cases ([626,Kronheimer & Mrowka,1995a,J. Differential Geom.], [314,Fintushel & Stern,1994a], [1012,Stipsicz & Szabó,1994a,Turkish J. Math.]), e.g. for a simply connected, elliptic surface without multiple fibers, the basic classes are nF where F is the fiber $(F \cdot F = 0)$ and $|n| \leq p_g - 1$ and $n \equiv p_g - 1 \pmod 2$ where p_g is the geometric genus.

The basic classes which are defined as those elements of $H^2(X;\mathbb{Z})$ which have non-zero Seiberg–Witten invariants, are conjectured to satisfy this equality (in which case the manifold is called simple in the Seiberg–Witten sense). Furthermore, the basic classes of Kronheimer–Mrowka and of Seiberg–Witten are conjectured to coincide for manifolds which are simple in both senses. Taubes has recently shown [1040,1995c,Math. Res. Lett.] that the Seiberg–Witten basic classes satisfy the equality if X^4 is a compact, oriented, symplectic 4-manifold.

(C) *If X is a minimal, algebraic surface of general type, does the following equality hold:*

$$D = 2^{c_1^2+3-(1-b_1+b_2^+)/2} exp(\frac{Q}{2}) cosh(K_X) \ \textit{for} \ b_2^+ \equiv 3 \pmod 4$$

with cosh replaced by sinh if $b_2^+ \equiv 1 \pmod 4$, where K_X is the canonical class?

Remarks: There are examples of such surfaces which are not deformation equivalent, but do have the same Donaldson series because they are homotopy equivalent by a homotopy which preserves the canonical class.

(D) *Given a class $\alpha \in H_2(X;\mathbb{Z})$, is there a smoothly imbedded surface Σ of genus g representing α which satisfies*

$$2g - 2 = \alpha \cdot \alpha + \max_s |K_s \cdot \alpha|?$$

Problem 4.129 *What is the Donaldson polynomial for $\mathbb{CP}^2$?*

Remarks: Kotschick & Lisca [613,1995] calculated all the $SO(3)$-invariants with non-trivial w_2 up to and including degree 16. In particular, they show that the invariants do not satisfy the simple type relation $D(x^2 z) = 4D(z)$ where x is the class of a point.

The $SU(2)$-invariants have only been evaluated on powers of the hyperplane class (no point class) up to and including degree 21, using stable bundle calculations by various people.

Problem 4.130 *If X^4 is irreducible, simply connected, admits an almost complex structure and satisfies $b_2^+ > 0$, does it have non-zero Donaldson invariants? Seiberg–Witten invariants?*

Remarks: All Kähler surfaces with $b_2^+ \geq 3$ have non-zero Donaldson invariants ([251,Donaldson,1990,Topology] for the 1-connected case, and [787,Morgan & Mrowka,1992,Internat. Math. Res. Notices, bound within Duke Math. J.] for a simple device to extend to the non-simply connected case). Some non-Kähler symplectic 4-manifolds are known to have non-zero Donaldson invariants [397,Gompf & Mrowka,1993,Ann. of Math.], [315,Fintushel & Stern,1994b,Invent. Math.].

Taubes [1037,1994,Math. Res. Lett.] has shown that if X^4 has a symplectic structure and $b_2^+ \geq 2$, then it has non-zero Seiberg–Witten invariants.

Problem 4.131 (Kronheimer & Mrowka) *Do all simply connected, closed, smooth 4-manifolds with $b_2^+ \geq 3$ have simple type?*

Remarks: Simple type means that the Donaldson polynomial q_k for a principal $SU(2)$-bundle with $c_2 = k$ satisfies

$$q_{k+1}(\sigma, \nu, \Sigma_1, \ldots, \Sigma_d) = 4q_k(\Sigma_1, \ldots, \Sigma_d)$$

where $\nu = \mu(1)$ ($\mu : H_0(X; \mathbb{Z}) \to H^4(\mathcal{M}_{k+1})$), $\Sigma_i \in H_2(X; \mathbb{Z})$, and $2d = \dim \mathcal{M}_k$. Manifolds which have simple type include:

- complete intersections,
- elliptic surfaces,
- any manifold with a Gompf nucleus,
- manifolds with a smoothly imbedded surface F satisfying

$$2(\text{genus}(F)) - 2 = F \cdot F > 0.$$

When $b_2^+ = 1$, some manifolds, e.g. $\mathbb{CP}^2$, $S^2 \times S^2$, $\mathbb{CP}^2 \# - \mathbb{CP}^2$, do not have simple type.

Problem 4.132 (Kotschick) Conjecture: *The parity of Donaldson polynomials is a homotopy invariant; that is, if $f : X \to Y$ is a homotopy equivalence, then $f^*(q_Y) = q_X$ in $Sym^d(H^2(X; \mathbb{Z}/2\mathbb{Z}))$.*

Remarks: We assume $b_2^+ \geq 3$, but the conjecture makes sense for the *absolute* invariants in the case $b_2^+ = 1$. In the definition of the polynomials, the μ-map has to be normalized to contain no unnecessary powers of 2, for otherwise the conjecture is trivially true. Many cases have been proved in [939,Ruan,1992,J. Differential Geom.], [313,Fintushel & Stern,1993,J. Amer. Math. Soc.] and [31,Akbulut, Mrowka, & Ruan,1995,Trans. Amer. Math. Soc.], [24,Akbulut,1995b], [25,Akbulut,1995c].

Problem 4.133 (Kotschick) (A) Conjecture: *There is no orientable, connected and simply connected, smooth 4-manifold for which Donaldson's polynomial invariants are defined and non-zero for both choices of orientation.*

Remarks: This refers only to the standard Donaldson polynomials for $SO(3)$ and $SU(2)$ with $b_2^+ \geq 3$. Conjecture (A) implies:

(B) Conjecture: *If two complex algebraic surfaces with finite fundamental groups are orientation reversing diffeomorphic (with respect to their complex orientations), then they are homeomorphic to a geometrically ruled surface, and in particular are simply connected.*

Remarks: Such a surface is a sphere bundle over a sphere. One can ask for diffeomorphism instead of homeomorphism.

Both conjectures would follow from the existence of certain imbedded spheres, or from the ACD property [607,Kotschick,1992b, Math. Ann.] .

Problem 4.134 (Conjecture) *If two complex surfaces are diffeomorphic then they are deformation equivalent.*

Remarks: Two complex structures on a smooth surface are deformation equivalent if their two J's on the underlying tangent bundle are homotopic through integrable J's. This is equivalent to requiring that there be a holomorphic map $h : V_1 \to V_2$ between two varieties which has maximal rank off an analytic subvariety of V_2 such that the two complex surfaces are pre-images of regular values.

Note that Problem 4.101 on Horikawa surfaces is a special case of this Problem and offers a possible counterexample.

Problem 4.135 (Ruan) The first betti number β_1 of a complex surface is even iff the surface is Kähler (see Update to Problem 4.37). A Kähler form is also a symplectic form. If two Kähler manifolds are deformation equivalent (see Problem 4.134 for the definition) then the associated symplectic manifolds are symplectic deformation equivalent.

Question: *Is the converse true?*

Remarks: Two symplectic structures are deformation equivalent if they belong to the same connected component of the moduli space of symplectic structures. The moduli space is constructed from the space of non-degenerate, closed 2-forms by dividing out by the orientation preserving diffeomorphisms; one gets a moduli space which is a cover of an open set in $H^2(X^4;\mathbb{R})$ (see the introduction to [940,Ruan,1993,

Geom. Funct. Anal.]). It may be an infinite cover [733,McDuff,1987, Invent. Math.].

Problem 4.136 (Ruan) Let X and Y be symplectic 4-manifolds with compatible almost complex structures. Assume that $c_1(X) = c_1(Y)$.

Conjecture: *X and Y are diffeomorphic iff $X \times S^2$ and $Y \times S^2$ (with the product symplectic structure) are symplectic deformation equivalent.*

Remarks: Neither implication is known in general. However the conjecture is true if X is an irrational, minimal surface and Y is a non-minimal surface [941,Ruan,1994,J. Differential Geom.], or if X is $E(n)$ with two logarithmic transforms of multiplicities p and q (see Problem 4.11) and Y is the same for p' and q' where $(p,q) = (p',q') = 1$ and $\{p,q\} \neq \{p',q'\}$ [942,Ruan & Tian,1995].

Problem 4.137 (Taubes) *A generic, self dual or ASD closed 2-form is a symplectic 2-form on the complement of a 1-manifold (in X^4). For such degenerate symplectic forms, formulate a theory of pseudo-holomorphic curves (smooth imbedded surfaces on which the symplectic form restricts to a volume form).*

Problem 4.138 (Gompf) (A) Conjecture*:* Any closed, minimal, symplectic 4-manifold is irreducible.

(B) *If X^4 is irreducible, simply connected, has $b_2^+ > 0$, and has an almost complex structure, then does it have a symplectic structure?*

Remarks: Minimality means that there is no symplectically imbedded 2-sphere of self-intersection -1.

(A) is nearly true. If X is symplectic (with 2-form compatible with the orientation of X) and $b_2^+ \geq 2$, then the first Chern class of the associated almost complex structure has non-zero Seiberg–Witten invariant [1037,Taubes,1994,Math. Res. Lett.]. This implies, as in the case of algebraic surfaces (see the Update to Problem 4.37) that if X decomposes as a smooth connected sum, then one of the summands has a negative definite (and therefore standard) intersection form and a fundamental group with no non-trivial finite quotients. This is close to saying that the summand is a connected sum of $-\mathbb{CP}^2$'s. The existence of such a connected summand in turn contradicts the minimality of X [1040, Taubes, 1995c, Math. Res. Lett.].

More recently, Kotschick [611, 1995a] proved that a minimal, symplectic, four-manifold with $b_2^+ > 1$, and with residually finite fundamental group, is irreducible (any summand of a connected sum decomposition is a smooth homotopy 4-sphere).

Problem 4.139 (A) (McDuff) *Let X^4 be a closed, minimal, symplectic 4-manifold containing a symplectic surface F^2 satisfying $c_1(F^2) > 0$. Does it follow that X must be rational or ruled?*

(B) (Gompf) *If X is closed and symplectic, and $3\sigma(X) > \chi(X) =$ ordinary Euler characteristic, does it follow that X is ruled?*

Remarks: Ruled surfaces are holomorphic $\mathbb{CP}^1$ bundles over Riemann surfaces. Rational means $\mathbb{CP}^2$ up to blowing up or down, which includes ruled surfaces over $\mathbb{CP}^1$. (A) is true for Kähler surfaces (see [205, Clemens, Kollar, & Mori, 1988]) and when $F^2 = S^2$ [734, McDuff, 1990, J. Amer. Math. Soc.], [735, McDuff, 1991, Ann. Inst. Fourier (Grenoble)].

Problem 4.140 (Symplectic Thom Conjecture) *In a symplectic 4-manifold X, does a symplectic 2-manifold F minimize genus in its homology class?*

Remarks: If X is a Kähler manifold then the answer is yes, at least for classes represented by holomorphic curves of non-negative self-intersection (see Update to Problem 4.36). The result extends to symplectic manifolds with $b_2^+ > 1$ (with the same proof), because Taubes [1037, 1994, Math. Res. Lett.] has shown that they have non-trivial Seiberg–Witten invariants with respect to the first Chern classes of the associated almost complex structures.

In a symplectic 4-manifold, the adjunction formula still holds. Thus, if F is a symplectically imbedded surface, then $2 \cdot \text{genus}(F) - 2 = F \cdot F - c_1(X)(F)$ where $c_1(X)$ is the first Chern class of T_X equipped with a $U(2)$ structure coming from the symplectic structure.

Thus the open cases are when $b_2^+ = 1$ or the self-intersection is negative.

Problem 4.141 (Eliashberg) An open symplectic 4-manifold M^4 is called convex if there exists a vector field X on M such that:

- X is contracting, i.e. $X\omega = -\omega$,
- X integrates to an $\mathbb{R}$-action which can have fixed points but must be complete,
- there is an exhaustion of M^4 by compact sets $K_1 \subset K_2 \subset \cdots \subset M^4$ with $\bigcup K_i = M$ such that all K_i are mapped into themselves under the positive flow of X.

(A) *Are there convex symplectic structures on all exotic $\mathbb{R}^4$'s?*

Remarks: A Stein manifold automatically has a convex symplectic structure; Gompf has shown that some exotic $\mathbb{R}^4$'s (they imbed in S^4) have Stein structures (see Problem 4.78).

(B) *Are there non-equivalent convex structures on the standard $\mathbb{R}^4$?*

Remarks: Stably (i.e. after crossing with $\mathbb{R}^4$ and, sometimes, $\mathbb{R}^2$) all convex structures are equivalent [281,Eliashberg & Gromov, 1991].

(C) *Is there a convex symplectic structure on $S^2 \times \mathbb{R}^2$?*

Remarks: If such a structure exists, it has to be topologically non-standard at infinity, i.e. the exhaustion mentioned above cannot be made by compact K_i which are diffeomorphic to $S^2 \times B^2$ [277,Eliashberg,1990a]. Gompf has shown that some exotic $S^2 \times \mathbb{R}^2$'s have Stein structures and hence convex symplectic structures.

(D) *Is the space Sympl of all standard-at-infinity symplectic structures on $\mathbb{R}^4$ contractible (or at least connected)?*

Remarks: This is equivalent to the contractibility of the group $\mathrm{Diff}(D^4, \partial D^4)$. Indeed, this group acts on *Sympl* and therefore we have an evaluation map

$$\mathrm{Diff}(D^4, \partial D^4) \to Sympl.$$

This is a fibration whose fiber is the group of symplectomorphisms of D^4 fixed at the boundary. This group is contractible according to [414,Gromov,1985,Invent. Math.].

Problem 4.142 (Eliashberg & Gompf) *Do all oriented 3-manifolds M^3 admit* tight *contact structures?* fillable *contact structures?*

Remarks: A contact structure is a 2-plane field on M which is nowhere integrable; it determines an orientation (let α be a 1-form such that $ker(\alpha)$ is the 2-plane field, and then contact means that $\alpha \wedge d\alpha \neq 0$ so $\alpha \wedge d\alpha$ gives an orientation) which must agree with the given one on M^3.

Such a field is *tight* if, given a smoothly imbedded B^2 in M, the 1-dimensional foliation (with singularities) on B^2 obtained from the intersections of contact planes and tangent planes to B^2 has no closed, non-singular, integral curves. Bennequin [76,1983,Astérisque] proved that the standard contact structure on $\mathbb{R}^3$ (planes equal to the kernel of $dz + xdy$) is tight, and constructed an example of a non-tight (overtwisted) contact structure on $\mathbb{R}^3$ which is, therefore, not diffeomorphic to the standard one.

There are at least three notions of *fillable*:

(1) M^3 is *holomorphically fillable* if M^3 is the oriented, strictly pseudoconvex boundary of a compact, complex surface (that is, using a Hermitian metric, at any $p \in M^3$ the outward normal n_p to M at p is taken by multiplication by i to a tangent vector to M at p, and then the orthogonal complex line should equal the contact plane at p).

(2) M^3 is *strongly symplectically fillable* if M^3 oriented bounds a compact, symplectic 4-manifold (W^4, ω) such that $\omega(P) \neq 0$ for all contact planes P, and there exists a vector field X near M, which is outward pointing and transverse to M at M, and has the property that its flow expands ω, i.e. $\mathcal{L}_X\omega = \omega$ (see Problem 4.141).

(3) M^3 is *symplectically fillable* if M^3 oriented bounds a symplectic 4-manifold (W^4, ω) such that $\omega(P) \neq 0$ for all contact planes P.

It is known that (1) $\Longrightarrow$ (2) $\Longrightarrow$ (3) $\Longrightarrow$ tight, but none of the converses are known. The Poincaré homology 3-sphere Σ with the orientation that does not bound the negative definite E_8 is a candidate (Gompf) for an oriented M^3 which has no tight contact structure, and then $\Sigma\# -\Sigma$ would have no tight contact structure with either orientation.

Tight structures which satisfy (3) but may not satisfy (1) *may* be given by tight structures on T^3 coming from certain cyclic covers of the cotangent S^1-bundle of T^2 [385, Giroux, 1994, Ann. Sci. École Norm. Sup. (4)].

This problem is a more detailed version of the old Problem 4.64 (B).

Problem 4.143 (Fukaya) *Do there exist simply connected, closed 4-manifolds (other than $\#n(S^2 \times S^2)$ or $\#p\mathbb{CP}^2\#q(-\mathbb{CP}^2)$) with metrics of positive scalar curvature which are either not spin or spin with zero signature?*

Remarks: (Kronheimer & Mrowka) It follows from [1117, Witten, 1994, Math. Res. Lett.] using the Seiberg–Witten monopole equation that any 4-manifold with $b^+ > 1$ and a nontrivial Seiberg–Witten class has no metric of positive scalar curvature. Thus complex surfaces of general type with $p_g > 0$ do not have metrics of positive scalar curvature.

Problem 4.144 (Suciu) *Consider a finite union $\mathcal{L}$ of complex lines in $\mathbb{C}^2$. Its complement, $M(\mathcal{L}) = \mathbb{C}^2 \setminus \mathcal{L}$, is a smooth 4-manifold. Does the combinatorial data determine the fundamental group of $M(\mathcal{L})$? the homotopy type of $M(\mathcal{L})$? The topological type? or even the diffeomorphism type?*

Remarks: Combinatorial data means the poset of lines, together with their points of intersection, ordered by reverse inclusion. A general reference is [860, Orlik & Terao, 1992].

The topological type of the complement determines the combinatorial data [541, Jiang & Yau, 1993, Bull. Amer. Math. Soc.], although the homotopy type does not [288, Falk, 1993, Invent. Math.]. If all the lines in $\mathcal{L}$ have real equations, stronger combinatorial data (the *oriented matroid*) determine the topological type of $M(\mathcal{L})$ [102, Björner & Ziegler, 1992, J. Amer. Math. Soc.]. G. Rybnikov has found two arrangements (with complex equations) that have the same intersection poset, but different fundamental groups.

Problem 4.145 (Suciu) Consider a finite union $\mathcal{H}$ of complex planes in $\mathbb{C}^3$ through the origin. We have the Milnor fibration

$$\mathbb{C}^3 \setminus \mathcal{H} \to \mathbb{C}^\star$$

given by taking the product of the linear forms which define the planes.

Questions: *Does the combinatorial data determine the homology of the Milnor fiber? the characteristic polynomial of the monodromy?*

Remarks: Combinatorial data means the poset of planes, together with their lines of intersection, and the origin. The cohomology ring of the complement itself is combinatorially determined [859, Orlik & Solomon, 1980, Invent. Math.].

The homology of the Milnor fiber has been computed [219, Cohen & Suciu, 1995, J. London Math. Soc.] using the presentation for $\pi_1(\mathbb{C}^3 \setminus \mathcal{H})$ found in [917, Randell, 1982, Invent. Math.] and [918, Randell, 1985, Invent. Math.] for real equations, and [47, Arvola, 1992, Topology] in general.

Problem 4.146 (Suciu) *Let $\mathcal{H}$ be an arrangement of planes in $\mathbb{C}^3$ through the origin. Does the combinatorial data determine whether the complement of $\mathcal{H}$ is a $K(\pi,1)$?*

Remarks: It is known that *simplicial* arrangements are $K(\pi,1)$'s [239, Deligne, 1972, Invent. Math.], but generic arrangements are not [466, Hattori, 1975, J. Fac. Sci. Univ. Tokyo Sect. IA Math.]. A long-standing conjecture of Saito (*free* arrangements are $K(\pi,1)$'s) has been disproved by Edelman & Reiner [269, 1995, Bull. Amer. Math. Soc.]. For a thorough discussion of the problem, and a new $K(\pi,1)$ test, see [289, Falk, 1995, Topology].

Problem 4.147 (Suciu) Consider a finite union $\mathcal{L}$ of projective lines in $\mathbb{CP}^2$, and an integer $n \geq 2$. Let $Y(\mathcal{L}, n)$ be the minimal desingularization of the branched cover of $\mathbb{CP}^2$ along $\mathcal{L}$, defined by the map $\pi_1(\mathbb{CP}^2 \setminus \mathcal{L}) \to H_1(\mathbb{CP}^2 \setminus \mathcal{L}; \mathbb{Z}/n\mathbb{Z})$.

Questions: *Does the combinatorial data determine the homotopy type of $Y(\mathcal{L}, n)$? the topological type? or even the diffeomorphism type?*

Remarks: The complex algebraic surfaces $Y(\mathcal{L}, n)$ were introduced in [495,Hirzebruch,1983]. See [489,Hironaka,1993] for further details.

Problem 4.148 (Boileau) Let $f, g : \mathbb{C}^2 \to \mathbb{C}$ be two polynomial maps. Call f and g topologically equivalent if the following diagram commutes:

$$\begin{array}{ccc} \mathbb{C}^2 & \xrightarrow{\ f\ } & \mathbb{C} \\ {\scriptstyle homeo}\downarrow & & \downarrow{\scriptstyle homeo} \\ \mathbb{C}^2 & \xrightarrow{\ g\ } & \mathbb{C} \end{array}$$

For any polynomial map $f : \mathbb{C}^2 \to \mathbb{C}$ there is a finite set of critical values Λ_f such that $f| : \mathbb{C}^2 - f^{-1}(\Lambda_f) \longrightarrow \mathbb{C} - \Lambda_f$ is a locally trivial fibration.

Question: *If f has only isolated singularities, does the homeomorphism type of the generic fiber and the topological monodromy of this locally trivial fibration determine the topological type of f?*

Remarks: The assumption of isolated singularities is necessary because the polynomials x^2y+x and $x(xy+1)^2$ both have critical value 0, generic fiber a twice-punctured 2-sphere, and the identity monodromy around the fiber over 0, but the first polynomial has no singularities whereas the second has non-isolated singularities.

Chapter 5

Miscellany

- Miscellaneous Problems 5.1–5.8 (1977), 5.9–5.29 (new).
- Groups, 5.1, 5.7, 5.9–5.11.
- Complexes, 5.2–5.5, 5.13, 5.14.
- Graphs, 5.8, 5.15–5.18.
- TQFT's, 5.19, 5.20, and 3.108.
- More dimensions, 5.21–5.29.

Problem 5.1 (M. Cohen) Let $\mathcal{P}$ and $\mathcal{P}'$ be finite presentations (with the same deficiency) of a given group π. Let $K_{\mathcal{P}}$ and $K_{\mathcal{P}'}$ be the 2-dim CW-complexes associated to these presentations. Consider the assertions:

(A) $K_{\mathcal{P}} \simeq K_{\mathcal{P}'}$ (homotopy equivalence),

(B) $K_{\mathcal{P}} \curvearrowright K_{\mathcal{P}'}$ (simple homotopy equivalence),

(C) $K_{\mathcal{P}} \underset{3}{\curvearrowright} K_{\mathcal{P}'}$ (simple homotopy equivalence by moves of dimension ≤ 3),

(D) $\mathcal{P}$ can be changed to $\mathcal{P}'$ by extended Andrews–Curtis moves [37, Andrews & Curtis, 1965, Proc. Amer. Math. Soc.] (i.e., we can change the presentation $\mathcal{P} = \{x_1, \ldots, x_n : R_1, \ldots, R_m\}$ in these ways

1. $R_i \to R_i^{-1}$,
2. $R_i \to R_i R_j$, $i \neq j$,
3. $R_i \to w R_i w^{-1}$, w any word,
4. add generator x_{n+1} and relation $w x_{n+1}$.)

Note: Redundant relations *cannot* be added. Then (D) $\Rightarrow$ (C) $\Rightarrow$ (B) $\Rightarrow$ (A) and (C) $\Rightarrow$ (D) [1121, Wright, 1975, Trans. Amer. Math. Soc.]. (A) fails for the trefoil group [257, Dunwoody, 1976, Bull. London Math. Soc.] and for many finite abelian groups [755, Metzler, 1976, J. Reine Angew. Math.].

Question: What other relations hold?

Update: (A) does not imply (B) [756, Metzler, 1990, J. Reine Angew. Math.], [686, Lustig & Moriah, 1991, Topology]. Whether (B) implies (C) is still open (generalized Andrews–Curtis Conjecture).

For a recent and thorough discussion of this problem, see Hog-Angeloni & Metzler's Chapters I and XII in [501, Hog-Angeloni, Metzler, & Sieradski, 1993]. This excellent book contains many other problems on the topics of Problems 5.1, 5.2 and 5.4.

Problem 5.2 (Lickorish) *Let K be a contractible finite 2-complex.*

(A) Zeeman Conjecture: *$K \times I$ collapses to a point [1138, Zeeman, 1964, Topology].*

(B) Conjecture: *K 3-deforms to a point, i.e., there exists a 3-complex L such that $K \swarrow\!\!\nearrow L \searrow pt$.*

(C) Conjecture: *The unique 5-dim regular neighborhood of K^2 in $\mathbb{R}^5$ is B^5.*

(D_0) Andrews–Curtis Conjecture: *Any presentation of the trivial group can be changed to the trivial presentation by Andrews–Curtis moves [37, Andrews & Curtis, 1965, Proc. Amer. Math. Soc.].*

Remarks: Conjecture (A) implies the Poincaré conjecture. Conjecture (C) is equivalent to knowing whether the boundary is S^4. (A) $\Rightarrow$ (B) $\Rightarrow$ (C) and (B) $\Leftrightarrow$ (D_0). The analogue of Conjecture (D_0) is false for nontrivial groups (see (D) of Problem 5.1). Possible counterexamples to (D_0) are $\{a, b : a^{-1}b^2a = b^3, b^{-1}a^2b = a^3\}$ and $\{a, b, c : [a, b]b = [b, c]c = [c, a]a = 1\}$. It is not known whether the regular neighborhoods in $\mathbb{R}^5$ of the corresponding 2-complexes are B^5.

Update: Still open. For a recent and thorough discussion of this problem, see Matveev & Rolfsen's Chapter XI in [501, Hog-Angeloni, Metzler, & Sieradski, 1993].

It follows from [622, Kreher & Metzler, 1983, Topology] that if K 3-deforms to a point, then K 2-expands to another 2-complex K' such that $K' \times I$ collapses to a point, (thus (B) implies a weakened version of (A)).

It is particularly interesting to note that the Zeeman conjecture for special polyhedra which are spines of compact 3-manifolds is equivalent to the Poincaré conjecture [380, Gillman & Rolfsen, 1983, Topology], and the Zeeman conjecture for special polyhedra which do not imbed in compact 3-manifolds is equivalent to the Andrews–Curtis Conjecture [720, Matveev, 1987, Sibirsk. Mat. Zh.] and [720, Matveev, 1987, Sibirsk. Mat. Zh.].

Another interesting presentation of the trivial group is $\{a, b : aba = bab, a^4 = b^5\}$ [29, Akbulut & Kirby, 1985, Topology]. (D_0) probably fails for this presentation, but an associated homotopy 4-sphere is shown to be standard by a judicious addition of a 2-,3-handle pair [392,Gompf, 1991a,Topology].

Problem 5.3 (R. Fenn) *Is there an acyclic 2-complex which does not imbed in $\mathbb{R}^4$?*

Remarks: All n-complexes with H^n cyclic imbed in $\mathbb{R}^{2n}$, $n \geq 3$.

Update: Using Freedman's work, Kranjc proved [617,Kranjc,1991,Pacific J. Math.] that every 2-complex with $H^1 = H^2 = 0$ imbeds topologically locally flat in $\mathbb{R}^4$. The PL case is still open.

Problem 5.4 (J. H. C. Whitehead) *Is every subcomplex of an aspherical 2-complex aspherical? Assume finiteness if you wish.*

Remarks: Aspherical means $\pi_k = 0$ for $k > 1$. For partial results, see [218,Cockroft,1954,Proc. London Math. Soc.], and [9,Adams,1955,J. London Math. Soc.].

Update: Still open. For a recent and thorough discussion of this problem, also known as Whitehead's Asphericity Question, see Bogley's Chapter X in [501,Hog-Angeloni, Metzler, & Sieradski,1993].

Problem 5.5 (A) (Lickorish) Conjecture*: Any linear subdivision of an n-simplex collapses simplicially.*

Remarks: True for $n \leq 3$ [200,Chillingworth,1967,Math. Proc. Cambridge Philos. Soc.] .

(B) (Goodrick) Conjecture*: Any linear subdivision of a star-like n-cell in $\mathbb{R}^n$ collapses simplicially.*

Remarks: True for $n < 3$. Note true for a triangulation of a topological n-cell [400,Goodrick,1968,Math. Proc. Cambridge Philos. Soc.] .

Update: Still open.

Problem 5.6 *What more can be said about nonsingular real algebraic varieties in $\mathbb{RP}^2$, $\mathbb{RP}^3$ or $\mathbb{RP}^4$?*

Remarks: A number of pretty results about algebraic invariants of such varieties have been established (by complexifying) in a series of papers by Gudkov, Arnold, Rohlin, Kharlamov and Zvonilov in *Functional analysis and its applications* during the 1970's; there is also Gudkov's survey [419,Gudkov,1974a,Успехи Мат. Наук].

Update: None attempted.

Problem 5.7 (Freedman) Let G be a nontrivial group, $G*\mathbb{Z}$ the free product, and $(G*\mathbb{Z})/r$ the free product with one relation r.

Conjecture (Kervaire): *$(G*\mathbb{Z})/r$ is nontrivial.*

Remarks: Any counterexample must satisfy: (i) G is perfect, (ii) the degree of t (= generator of $\mathbb{Z}$) in the relation r is ± 1 (proof: abelianize).

The conjecture is true if G has a normal subgroup of finite index [373,Gerstenhaber & Rothaus,1962,Proc. Nat. Acad. Sci. U.S.A.].

It seems rare that both natural maps $A \to A*B/r$ and $B \to A*B/r$ have nontrivial kernels. This happens if $A = \mathbb{Z}/2\mathbb{Z}$, $B = \mathbb{Z}/3\mathbb{Z}$ and $r = ab$. Are there any other examples of a different flavor? Specifically if A is torsion-free and B nontrivial, is either A or B into $A*B/r$ always an injection?

Here is a relation to knot theory. Let M^3_K be the result of surgery on a knot K with 0-framing. There is a natural map $f_K : M_K \to S^1 \times S^2$ and the closer K is to being trivial, the closer f_K is to a homotopy equivalence, e.g., f_K always induces an isomorphism on integral homology, and induces a $\mathbb{Z}[\mathbb{Z}]$-homology isomorphism exactly when the Alexander polynomial is trivial. $H_2(\pi_1(S^3 - K)/l) = 0$ iff M_K is diffeomorphic to $S_1 \times S^2 \# H^3$ where H^3 is a homology 3-sphere (l represents the longitude of K). Since $\pi_1(M_K) = \pi_1(S^3 - K)/l$ is normally generated by a meridian, the conjecture would imply that H^3 above could be replaced by a homotopy 3-sphere. (See Problems 1.16 and 1.17.)

Update: Klyachko [588,Klyachko,1993,Comm. Algebra.] has recently proved the conjecture in the case where G is torsion free; indeed, he shows here that G injects into $(G*\mathbb{Z})/r$ if (ii) holds.

It should be noted that in the final paragraph, the relevance of the Kervaire conjecture is now moot because Gabai has proved [355,1987a,

J. Differential Geom.] that if M_K has an $S^1 \times S^2$ summand, then K is the unknot, in which case $M_K = S^1 \times S^2$.

Problem 5.8 *Does there exist a graph G such that for any imbedding $f : G \to \mathbb{R}^3$, $f(G)$ contains a nontrivial knot?*

Remarks: It suffices to consider $G = C_n$ = complete graph on n-vertices. (Added in proof. Yes, for $n = 7$, John Conway.)

Update: Yes, for C_7 [226,Conway & Gordon,1983,J. Graph Theory]. See Problem 5.15 for further material.

NEW PROBLEMS

Problem 5.9 (A) *Consider the closure of groups of subexponential growth under the operations of extensions and direct limits. Is the resulting class the class of all amenable groups?*

Remarks: Freedman & Teichner [339,1995], [340,1996], have shown that 4-dimensional topological surgery works when the fundamental group belongs to the above class (see old Problem 4.6).

Grigorchuk [411,1985,Math. USSR-Izv.] has constructed (uncountably many) finitely generated groups of intermediate growth, i.e., their growth function (see below) is eventually bigger than any polynomial but smaller than any exponential function. This implies that the class of amenable groups (see below) is larger than the closure of finite and abelian groups under extension and direct limits, since no group in this closure has intermediate growth [204, Chou,1980,Illinois J. Math.].

A locally compact group is called *amenable* if it has a left invariant mean (which is basically a finite Haar measure). Using the so called Følner conditions (see e.g. [869,Paterson,1988]) one shows that finitely generated groups of subexponential growth are amenable.

Given a finite set of generators of a group, one can define the corresponding *growth function* $\mathbb{N} \to \mathbb{N}$ by sending a natural number r to the number of distinct group elements which can be written as words of length $\leq r$ in the generators and their inverses. (In the

Cayley graph with the word metric, this number is the cardinality of the ball of radius r.)

There is another definition of amenable which intuitively says that the quotient of the volume of the boundary of the ball of radius r by the volume of the ball itself goes to zero as r goes to infinity; but balls are not quite general enough to get all amenable groups. Amenability is equivalent to the *Følner condition*: Given $\epsilon > 0$ and a compact subset C of the group G, there exists a non-null compact subset K of G such that

$$\lambda(x \cdot K \Delta K)/\lambda(K) < \epsilon \quad \forall x \in C.$$

Here λ is a Haar measure on G and Δ denotes the symmetric difference.

(B) *Can a finitely presented group have intermediate growth?*

Remarks: Grigorchuk's examples above are finitely generated but not finitely presented.

Problem 5.10 (G. Martin) *Are finitely generated convergence groups (in Homeo(S^n)) accessible?*

Remarks: A group G is *accessible* if every sequence of non-trivial algebraic splittings of G as free products with amalgamation along finite subgroups or as HNN extensions along finite subgroups, is finite, or, equivalently, G is the fundamental group of a graph of groups in which all edge groups are finite and every vertex group has at most one end. (For the definition of convergence group, see Problem 3.71.)

Wall [1096,1971,J. Pure Appl. Algebra] conjectured that all finitely generated groups are accessible, but this is not true [259,Dunwoody, 1993]

A finitely generated G *is* accessible if G has uniformly bounded torsion [662,Linnell,1983,J. Pure Appl. Algebra] or if G is finitely presented [258,Dunwoody,1985,Invent. Math.]. It follows from Grushko's theorem that a finitely generated torsion free group is accessible; since Mobius groups are virtually torsion free (Selberg's Lemma), they are also accessible.

Recently Dunwoody [260,Dunwoody,1995] has shown that a finitely generated G is accessible unless it has an infinite torsion subgroup. Martin & Skora have shown [699,1989,Amer. J. Math.] that a convergence group on S^2 has no infinite torsion subgroup, so those groups are accessible. (See Problem 3.71.)

Problem 5.11 (Mess) Let X be a finite, aspherical complex.

(A) *Is the center $Z(\pi_1)$ of $\pi_1(X)$ finitely generated?*

Remarks: This problem appears in [223,Conner & Raymond, 1977,Bull. Amer. Math. Soc.].

The center must be a group of finite rank, i.e. a subgroup of Q^n where Q is the rationals. The answer to (A) is yes if X is a 2-complex [86,Bieri,1976,J. Pure Appl. Algebra], or if $\pi_1(X)$ has a faithful, finite dimensional, linear, representation [87,Bieri,1980, Math. Zeit.].

(A) is also interesting if X is only assumed to have finitely many cells in each dimension.

(B) *Is it possible that $\pi_1(X)/Z(\pi_1(X))$ has nontrivial rational cohomology in infinitely many dimensions? This is open in the simple case of the center being $\mathbb{Z}$.*

(C) *Is there an example of an X with simple fundamental group?*

(D) *Can a finite dimensional, aspherical complex have a fundamental group which is of intermediate growth (subexponential but faster than polynomial)? Can non-elementary amenable groups occur?*

Problem 5.12 (Short & Neumann) Let $\mathcal{P}$ be the set of all polynomials in one variable with integer coefficients other than the cyclotomic polynomials. Let $\rho : \mathcal{P} \to \mathbb{R}$ be the function defined by letting $\rho(P)$ be the product of the modulus of all complex roots which lie strictly outside the unit circle, that is:

$$\rho(P) = \prod_{roots\ r_i, |r_i|>1} |r_i|.$$

(ρ is also know as the *Mahler measure* of P [647,Lehmer,1933,Ann. of Math.].)

Conjecture (Lehmer) *[ibid.]: ρ has a minimum, and it is 1.176780821... which is achieved by the degree 10 polynomial*

$$x^{10} - x^9 + x^7 - x^6 + x^5 - x^4 + x^3 - x + 1.$$

Remarks: The above polynomial has one root outside the unit circle, listed above. Lehmer conjectures a lower bound for polynomials with a non-real root outside the unit circle, namely 1.2013961862..., which is realized by the polynomial

$$x^{18} + x^{17} + x^{16} - x^{13} - x^{11} - x^9 - x^7 - x^5 + x^2 + x^1$$

which has two complex conjugate roots outside the unit circle.

These polynomial are related to the lengths of geodesics in arithmetic manifolds: the first (second) Lehmer conjecture implies that a geodesic in an arithmetic 2-orbifold (3-orbifold) has length at least 0.162357614... (0.09174218...) [838,Neumann & Reid,1992a].

This polynomial is the Alexander polynomial of the (2,3,7)-pretzel knot, which has three surgeries giving finite π_1.

Problem 5.13 (Habegger) *Suppose M and N are PL-homeomorphic cubulated n-manifolds. Are they related by the following set of moves: excise B and replace it by B', where B and B' are complementary balls (unions of n-cubes which are homeomorphic to B^n) in the boundary of the standard $(n+1)$-cube?*

Remarks: A similar statement is true for combinatorial triangulations [864,Pachner,1991,European. J. Combin.].

Problem 5.14 (Zhong-mou Li) *The standard CW-complex of a presentation*

$$\langle a_1, a_2, \cdots, a_k | R_1, R_2, \cdots, R_k \rangle$$

is called a generalized dunce hat *if for each i, $1 < i > K$, the word R_i cancels to a_i.*

Question: *Is the Zeeman Conjecture true for generalized dunce hats?*

Remarks: Possibly the Zeeman Conjecture (see Problem 5.2 [1138, Zeeman, 1964, Topology]) reduces to this special case. A generalized dunce hat collapses after crossing with an appropriate tree [1099, Wedel, 1994].

Problem 5.15 (Flapan & Gordon) *Characterize all graphs G that have the following Property: every imbedding of G in $\mathbb{R}^3$ contains a subgraph which is a knotted, simple closed curve, (this generalizes Problem 5.8).*

Remarks: It is known (Robertson & Seymour) by that there is some finite *obstruction set* of graphs, $\mathcal{F}$, such that a graph G has an imbedding in $\mathbb{R}^3$ with no nontrivial knotted subgraph iff G has no minor in $\mathcal{F}$. But the graphs in $\mathcal{F}$ are not known.

Graphs are assumed to be finite, and loops and multiple edges are allowed. By definition, H is a minor of G if H can be obtained from a subgraph of G by contracting edges.

K_7, the complete graph on 7 vertices, has the Property [226, Conway & Gordon, 1983, J. Graph Theory], as does $K_{5,5}$ [993, Shimabara, 1988, Tokyo J. Math.]. An interesting open case is $K_{3,3,1,1}$ [598, Kohara & Suzuki, 1992]. ($K_{m_1,\ldots,m_k}$, for $k > 1$, denotes the graph obtained from k sets of vertices, the ith set having m_i vertices, by attaching an edge to any two vertices not belonging to the same set.)

The case of links (rather than knots) has been completely solved [930, Robertson, Seymour, & Thomas, 1993, Bull. Amer. Math. Soc.]. They prove that a graph G has an imbedding containing no non-trivial link (of one or more components) iff G has no minor in the Petersen family. The Petersen family consists of the seven graphs which are K_6 and the six graphs which can be obtained from K_6 by a sequence of exchanges of a Y (a neighborhood of a 3-valent vertex) with a triangle (complete graph on 3 vertices).

Problem 5.16 (Kinoshita & Mikasa) Fix a plane $\mathbb{R}^2$ in $\mathbb{R}^3$, and let $proj(K)$ be the subcomplex in $\mathbb{R}^2$ obtained by projecting a PL knot K from $\mathbb{R}^3$ to $\mathbb{R}^2$ (assume always that K is in general position with respect to the projection). Call K and K' equivalent, $K \sim K'$, if there exists a (perhaps orientation reversing) homeomorphism of $\mathbb{R}^3$ carrying K to K'. Define $Proj(K) = \{proj(K') \mid K' \sim K\}$.

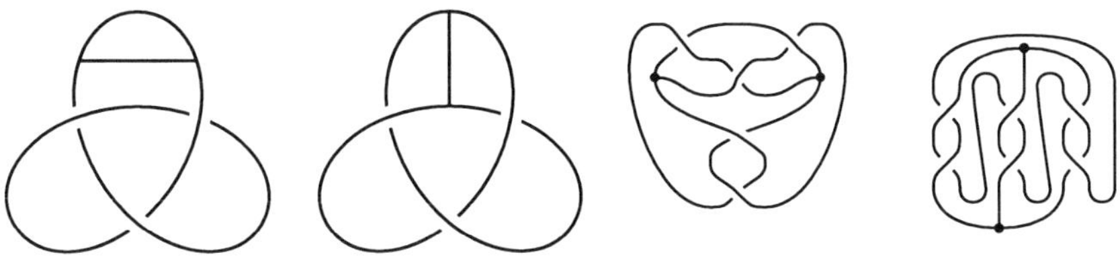

3_1^* 3_1 5_1 6_1

Figure 5.16.1.

(A) *Is it true that if $Proj(K_1) = Proj(K_2)$, then $K_1 \sim K_2$?*

The same definitions can be made for Θ-curves in $\mathbb{R}^3$ (a Θ-curve is the image of a PL imbedding of the graph with two vertices and three edges between them).

(B) *Is it true that if $Proj(\Theta_1) = Proj(\Theta_2)$, then $\Theta_1 \sim \Theta_2$?*

A Θ-curve is called trivial if it is equivalent to a Θ-curve in $\mathbb{R}^2$. A non-trivial Θ-curve Θ is called an *atom* if, for any Θ-curve Θ' with $Proj(\Theta') \supset Proj(\Theta)$, then $\Theta' \sim \Theta$ or Θ' is trivial. It is known that $3_1^*, 3_1, 5_1, 6_1$ are atoms (see Figure 5.16.1) [761, Mikasa, 1993]. (For knots, the trefoil knot is the only atom [1035, Taniyama, 1989, Tokyo J. Math.].)

(C) *How many atoms are there for Θ-curves?*

A Θ-curve is called *strongly almost trivial* if it has some projection to $\mathbb{R}^2$ which does not contain a projection of any non-trivial knot. The Θ-curves 5_1 and 6_1 in Figure 5.16.1 are strongly almost trivial. A Θ-curve is called *almost trivial* if its three constituent knots are trivial. Strongly almost trivial implies almost trivial.

(D) *Does there exist a Θ-curve which is almost trivial but not strongly almost trivial?*

Problem 5.17 (Freedman) *Given a finite set of points X in ∂B^3, let T be a tree in B^3 of minimal length with $\partial T = X$. Is T unknotted, that is, is there a PL imbedded 2-ball in B^3 containing T?*

Remarks: Krystyna Kuperberg proposes the following knotted tree which, if indeed minimal, would mean a negative answer to the problem.

Assume the following notation:

$S^2 = \{(x,y,z) \in \mathbb{R}^3 | x^2 + y^2 + z^2 = 1\}$,

$P = \{(x,y,z) \in S^2 | y = 0\}$,

$Q = \{(x,y,z) \in S^2 | z = 0\}$.

The set X of points on S^2 will depend on four small constants ε_1, ε_2, δ_1, and δ_2.

Let A_1, B_1, C_1, A_2, B_2, and C_2 be points on P close to the vertices of a regular hexagon inscribed in P such that: $B_1 = (-1,0,0)$, $B_2 = (1,0,0)$; A_1 and A_2 are at ε_1 distance and above the points $(-\frac{1}{2}, 0, \frac{\sqrt{3}}{2})$ and $(\frac{1}{2}, 0, \frac{\sqrt{3}}{2})$; C_1 and C_2 are at ε_1 distance and below the points $(-\frac{1}{2}, 0, -\frac{\sqrt{3}}{2})$ and $(\frac{1}{2}, 0, -\frac{\sqrt{3}}{2})$. The minimal tree of each of the triples $\{A_i, B_i, C_i\}$ is a triod as shown in Figure 5.17.1.

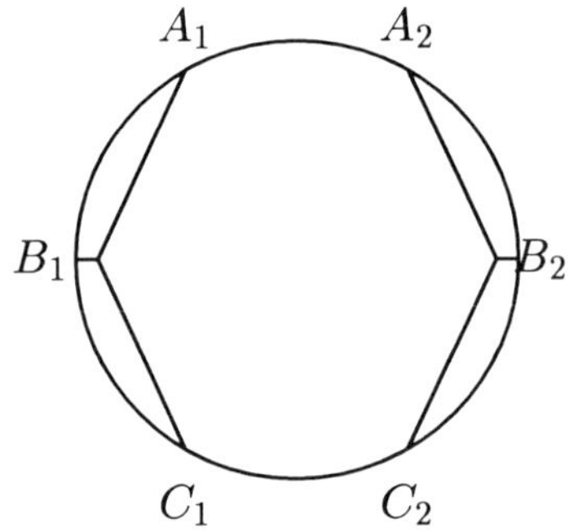

Figure 5.17.1.

Split the point C_1 $[A_2]$ into two points E_1 and F_1 $[E_2$ and $F_2]$ which are ε_2 apart and symmetric with respect to the xz-plane so that they are farther from the point $(\frac{1}{2}, 0, -\frac{\sqrt{3}}{2})$ $[(-\frac{1}{2}, 0, \frac{\sqrt{3}}{2})]$ than from the point B_1 $[B_2]$.

Let L be an arc in the δ_1-neighborhood N of Q *going around* N one-and-a-half times and connecting two points D_1 and D_2 in different boundary components of N, close to B_1 and B_2 respectively. Choose a sequence of points $x_1 = D_1, x_2, \ldots, x_n = D_2 \in L$ with the line order of

L and such that x_i and x_{i+1} are less than δ_2 apart.

Choose the constants ε_2, ε_1, δ_1, and δ_2 (in this order) so that the minimal tree T (drawn in Figure 5.17.2) of the set

$$X = \{A_1, D_1, E_1, F_1, E_2, F_2, D_2, C_2\} \cup \{x_1, ...x_n\}$$

is knotted (Figure 5.17.3 shows a subtree of T which is clearly knotted).

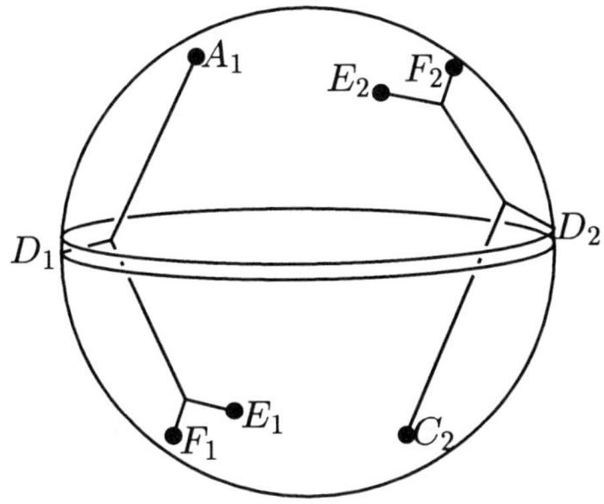

Figure 5.17.2.

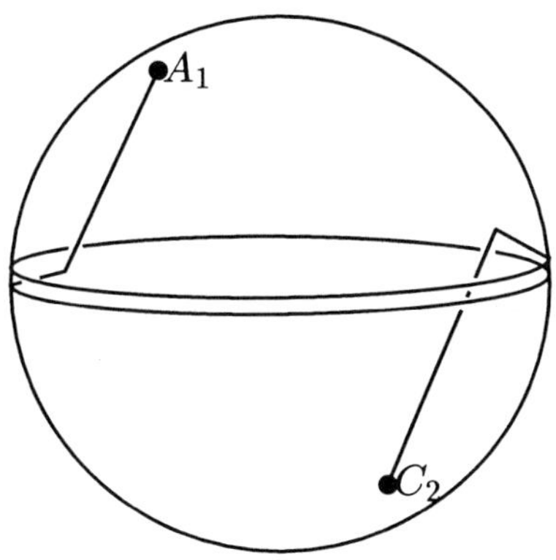

Figure 5.17.3.

Problem 5.18 Let P be a polygon, homeomorphic to S^1, in the plane $\mathbb{R}^2$; consider motions of P which allow bending at the vertices, but which preserve edge length (and straightness) and imbeddedness.

(A) Question (S. Schanuel): *Can any P be moved to a convex polygon by such motions?*

(B) (G. Bergman) *Same question as (A) except that P is assumed to be homeomorphic to an interval $[0,1]$, and should be moved into a line.*

Remarks: Yes to (A) implies yes to (B) (just close up the arc with any disjoint polygonal arc). The polygonal arc in Figure 5.18.1 was proposed as a possible counterexample to (B), but in fact it can be moved into a line.

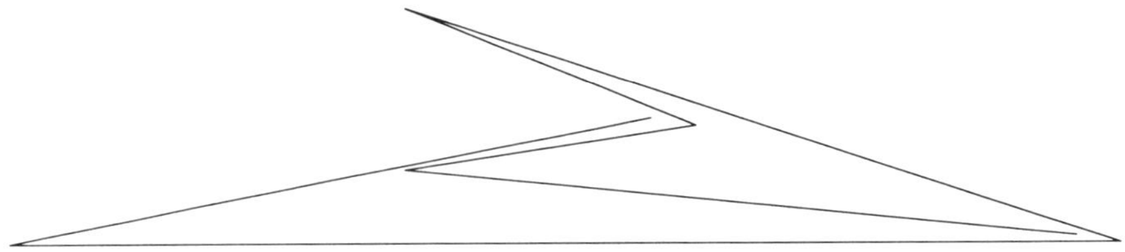

Figure 5.18.1.

Problem 5.19 (Kontsevich) *Compute the cohomology $H^*(C_n;\mathbb{Q})$ of the graph complex C_n for each $n > 0$.*

Remarks: The graph complex C_n is defined as follows (see [603, Kontsevich, 1994]): consider the pairs (Γ, or) where Γ is a finite, connected graph whose vertices have valence ≥ 3, and or is an orientation of $H^1(\Gamma;\mathbb{R})$ together with an ordering of the edges of Γ (or equivalently, using the exact sequence

$$0 \to H_1(\Gamma) \to C_1(\Gamma) \to C_0(\Gamma) \to H_0(\Gamma) \to 0,$$

an ordering of the vertices and an orientation of each edge). Let C_n be the finite dimensional vector space over $\mathbb{Q}$ generated by the pairs (Γ, or) for which $n+1 = rank(H^1(\Gamma;\mathbb{Q}))$, subject to the equivalence relation $(\Gamma, -or) = -(\Gamma, or)$, (this implies that $(\Gamma, or) = 0$ for any Γ containing a loop (an edge whose endpoints are identified)).

Decompose C_n into C_n^k, $k \geq 0$, where $k = 3V - 2E$ and E is the number of edges and V is the number of vertices. Thus C_n^0 consists of graphs whose vertices are all 3-valent, C_n^1 consists of graphs which have exactly one 4-valent vertex and the rest 3-valent, and so on. Define a boundary homomorphism $d : C_n^k \to C_n^{k+1}$ by

$$d(\Gamma, or) = \sum_{all\ edges\ e_i} (\Gamma/e_i\ ,\ \text{induced orientation}),$$

where the induced orientation is the same orientation on $H^1(\Gamma/e_i) \cong H^1(\Gamma)$ and the ordering of the edges of $H^1(\Gamma/e_i)$ is $(-1)^{i+1}e_1 \ldots \hat{e}_i \ldots e_E$. Clearly $dd = 0$. Then $H^*(C_n;\mathbb{Q})$ is the cohomology of the chain complex $\{C_n^k, d\}$.

Little is known about $H^p(C_n;\mathbb{Q})$ for $p > 0$, but it can be non-zero, e.g. rank$H^3(C_5;\mathbb{Q}) = 1$ (Bar-Natan). The case of real interest, however, is $H^0(C_n)$ which is closely related to low-dimensional problems [Kontsevich, ibid.], [1038, Taubes, 1995a], [231, Culler & Vogtmann, 1986, Invent. Math.] and the next Problem 5.20. Bar-Natan has shown that rank$H^0(C_n;\mathbb{Q})$ is $1, 1, 1, 2, 2, 3, 4, 5$ for $n = 1, \ldots, 8$.

Problem 5.20 (Kirby & Melvin) Let M be a closed 3-manifold with a spin structure, and let τ be a framing (trivialization) of the tangent bundle which extends the spin structure. Fix a point $p \in M$, let Δ be the diagonal in $M \times M$, and let

$$\Theta = (M \times p) \cup (p \times M) \cup \Delta.$$

If N is a suitably chosen neighborhood of Θ in $M \times M$ then $\partial N = \partial(M \times M - intN)$ is a 3-manifold bundle over S^2 with fiber equal to the connected sum of M, M and $-M$.

To obtain N, first choose a coordinate chart $\mathbb{R}^3$ centered at p and then let $f : \mathbb{R}^3 \times \mathbb{R}^3 \to \mathbb{R}^3$ be defined by $f(x, y) = |x|^2 y - x|y|^2$. Extend f over a neighborhood of Θ by using the projection maps $M \times \mathbb{R}^3 \to \mathbb{R}^3$ and $\mathbb{R}^3 \times M \to \mathbb{R}^3$, and using the framing of M which frames the normal bundle of the diagonal Δ. Then N is f^{-1} of a suitable 3-ball, say B^3 in $\mathbb{R}^3$ (for a careful description of this construction using singular framings, see Section 1 in [1038, Taubes, 1995a]).

The volume 2-form on S^2, or its integral cohomology class, pulls back to a 2-form $\omega_{\partial N}$, or class, on ∂N. If M is an integral homology 3-sphere, then the 2-form or integral class extends to a 2-form ω, or integral cohomology class $[\omega]$, on $(M \times M) - intN$ (if M is only a rational homology 3-sphere, then only a multiple (by $|H_1(M;\mathbb{Z})|$) of ω

extends, and for non-rational homology 3-spheres, more must be deleted from $M \times M$).

The integral

$$I(M,\tau) = \int_{(M\times M)-intN} \omega \wedge \omega \wedge \omega$$

is an invariant of the framed M which has been shown to be zero for the *canonical framing* on homology 3-spheres by Taubes [1038,1995a]. Moreover, Taubes [1039,1995b] has shown the following: let M and M' be rational homology spheres, let W be an oriented, spin bordism between M and M' with the following properties: (1) the intersection form of W is equivalent to a sum of metabolic pairs, and (2) the inclusions of both M and M' into W induce injective maps on the first homology with $\mathbb{Z}/2\mathbb{Z}$ coefficients; then the invariants of M and of M' are the same (where W induces the same framing on M and M').

In this subject, framings of the tangent bundle T_M, of $T_M \oplus \epsilon$, or of $T_M \oplus T_M$ [51,Atiyah,1990,Topology], can be interesting; in the case at hand, we use the framing of T_M because of its use in defining N. One can define a function $h : (M,\tau) \to \mathbb{Z}$ by $h(M,\tau) = p_1(Y,\tau) - 3\sigma(Y)$ where $\partial Y^4 = M$. One wishes to define the canonical framing as the τ for which $h = 0$; however

$$h(M,\tau) \equiv 2(\text{rank}(H_1(M;\mathbb{Z}/2\mathbb{Z})) + 1) + \mu \pmod 4$$

where μ is the μ-invariant of M with the spin structure defined by τ [583,Kirby & Melvin,1995]. Since $h(S^3) \equiv 2 \pmod 4$, on homology 3-spheres one uses, in effect, the framing τ for which $h(M,\tau) = 2$ [Taubes, ibid.]. If the framing changes by the generator $1 \in \pi_3(SO(3) = \mathbb{Z}$, then $I(M,\tau)$ changes by 1.

Given a finite, oriented graph Γ with no loops (see previous Problem 5.19) whose vertices are all 3-valent, one can also define a number $I(M,\tau,\Gamma)$ as follows: the orientation orders the vertices and orients the edges, so we denote an edge by e_{ij} if it joins the ith vertex to the jth vertex. This defines a 2-form ω_{ij} on X_V, the cartesian product of V copies of M, by setting:

$$\omega_{ij} = p_{ij}^*(\omega)$$

where p_{ij}^* is projection of X_V onto its ith and j^{th} factors, (V and E are the number of vertices and edges, and $3V = 2E$). Now let

$$I(M,\tau,\Gamma) = \int_{X_V} \wedge_{all\ edges\ e_{ij}} \omega_{ij}.$$

This integral is well defined and gives an invariant on M with values in $Hom(H^0(C_n;\mathbb{Q}),\mathbb{R})$ (see Problem 5.19) [603,Kontsevich,1994] [55, Axelrod & Singer,1992], and is equal to $I(M,\tau)$ when Γ is the 3-valent graph with 2 vertices and 3 edges, Γ_Θ. Also, there are further generalizations, in particular to 2-forms with values in a flat bundle of Lie algebras [Kontsevich, ibid.] [Axelrod & Singer, ibid.], and these should be closely related to the perturbative Chern–Simons invariants of 3-manifolds [Axelrod & Singer, ibid.], [1116,Witten,1989,Comm. Math. Phys.].

(A) *Compute* $I(M,\tau,\Gamma)$ *for* $\Gamma \in H^0(C_n;\mathbb{Q})$

Remarks: As mentioned above, Taubes [ibid.] proved that $I(M,\tau,\Gamma_\Theta) = 0$ for the canonical framing on M.

One approach is to reinterpret the integral in terms of intersection theory using Poincaré duality. For example in the case $\Gamma = \Gamma_\Theta$ and $M = S^3$, then

$$(S^3 \times S^3) - N = S^2 \times B^4.$$

A convenient Poincaré dual to $[\omega]$ (using what must turn out to be the canonical framing) is (north pole) $\times B^4$ which we call B_{np}. But it may be necessary to choose some other Poincaré dual which we call B. In general, for a homology sphere M, the Poincaré dual to $[\omega]$ is an oriented 4-manifold W^4 with $\partial W \subset \partial N$ determined by the framing on T_M. Taubes' calculation can be reinterpreted as calculating the triple intersection of three copies of W, where the copies of ∂W are first made transverse (in fact, disjoint) using the framing and then, rel boundary, the rest of the W's are made transverse.

(B) *Can the integral* $I(M,\tau,\Gamma)$ *be reinterpreted, as in this simple example, in terms of intersection theory; one should count the points of intersection of* E *transverse, codimension-two manifolds in* X_V *(each edge* e_{ij} *determines a* W_{ij} *in the* i^{th} *and* j^{th} *copies of* M *in* X_V*)?*

Remarks: The difficulty seems to lie in understanding the correct way to make these codimension-two manifolds transverse at their boundaries (see [138, Bott & Taubes, 1994, J. Math. Phys.] and [351, Fulton & MacPherson, 1994, Ann. of Math.] for a discussion of boundary issues).

One can understand perturbative Chern–Simons invariants of knots in S^3 in terms of the 2-form ω and similar integrals [Kontsevich, ibid.], [61, Bar-Natan, 1991], [62, Bar-Natan, 1995a, Topology], [138, Bott & Taubes, 1994, J. Math. Phys.]. One uses trivalent graphs with boundaries lying on a circle to organize wedges of 2-forms to be integrated over the analogue of X_V. For example, using the graph Γ in Figure 5.20.1, an integral $I(K,\Gamma)$ can be defined as follows:

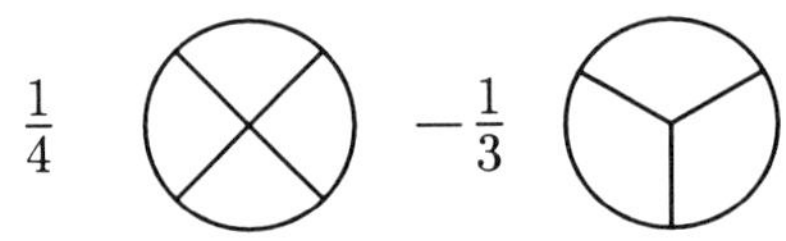

Figure 5.20.1.

$$I(K,\Gamma) =$$

$$\frac{1}{4}\{\int_{t_1<t_2<t_3<t_4} v(t_1t_3)\wedge v(t_2t_4)\} - \frac{1}{3}\{\int_{t_1<t_2<t_3,B^3} v(t_1x)\wedge v(t_2x)\wedge v(t_3x)\}$$

for the volume form v on S^2, $t_i \in K$ parameterized by $[0,2\pi]$, $x \in B^3 - K$, and t_it_j the unit vector from t_i to t_j (same for t_ix) [61, Bar-Natan, 1991].

For $M = S^3$, B_{np} can be understood as the pairs of points in $B^3 = S^3 -$ (neighborhood of p) whose first component lies north of the second, and B can be understood as *defining* all pairs of points with one component *above* the other. Aspects of knot theory using projections of the knot to the plane and counting overcrossings should be translatable into intersection theory using the knots and B_{np} or maybe B. For example, the linking number of knots K_1 and K_2 in B^3 is given by the transverse intersection

$$(K_1 \times B^3) \cap (B^3 \times K_2) \cap B_{np}$$

in $S^2 \times B^4 \subset S^3 \times S^3$.

(C) *Can the integral $I(K,\Gamma)$ for the Γ in Figure 5.20.1 be understood by counting points with sign of the transverse intersection in X_4 of $(K \times B^3 \times B^3 \times B^3) \cap (B^3 \times K \times B^3 \times B^3) \cap$ $\cap (B^3 \times B^3 \times K \times B^3) \cap (B^3 \times B^3 \times B^3 \times K) \cap B_{13} \cap B_{24}$, where*

$$B_{ij} = \{(x_1, x_2, x_3, x_4) \in X_4 | (x_i, x_j) \in B \subset X_2 = S^2 \times B^4\}?$$

Can this be generalized to other graphs? Again, the difficulty apparently lies in making sense of transversality near the boundaries.

For more general 3-manifolds M, at least including integral homology 3-spheres, W is the analogue of B_{np} or B for S^3; one could say that W consists of all pairs of points in M^3 for which the first point is *above* the second.

(D) *Can the perturbative Chern–Simons knot invariants of (C) be extended to knots in more general 3-manifolds, particularly through the use of an appropriate W^4 and appropriate rules for making manifolds transverse at their boundaries?*

(E) *If the answer to (D) is yes, can a Dehn surgery formula be found relating the 3-manifold invariants of (A) and (B) to the knot invariants of (C) and (D) applied to a framed link description of the 3-manifold?*

Problem 5.21 (Yau) Conjecture: *All smooth 2k-manifolds, $k > 2$, with a given almost complex structure, are actually complex manifolds (the complex structure may not deform to the given almost complex structure).*

Remarks: For $k = 1$, the conjecture is true since all orientable 2-manifolds are complex. For $k = 2$, the conjecture fails dramatically because, assuming simply connectedness, X^4 is almost complex iff b_2^+ is odd, whereas there are many restrictions on the characteristic classes of a complex surface, e.g. $c_1^2 \leq 3c_2$ ([774, Miyaoka, 1977, Invent. Math.] and [1132, Yau, 1977, Proc. Nat. Acad. Sci. U.S.A.]), so that the connected sum of three copies of $\mathbb{CP}^2$ is not complex. Note that [1131,

Yau,1976,Topology] contains an example of a parallelizable 4-manifold which is not complex.

Furthermore, it follows from the Kodaira classification of surfaces that the fundamental groups of complex surfaces are very restricted, e.g. Abelian groups of odd rank bigger than 2 are not possible; on the other hand, Kotschick [606,Kotschick,1992a,Bull. London Math. Soc.] showed that every finitely presented group is the fundamental group of a closed, almost complex 4-manifold.

S^6 is almost complex, but whether it is complex is not known (various papers are not convincing, e.g. [10,Adler,1969,Amer. J. Math.]), so this is a good low(?)-dimensional case to try. An orientable, smooth, closed 6-manifold has an almost complex structure iff $W_3 = 0$, and almost complex structures are classified by integral lifts of w_2 (since $\pi_k(SO(6)/U(3) = \mathbb{CP}^3) = 0$ for $k = 0, 1, 3, 4, 5, 6,$ and $= \mathbb{Z}$ for $k = 2$) [1122,Wu,1952,Publ. Inst. Math. Univ. Strasbourg], (also see section 7 in [1095,Wall,1966,Invent. Math.]).

Problem 5.22 (Yau) *What finite groups can act freely and linearly on a complex 3-dimensional complete intersection in the product of weighted projective spaces?*

Remarks: The action may be holomorphic for one complex structure, but not be holomorphic after a deformation.

Problem 5.23 (Yau) *Is every Kähler manifold diffeomorphic to a complex algebraic manifold?*

Remarks: This is true in complex dimensions 1 and 2, but is unknown in complex dimension 3, even for simply connected manifolds.

Problem 5.24 (Spanier) *Given a continuous function $f : S^3 \to \mathbb{R}^2$ and three arbitrary points $x_1, x_2, x_3 \in S^3$, does there exist a rotation $\rho \in SO(4)$ such that*

$$f\rho(x_1) = f\rho(x_2) = f\rho(x_3)?$$

Remarks: This is a special case of the more general Knaster conjecture which is now known to be false in some cases.

Knaster Conjecture: If $f : S^n \to \mathbb{R}^m$ is continuous and $n-m+2$ arbitrary points, $x_1, , , x_{n-m+2}$, are given, then there exists a rotation $\rho \in SO(n+1)$ such that

$$f\rho(x_1) = \cdots = f\rho(x_{n-m+2}).$$

This conjecture is known to fail for a polynomial $f : S^6 \to \mathbb{R}^3$ (and in other cases) [56,Babenko & Bogatyi,1989,Математисческие Заметки]. Note that if the Knaster conjecture fails for some setting, then it fails for a polynomial map as a consequence of the Stone–Weierstrass approximation theorem. More and simpler counterexamples have recently been found by W. Chen [196,1995].

The case $n = m$ was proved by Hopf [502,Hopf,1944,Portugal. Math.] generalizing the Borsuk–Ulam theorem. In the case $m = 1$, the conjecture is true if the $n+1$ points form a frame [1128,Yamabe & Yujobo,1950,Osaka Math. J.]. Also, for $f : S^2 \to \mathbb{R}$ the conjecture holds for four points if they are endpoints of two diameters [667,Livesay, 1954,Ann. of Math.].

In the case $n = m+1$, the conjecture holds if the three points are vertices of an equilateral triangle in $S^n \subset \mathbb{R}^{n+1}$ [1130,Yang,1957,Amer. J. Math.]. The case $f : S^2 \to \mathbb{R}^1$ for 3 arbitrary points was proved in [320,Floyd,1955,Proc. Amer. Math. Soc.].

Problem 5.25 (G. Martin) Sullivan [1017,Sullivan,1979] showed that any topological m-manifold M, $m \geq 5$ (or $m \geq 6$ if $\partial M \neq \emptyset$), has a quasiconformal structure. Let

$$K(M) = \inf_K \{\text{M has a K quasiconformal atlas}\};$$

it is a conformal invariant. If M is smooth, then $K(M) = 1$ by an elementary argument using arbitrarily small charts.

(A) Does there exist an M^m with $K(M) > 1$?

(B) Does $K(M) = 1$ imply that M is smooth?

Problem 5.26 (Hopf) *Given a closed, orientable M^m, is every map $f : M \to M$ of degree ± 1 a homotopy equivalence?*

Remarks: Following [467,Hausmann,1987], the problem splits into two parts:

(A) is $\pi_1(f) : \pi_1(M) \to \pi_1(M)$ injective?

(B) if so, is $f_* : H_*(M;\mathbb{Z}\pi) \to H_*(M;\mathbb{Z}\pi)$ an isomorphism ($\pi = \pi_1(M)$).

Note that $\pi_1(f)$ is always surjective. (B) is true for $m \leq 4$ and (A) fails if the degree of f is not ± 1 and $m \geq 6$ [ibid.].

Problem 5.27 *Let M^m be a compact m-manifold. Is the topological group* Homeo(M) *an ANR?*

Remarks: Ferry [301,1977,Ann. of Math.] proved this when M is a compact Hilbert cube manifold. It is true when $m = 2$ [684,Luke & Mason,1972,Trans. Amer. Math. Soc.], but is unknown in higher dimensions. The problem reduces to the case of showing that $\mathrm{H}omeo_\partial(B^m)$ (homeomorphisms fixing ∂B^m) is an ANR. By work of Geoghegan [370, 1972,Topology] and Toruńczyk [1059,1973,Fund. Math.], a positive answer would imply that Homeo(M) is an infinite-dimensional manifold modeled on ℓ_2, the separable Hilbert space of square summable sequences.

Recall that Y is an ANR (absolute neighborhood retract) if it is T_0 and any function $f : A \to Y$, where A is a subset of a T_0 space X, extends to a function $F : X \to Y$.

Problem 5.28 (Browder) *Is every finite dimensional H-space homotopy equivalent to an orientable closed m-manifold?*

Remarks: An H-space Y is a topological space with a multiplication $Y \times Y \to Y$ with a unit. Browder [150,1961,Ann. of Math.] proved that every finite dimensional H-space is a Poincaré complex. Not all

H-spaces are Lie groups [488,Hilton & Roitberg,1969,Ann. of Math.], [1135,Zabrodsky,1970,Topology]. The answer is yes if $\pi_1(Y)$ is an odd p-group, or infinite with at most cyclic 2-torsion [184,Cappell & Weinberger,1988,Topology]. Does it help to assume dimension ≤ 5?

Problem 5.29 (Ferry) (A) (Borel Conjectures)

Existence: *Given a Poincaré duality group π, its $K(\pi, 1)$, K, is homotopy equivalent to a closed, topological m-manifold M.*

Remarks: Note that Problem 3.77 is a 3-dimensional version of this Conjecture.

By definition, π is a Poincaré duality group if K satisfies Poincaré duality over $\mathbb{Z}\pi$ with respect to a fundamental class in $H_m(K;\mathbb{Z}) = \mathbb{Z}$. An attempt at the Conjecture naturally breaks up into three beginning steps:

(Step 1) Prove that π is finitely presented.

If this is true, Browder [151,1972,Invent. Math.] and Brown [152,1982] show that K is dominated by a finite complex.

(Step 2) Prove that K can be chosen to be a finite complex.

The obstruction to K being homotopy equivalent to a finite complex is in $\widetilde{K_0}(\mathbb{Z}\pi)$, which vanishes if (C) below is true since $\widetilde{K_0}(\mathbb{Z}\pi)$ is a summand of $Wh(\mathbb{Z}[\pi \times \mathbb{Z}])$.

Poincaré duality gives a $\mathbb{Z}\pi$ homomorphism (cap product) between the based chain complexes C^k and C_{m-k} which is a $\mathbb{Z}\pi$ homology equivalence and thus has Whitehead torsion in $Wh(\mathbb{Z}\pi)$. K is called a *simple* PD space if this torsion is zero.

(Step 3) Prove that K is simple.

Note that all closed, compact manifolds have finitely presented π_1, are homotopy equivalent to finite complexes [585,Kirby & Siebenmann,1977], and are simple, and sometimes these properties are assumed in the Conjecture. With these properties, one is ready to apply the surgery exact sequence described below.

Uniqueness: *If $f : M^m \to N^m$ is a homotopy equivalence between closed, aspherical manifolds, then f is homotopic to a homeomorphism.*

Remarks: This is a topological analog of Mostow rigidity and is true in dimensions ≥ 5 in case M (but not necessarily N) is a non-positively curved Riemannian manifold [294, Farrell & Jones, 1989, J. Amer. Math. Soc.]. Note the relation with Problem 4.83.

(B) *Under the assumptions of the Uniqueness conjecture, is f a tangential equivalence, i.e. is $f^*(T_N)$ stably isomorphic to T_M?*

Remarks: This is a version of the integral Novikov conjecture for $\pi_1(M)$.

(C) (Hsiang) *If Γ is a torsion-free, finitely presented group, then is $Wh(\mathbb{Z}\Gamma) = 0$?* .

Remarks: (C) may be a step in proving (B), for if f is homotopic to a homeomorphism, it must be homotopic to a simple homotopy equivalence which implies that $Wh(\mathbb{Z}\pi_1(M)) = 0$. Note that M aspherical implies that $\pi_1(M)$ is torsion free (and of course finitely presented). The answer to (C) is yes when Γ is π_1 of a non-positively curved polyhedron [515, Hu, 1993, J. Differential Geom.].

The old Problem 3.32 is a special case of this conjecture which was proposed in [511, Hsiang, 1984].

(D) *Is there a closed, aspherical, ANR, homology manifold with Quinn index $\neq 1$?*

Remarks: Compare Problem 4.69. If we do not require the homology manifold to be aspherical, then these exist; in fact there are ones which are homotopy equivalent to S^m for each $m \geq 6$. A *yes* answer to (D) implies that either the Borel existence conjecture or the integral Novikov conjecture (see below) fails. For if H is the homology manifold asked for in (D), and M is the manifold conjectured in (A), then, using a version of the material below which encompasses homology manifolds (see [1103, Weinberger, 1994]), H and M (which live in $\mathcal{S}(X)$) go to different elements in $H_m(B\pi; \mathbb{L}_0)$ but the same element in $L^s_m(\mathbb{Z}\pi)$.

Further remarks: All these problems are versions assuming asphericity of classical problems in surgery theory. We give a very brief sketch below, but an excellent source for this material is [1103, Weinberger, 1994].

Associated to an m-manifold M^m is a map $f : M \to B\pi$ which classifies the universal cover of M ($\pi = \pi_1(M)$). Let $\alpha \in H^k(B\pi;\mathbb{Q})$. Then the rational Novikov conjecture for α is that $\langle L_{4i}(M) \cup f^*(\alpha), \mu_M\rangle$ is a homotopy invariant.

Any α provides a map $g : (B\pi \times B^l, \partial) \to (S^{k+l}, *)$ for which $\alpha = g^*(1)$, $1 \in H^{k+l}(S^{k+l};\mathbb{Q})$. After making $g(f \times id) : (M \times B^l, \partial) \to (S^{k+l}, *)$ transverse to a point $p \neq *$, then the preimage of p is a $4i$-manifold in $M \times B^l$ whose signature equals $\langle L_{4i}(M) \cup f^*(\alpha), \mu_M\rangle$. Thus the rational Novikov conjecture asks whether the signatures of certain submanifolds of M (or $M \times B^l$) depending on elements of $H^k(B\pi;\mathbb{Q})$ are homotopy invariants.

The modern way to attack this conjecture is, however, not via the above description. Instead, recall the surgery exact sequence: if X^m is a Poincaré complex with Spivak normal fibration given by $X^m \xrightarrow{\nu} BG$, then suppose, to get started, that ν has a lift to $X^m \xrightarrow{\nu'} BTOP$. Then ν' provides a basepoint in $\mathcal{N}(X)$ (the set of liftings of $X^m \xrightarrow{\nu} BG$) and $[X, G/TOP]$ acts simply transitively on $\mathcal{N}(X)$. Then the surgery exact sequence is

$$\to L^s_{m+1}(\mathbb{Z}\pi) \to \mathcal{S}(X) \to \mathcal{N}(X) \xrightarrow{\theta} L^s_m(\mathbb{Z}\pi)$$

where, (1) an element (X, ν') of $\mathcal{N}(X)$ provides a normal map $g : N^m \to X$ (N is a manifold, g is degree one, and g is covered by a bundle map from the stable normal bundle of N to the bundle over X induced by ν'), and $\theta(X, \nu')$ is the obstruction to finding a normal bordism to a manifold M simple homotopy equivalent to X, and (2) $\mathcal{S}(X)$ is the equivalence classes of pairs (M, g), with $M \xrightarrow{g} X$ being a simple homotopy equivalence, and (M', g') equivalent to (M, g) if there exists a homeomorphism of M to M' making the obvious diagram homotopy commute.

The surgery exact sequence is now studied via the following commutative diagram:

$$\begin{array}{ccccc}
\mathcal{N}(X) & \xrightarrow{\theta} & L^s_m(\mathbb{Z}\pi) & & \\
\downarrow & & \uparrow A_\pi \ \text{assembly map} & & \\
\pi_m(X_+ \wedge \mathbb{L}_0) = H_m(X;\mathbb{L}_0) & \longrightarrow & H_m(B\pi;\mathbb{L}_0) & = & \pi_m(B\pi_+ \wedge \mathbb{L}_0)
\end{array}$$

where $\mathbb{L}_0$ is the quadratic L-theory spectrum of the trivial group; thus

$$\pi_k(\mathbb{L}_0)) = \begin{cases} 0 & k < 0 \\ \mathbb{Z} & k \equiv 0 \pmod 4 \\ 0 & k \equiv 1 \pmod 4 \\ \mathbb{Z}/2\mathbb{Z} & k \equiv 2 \pmod 4 \\ 0 & k \equiv 3 \pmod 4 \end{cases}$$

Very roughly, one can break up the surgery obstruction for $N^m \to X$ into the pieces over each simplex of X (since these simplices are simply connected, one is led to $\mathbb{L}_0$ and the simply connected surgery groups, $\mathbb{Z}$ (signature), 0, $\mathbb{Z}/2\mathbb{Z}$ (Kervaire invariant), 0); putting the pieces together leads to the assembly map A_π. (The real assembly map, at the level of spectra, is $B\pi_+ \wedge \mathbb{L}_0 \to \mathbb{L}_0(\mathbb{Z}\pi)$ where the latter spectrum is that for which $\pi_m(\mathbb{L}_0(\mathbb{Z}\pi)) = L^s_m(\mathbb{Z}\pi)$).

In this language, we have the following versions of the Novikov and Borel Conjectures (which are purely algebraic):

- **Integral Novikov Conjecture**: *If π is torsion free, then A_π is injective.*
- **Rational Novikov Conjecture:** *$A_\pi \otimes \mathbb{Q}$ is injective.*
- **Borel Conjecture for π:** *If π is a Poincaré duality group, then A_π is an isomorphism. (This is false for many other π). This Conjecture implies the Borel Existence and Uniqueness Conjectures above.*
- **Modern Borel Conjecture:** *The L–theory and K–theory assembly maps are all isomorphisms for Poincaré duality groups.*

Index of Conjectures

Index

L

W

Z

Old Problem Lists

Adams, J. F. (1989). Problem session for homotopy theory. In G. Carlson, R. L. Cohen, H. R. Miller, & D. C. Ravenel (Eds.), *Algebraic Topology*, Volume 1370 of *Lecture Notes in Math.*, pp. 438–456. Springer-Verlag. Proc. of Arcata 1986.

Alexander, J. W. (1932). Some problems in topology. In *Proc. Intern. Congress of Mathematicians*, Volume 1, Zürich, pp. 249–257.

Arnold, V. I. (1976). Некоторые нерешенные задачи теории особенностей. *Труды семинара академика С. Л. Соболева 1*, 5–15. Some unsolved problems in singularity theory.

Arnold, V. I. (1981a). On some problems in singularity theory. *Proc. Indian Acad. Science (Math. Sci.) 90*, 1–9. Also published as [Arnold1981b] and further comments on [Arnold1983].

Arnold, V. I. (1981b). On some problems in singularity theory. In *Geometry and Analysis: Papers dedicated to the memory of V. K. Patodi*, pp. 1–9. Springer-Verlag. Also published as [Arnold1981a] and further comments on [Arnold1983].

Arnold, V. I. (1983). Some open problems in the theory of singularities. In P. Orlik (Ed.), *Singularities*, Volume 40 of *Proc. Sympos. Pure Math.*, pp. 57–69. American Math. Soc. Expanded translation of [Arnold1976] and comments on the previous list published in [Arnold1981a] and [Arnold1981b].

Arnold, V. I. (1990a). Ten problems. In V. I. Arnold (Ed.), *Theory of Singularities and its Applications*, Volume 1 of *Adv. Soviet Math.*, pp. 1–8. American Math. Soc.

Arnold, V. I. (1993). Problems on singularities and dynamical systems. In V. Arnold & M. Monastyrsky (Eds.), *Developments in mathematics: the Moscow school*, pp. 251–274. Chapman & Hall.

Arnold, V. I. (1994). Sur quelques problèmes de la théorie des systèmes dynamiques. Preprint.

Bing, R. H. & Bean, R. J. (1966a). Additional questions on 3-manifolds. In R. H. Bing & R. J. Bean (Eds.), *Topology Seminar, Wisconsin 1965*, Volume 60 of *Ann. of Math. Stud.*, pp. 81–82. Princeton Univ. Press.

Birman, J. S. (1974b). Research problems. In J. S. Birman (Ed.), *Braids, Links, and Mapping Class Groups*, Volume 82 of *Ann. of Math. Stud.*, pp. 216–220. Princeton Univ. Press.

Bogley, W. H. & Howie, J. (1993). J. H. C. Whitehead's Asphericity Question - Open Problems. In C. Hog-Angeloni, W. Metzler, & A. J. Sieradski (Eds.), *Two-dimensional Homotopy and Combinatorial Group Theory*, Volume 197 of *London Math. Soc. Lect. Note Ser.*, pp. 333–334. Cambridge Univ. Press.

Bott, R., Mostow, M., & Schweitzer, P. A. (1978). Foliations problem session. In R. J. Milgram (Ed.), *Algebraic and Geometric Topology (Stanford, 1976)*, Volume 32, part II of *Proc. Sympos. Pure Math.*, pp. 269–271. American Math. Soc.

Browder, F. E. (1976b). Problems of present day mathematics. In F. E. Browder (Ed.), *Mathematical developments arising from Hilbert problems, Northern Illinois University, 1974*, Volume 28 of *Proc. Sympos. Pure Math.*, pp. 35–79. American Math. Soc.

Browder, W. & Hsiang, W.-C. (1978). Homotopy theory, manifolds and transformation groups. In R. J. Milgram (Ed.), *Algebraic and Geometric Topology (Stanford, 1976)*, Volume 32, part II of *Proc. Sympos. Pure Math.*, pp. 251–268. American Math. Soc.

Duncan, A. J., Gilbert, N. D., & Howie, J. (1994). Problem session. In *Combinatorial and Geometric Group Theory*, Volume 204 of *London Math. Soc. Lect. Note Ser.*, pp. 322–325. Cambridge Univ. Press. Proc. of ICMS Workshop held at Heriot-Watt Univ., 1993.

Eliashberg, Y. (1988). Three lectures on symplectic topology in Cala Gonone: Basic notions, problems, and some methods. *Rendiconti del Seminario della Facoltà di Scienze dell' Univerità di Cagliari, Supplemento al 58*, 27–49.

Eliashberg, Y. (1995). Contact 3-manifolds: twenty years since J. Martinet's work. Problem section at the end of the Preprint.

Fenn, R. A. (1985a). Conference problem list. In R. A. Fenn (Ed.), *Low Dimensional Topology (Sussex, 1982)*, Volume 95 of *London Math. Soc. Lect. Note Ser.*, pp. 256–258. Cambridge Univ. Press.

Ferry, S. C., Ranicki, A. A., & Rosenberg, J. M. (1995a). Annotated problem list. In S. C. Ferry, A. A. Ranicki, & J. M. Rosenberg (Eds.), *Novikov Conjectures, Index Theorems and Rigidity, Volume 1 (Oberwolfach, 1993)*, Volume 226 of *London Math. Soc. Lect. Note Ser.*, pp. 67–78. Cambridge Univ. Press.

Fox, R. H. (1962a). Knots and periodic transformations. In M. K. Fort, Jr. (Ed.), *Topology of* 3*-Manifolds and Related Topics (Georgia, 1961)*, pp. 177–182. Prentice-Hall.

Fox, R. H. (1962b). Some problems in knot theory. In M. K. Fort, Jr. (Ed.), *Topology of* 3*-Manifolds and Related Topics (Georgia, 1961)*, pp. 168–176. Prentice-Hall.

Gabai, D. (1995). Eight problems in the geometric theory of foliations and laminations on 3-manifolds. In W. H. Kazez (Ed.), *Geometric Topology, Proc. of the 1993 Georgia International Topology Conference*, pp. 522–553. International Press.

Gersten, S. M. (1987). Selected problems. In S. M. Gersten & J. R. Stallings (Eds.), *Combinatorial Group Theory and Topology*, Volume 111 of *Ann. of Math. Stud.*, pp. 545–551. Princeton Univ. Press.

Gersten, S. M. (1995). Questions on geometric group theory for the Max Dehn seminar. Published electronically at the server **ftp.math.utah.edu** in the file `/u/ma/gersten/MaxDehnSeminar/hyp-quest.dvi`.

Gompf, R. E. (1990). Problems in 4-manifold topology. *Topology Proceedings 15*, 29–38.

Gordon, C. M. (1978). Problems. In J.-C. Hausmann (Ed.), *Knot Theory*, Volume 685 of *Lecture Notes in Math.*, pp. 309–311. Springer-Verlag. Proc. of a Conf. in Plan-sur-Bex, 1977.

Hass, J. (1995). Problems in minimal surfaces and low dimensional topology. To appear.

Hauser, H. & Randell, R. C. (1989). Report on the problem session. In R. C. Randell (Ed.), *Singularities*, Volume 90 of *Contemp. Math.*, pp. 119–134. American Math. Soc. Proc. of the IMA Conference - 1986.

Hempel, J. P. (1976b). Open problems. In J. P. Hempel (Ed.), *3-Manifolds*, Volume 86 of *Ann. of Math. Stud.*, pp. 169–184. Princeton Univ. Press.

Hillman, J. A. (1989). *2-Knots and their groups*, Volume 5 of *Australian Math. Soc. Lect. Series.* Cambridge Univ. Press. Pages 147-149.

Hillman, J. A. (1994). Problems. In *The Algebraic Characterization of Geometric 4-manifolds*, Volume 198 of *London Math. Soc. Lect. Note Ser.*, pp. 158–159. Cambridge Univ. Press.

Hirzebruch, F. (1954). Some problems in differentiable and complex manifolds. *Ann. of Math. 60*, 213–236. Republished in [Hirzebruch1987b] and added comments on the problems in [Hirzebruch1987a].

Hirzebruch, F. (1987a). Kommentare on "Some problems in differentiable and complex manifolds". In *Gesammelte Abhandlungen - Collected Papers*, Volume 1, pp. 762–784. Springer-Verlag. Update to the problem list published in [Hirzebruch1954] and [Hirzebruch1987b].

Hirzebruch, F. (1987b). Some problems in differentiable and complex manifolds. In *Gesammelte Abhandlungen - Collected Papers*, Volume 1, pp. 104–127. Springer-Verlag. Republication of [Hirzebruch1954] and with additional comments on the problems in [Hirzebruch1987a].

Hog-Angeloni, C. & Metzler, W. (1993). Geometric aspects of two-dimensional complexes - Three conjectures and further problems. In C. Hog-Angeloni, W. Metzler, & A. J. Sieradski (Eds.), *Two-dimensional Homotopy and Combinatorial Group Theory*, Volume 197 of *London Math. Soc. Lect. Note Ser.*, pp. 44–50. Cambridge Univ. Press.

Hollingsworth, J. G. (1979). A short list of problems. In J. C. Cantrell (Ed.), *Geometric Topology (Georgia, 1977)*, pp. 685–692. Academic Press.

Kirby, R. C. (1978). Problems in low dimensional manifold theory. In R. J. Milgram (Ed.), *Algebraic and Geometric Topology (Stanford, 1976)*, Volume 32, part II of *Proc. Sympos. Pure Math.*, pp. 273–312. American Math. Soc.

Kirby, R. C. (1984). 4-manifold problems. In C. M. Gordon & R. C. Kirby (Eds.), *Four-Manifold Theory (Durham, NH, 1982)*, Volume 35 of *Contemp. Math.*, pp. 513–528. American Math. Soc.

Kirby, R. C. (1988). Problems in 4-manifolds. Proceedings (Tagungsberichte) Oberwolfach meeting on 4-manifolds.

Kister, J. M. (1962). Questions on isotopies in manifolds. In M. K. Fort, Jr. (Ed.), *Topology of 3-Manifolds and Related Topics (Georgia, 1961)*, pp. 229–230. Prentice-Hall.

Lashof, R. K. (1965). Problems in differential and algebraic topology. Seattle Conference, 1963. *Ann. of Math. 81*, 565–591. Russian commented translation by S. P. Novikov in [Novikov1965a] and translation back to English of the commented version in [Novikov1965b].

Lê Dũng Tráng & Teissier, B. (1983). Report on the problem session. In P. Orlik (Ed.), *Singularities*, Volume 40 of *Proc. Sympos. Pure Math.*, pp. 105–116. American Math. Soc.

Lyndon, R. (1993). Problems in combinatorial group theory. In C. Hog-Angeloni, W. Metzler, & A. J. Sieradski (Eds.), *Two-dimensional Homotopy and Combinatorial Group Theory*, Volume 197 of *London Math. Soc. Lect. Note Ser.*, pp. 3–33. Cambridge Univ. Press.

Matveev, S. V. & Rolfsen, D. P. (1993). Zeeman's collapsing conjecture - open problems. In C. Hog-Angeloni, W. Metzler, & A. J. Sieradski (Eds.), *Two-dimensional Homotopy and Combinatorial Group Theory*, Volume 197 of *London Math. Soc. Lect. Note Ser.*, pp. 363–364. Cambridge Univ. Press.

Mauldin, R. D. (1981). *The Scottish book: Mathematics from the Scottish Cafe.* Birkhäuser.

Meeks, III, W. H. (1978). *Some outstanding problems on the existence, topology and geometry of minimal surfaces in* $\mathbb{R}^3$. Universidade Federal do Ceará. Escola de Geometria Diferencial, Fortaleza, 17-28 de Julho de 1978, Pages 74-98.

Miyazaki, K. (1992). A collection of problems. *Sūrikaisekikenkyūsho Kōkyūroku 813*, 155–168.

Neuwirth, L. P. (1965). Problems. In *Knot Groups*, Volume 56 of *Ann. of Math. Stud.*, pp. 99–102. Princeton Univ. Press. See update in [Neuwirth1974].

Neuwirth, L. P. (1974). The status of some problems related to knot groups. In R. F. Dickman & P. Fletcher (Eds.), *Topology Conference*, Volume 375 of *Lecture Notes in Math.*, pp. 209–230. Springer-Verlag. Proc. of the Virginia Polytechnic Institute and State Univ. Topology Conf. 1973, Update to the list published in [Neuwirth1965].

Niblo, G. A. (1993). Geometric group theory 1991 problem list. In G. A. Niblo & M. A. Roller (Eds.), *Geometric Group Theory I (Sussex, 1991)*, Volume 181 of *London Math. Soc. Lect. Note Ser.*, pp. 208–212. Cambridge Univ. Press.

Novikov, S. P. (1965a). О летнем топологическом институте. *Успехи Матем. Наук 20*, 147–170. Commented translation by S. P. Novikov of [Lashof1965]. This article was later translated back to English in [Novikov1965b].

Novikov, S. P. (1965b). The topology summer institute (Seattle, USA, 1963). *Russian Math. Surveys 20*, 145–167. This is a translation back to English of the commented translation by S. P. Novikov of [Novikov1965a]. The original is [Lashof1965].

Palis, Jr., J. & Pugh, C. C. (1975). Fifty Problems in Dynamical Systems. In A. Manning (Ed.), *Dynamical Systems – Warwick 1974*, Volume 468 of *Lecture Notes in Math.*, pp. 345–353. Springer-Verlag. Proc. of a Symp. at University of Warwick, 1973/74.

Rolfsen, D. P. (1985b). Problems. In D. P. Rolfsen (Ed.), *Knot Theory and Manifolds, Proc. Vancouver 1983*, Volume 1144 of *Lecture Notes in Math.*, pp. 160–163. Springer-Verlag.

Schultz, R. (1985). Problems submitted to the AMS Summer Research Conference on Group Actions on Manifolds. In R. Schultz (Ed.), *Group actions on Manifolds*, Volume 36 of *Contemp. Math.*, pp. 513–568. American Math. Soc.

Shaneson, J. L. (1973). Some problems in hermitian K-theory, III. In H. Bass (Ed.), *Algebraic K-theory*, Volume 343 of *Lecture Notes in Math.*, pp. 41–51. Springer-Verlag. Proc. of a Seattle conference, 1972.

Smale, S. (1991). Dynamics retrospective: great problems, attempts that failed. *Physica D 51*, 267–273.

Sumners, D. W. (1988). Some problems in applied knot theory, and some problems in geometric topology. *Topology Proceedings 13*, 163–176.

Turaev, V. G. (1994). *Quantum Invariants of Knots and 3-manifolds*, Volume 18 of *de Gruyter Studies in Mathematics.* W. de Gruyter. Pages 571-572.

van Mill, J. & Reed, G. M. (1990). *Open Problems in Topology.* North-Holland and Elsevier Science Pub. Co. See also the periodic updates published in *Topology Appl.*, e.g. Vol. 62 (1995) 93-99.

Waldhausen, F. (1978). Some problems on 3-manifolds. In R. J. Milgram (Ed.), *Algebraic and Geometric Topology (Stanford, 1976)*, Volume 32, part II of *Proc. Sympos. Pure Math.*, pp. 313–322. American Math. Soc.

Wall, C. T. C. (1979b). List of problems. In C. T. C. Wall (Ed.), *Homological Group Theory*, Volume 36 of *London Math. Soc. Lect. Note Ser.*, pp. 369–394. Cambridge Univ. Press. Proc. of a symposium, held at Durham in 1977.

Weber, C. (1983). Problems. In C. Weber (Ed.), *Nœuds, Tresses, et Singularités*, Volume 31 of *Monographie L'Enseignement Math.*, pp. 247–259. Université de Genève. Comptes Rendus du Séminaire tenu aux Plans-sur-Bex - Suisse, 1982.

Yau, S.-T. (1982). Problem section. In S.-T. Yau (Ed.), *Seminar on Differential Geometry*, Volume 102 of *Ann. of Math. Stud.*, pp. 669–706. Princeton Univ. Press.

Yau, S.-T. (1993). Open problems in geometry. In R. Greene & S.-T. Yau (Eds.), *Differential Geometry*, Volume 54 part 3 of *Proc. Sympos. Pure Math.*, pp. 1–28. American Math. Soc.

Bibliography

[1] Abhyankar, S. S. & Moh, T. T. (1975). Embeddings of the line in the plane. *J. Reine Angew. Math. 276*, 148–166. Problem: 1.102.

[2] Adams, C. C. (1987). The noncompact hyperbolic 3-manifold of minimal volume. *Proc. Amer. Math. Soc. 100*, 601–606. Problem: 3.60.

[3] Adams, C. C. (1988). Volumes of N-cusped hyperbolic 3-manifolds. *J. London Math. Soc. 38*, 555–565. Problem: 3.60.

[4] Adams, C. C. (1991). Limit volumes of hyperbolic three-orbifolds. *J. Differential Geom. 34*, 115–141. Problem: 3.60.

[5] Adams, C. C. (1992a). Noncompact hyperbolic 3-orbifolds of small volume. In B. N. Apanasov, W. D. Neumann, A. W. Reid, & L. C. Siebenmann (Eds.), *Topology 90, (Ohio St. U., 1990)*, pp. 1–16. W. de Gruyter. Problem: 3.60.

[6] Adams, C. C. (1992b). Volumes of hyperbolic 3-orbifolds with multiple cusps. *Indiana Univ. Math. J. 41*, 149–172. Problem: 3.60.

[7] Adams, C. C. (1995). Unknotting tunnels in hyperbolic 3-manifolds. To appear in *Math. Ann.*. Problems: 1.71, 1.72, 3.94.

[8] Adams, C. C. & Reid, A. W. (1994). Unknotting tunnels in 2-bridge knot and link complements. Preprint. Problem: 1.71.

[9] Adams, J. F. (1955). A new proof of a theorem of W. H. Cockcroft. *J. London Math. Soc. 30*, 482–488. Problem: 5.4.

[10] Adler, A. (1969). The second fundamental forms on S^6 and $P^n(\mathbb{C})$. *Amer. J. Math. 91*, 657–670. Problem: 5.21.

[11] Ahlfors, L. (1966). Fundamental polyhedrons and limit sets of Kleinian groups. *Proc. Nat. Acad. Sci. U.S.A. 55*, 251–254. Problem: 3.52.

[12] Aitchison, I. R. & Rubinstein, J. H. (1984). Fibered knots and involutions on homotopy spheres. In C. M. Gordon & R. C. Kirby (Eds.), *Four-Manifold Theory (Durham, NH, 1982)*, Volume 35 of *Contemp. Math.*, pp. 1–74. American Math. Soc. Problem: 4.13.

[13] Aitchison, I. R. & Rubinstein, J. H. (1992). Combinatorial cubings, cusps, and the dodecahedral knots. In B. N. Apanasov, W. D. Neumann, A. W. Reid, & L. C. Siebenmann (Eds.), *Topology 90, (Ohio St. U., 1990)*, pp. 17–26. W. de Gruyter. Problem: 3.64.

[14] Aitchison, I. R. & Rubinstein, J. H. (1993). Incompressible surfaces and the topology of 3-dimensional manifolds. *J. Austral. Math. Soc. Ser. A 55*, 1–22. Problem: 3.2.

[15] Akbulut, S. (1977). On 2-dimensional homology classes of 4-manifolds. *Math. Proc. Cambridge Philos. Soc. 82*, 99–106. Problem: 1.41.

[16] Akbulut, S. (1984). A fake 4-manifold. In C. M. Gordon & R. C. Kirby (Eds.), *Four-Manifold Theory (Durham, NH, 1982)*, Volume 35 of *Contemp. Math.*, pp. 75–142. American Math. Soc. Problems: 4.14, 4.15, 4.51, 4.85.

[17] Akbulut, S. (1985). On a fake $S^3 \tilde{\times} S^1 \# S^2 \times S^2$. In J. R. Harper & R. Mandelbaum (Eds.), *Combinatorial Methods in Topology and Algebraic Geometry (Rochester, 1982)*, Volume 44 of *Contemp. Math.*, pp. 281–286. Problem: 4.15.

[18] Akbulut, S. (1988). Constructing a fake 4-manifold by Gluck construction to a standard 4-manifold. *Topology 27*, 239–243. Problems: 4.15, 4.45, 4.51.

[19] Akbulut, S. (1991a). A fake compact contractible 4-manifold. *J. Differential Geom. 33*, 335–356. Problems: 4.2, 4.16, 4.105.

[20] Akbulut, S. (1991b). A solution to a conjecture of Zeeman. *Topology 30*, 513–515. Problems: 1.31, 4.21.

[21] Akbulut, S. (1991c). An exotic 4-manifold. *J. Differential Geom. 33*, 357–361. Problems: 4.16, 4.45, 4.90.

[22] Akbulut, S. (1994). On quotients of complex surfaces under complex conjugation. *J. Reine Angew. Math. 447*, 83–90. Problem: 4.104.

[23] Akbulut, S. (1995a). A note on a homology sphere. Preprint. Problem: 4.28.

[24] Akbulut, S. (1995b). On universal relations in gauge theory. To appear in *Trans. Amer. Math. Soc.*. Problem: 4.132.

[25] Akbulut, S. (1995c). Torsion in gauge theory. In S. Akbulut, T. Onder, & R. J. Stern (Eds.), *Proc. of Gökova Topology–Geometry Conference, 1994*, pp. 49–59. Scientific and Technical Research Council of Turkey. Problem: 4.132.

[26] Akbulut, S. & King, H. C. (1992). Algebraicity of immersions in $\mathbb{R}^n$. *Topology 31*, 701–712. Problem: 4.61.

[27] Akbulut, S. & Kirby, R. C. (1979). Mazur manifolds. *Michigan Math. J. 26*, 259–284. Problem: 4.2.

[28] Akbulut, S. & Kirby, R. C. (1980). Branched covers of surfaces in 4-manifolds. *Math. Ann. 252*, 111–131. Problems: 4.18, 4.113.

[29] Akbulut, S. & Kirby, R. C. (1985). A potential smooth counterexample in dimension 4 to the Poincaré conjecture, the Schoenflies conjecture, and the Andrews–Curtis conjecture. *Topology 24*, 375–390. Problems: 4.45, 5.2.

[30] Akbulut, S. & McCarthy, J. D. (1990). *Casson's Invariant for Oriented Homology 3-Spheres*, Volume 36 of *Math. Notes*. Princeton Univ. Press. Problems: 3.1, 3.43, 4.72.

[31] Akbulut, S., Mrowka, T. S., & Ruan, Y. (1995). Torsion classes and a universal constraint on Donaldson invariants for odd manifolds. *Trans. Amer. Math. Soc. 347*, 63–76. Problem: 4.132.

[32] Alexander, J. W. (1924). On the subdivision of a 3-space by a polyhedron. *Proc. Nat. Acad. Sci. U.S.A. 10*, 6–8. Problem: 3.11.

[33] Allenby, R. B. J. T., Boler, J., Evans, B., Moser, L. E., & Tang, C. Y. (1979). Frattini subgroups of 3-manifold groups. *Trans. Amer. Math. Soc. 247*, 275–300. Problem: 3.33.

[34] Ancel, F. D. & Guilbault, C. R. (1995a). Compact contractible n-manifolds have arc spines ($n \geq 5$). *Pacific J. Math. 169*, 1–10. Problem: 4.73.

[35] Ancel, F. D. & Guilbault, C. R. (1995b). Mapping swirls and pseudo-spines of compact 4-manifolds. Preprint. Problem: 4.73.

[36] Anderson, J. (1991). On the finitely generated intersection property for Kleinian groups. *Complex Variables Theory Appl. 17*, 111–112. Problem: 3.52.

[37] Andrews, J. J. & Curtis, M. L. (1965). Free groups and handlebodies. *Proc. Amer. Math. Soc. 16*, 192–195. Problems: 5.1, 5.2.

[38] Anosov, D. V. (1967). Геодезические потоки на замкнутых римановых многообразиях отрицательной кривизны. *Труды Мат. ин-та. им. Стеклова 90*, 209pp. Translation to English in [39].

[39] Anosov, D. V. (1969). Geodesic flows on closed Riemann manifolds with negative curvature. *Proc. Steklov Inst. Math. 90*, 1–235. Translation of [38]. Problem: 3.53.

[40] Anstee, R. P., Przytycki, J. H., & Rolfsen, D. P. (1989). Knot polynomials and generalized mutation. *Topology Appl. 32*, 237–249. Problem: 1.60.

[41] Antonelli, P. L., Burghelea, D., & Kahn, P. J. (1972). The non-finite homotopy type of some diffeomorphism groups. *Topology 11*, 1–49. Problem: 4.126.

[42] Apéry, F. (1986). La surface de Boy. *Adv. Math. 61*, 185–266. Problem: 4.61.

[43] Apéry, F. (1987). *Models of the Real Projective Plane.* Vieweg. Problem: 4.61.

[44] Artal-Bartolo, E. (1991). Forme de Seifert des singularités de surface. *C.R. Acad. Sci. Paris Sér. I Math. 313*, 689–692. Problem: 3.29.

[45] Artal-Bartolo, E. (1993). Combinatoire et type topologique des applications polynomials de $\mathbb{C}^2$ dans $\mathbb{C}$. *L'Enseign. Math. 39*, 211–224. Problem: 1.100.

[46] Artin, E. (1925). Zur Isotopie zweidimensionaler Flächen in R_4. *Abh. Math. Sem. Univ. Hamburg 4*, 174–177. Problem: 1.49.

[47] Arvola, W. A. (1992). The fundamental group of the complement of an arrangement of complex planes. *Topology 31*, 757–765. Problem: 4.145.

[48] Asanao, K. (1978). Homeomorphisms of prism manifolds. *Yokohama Math. J. 26*, 19–25. Problem: 3.95.

[49] Asimov, D. (1975). Round handles and non-singular Morse-Smale flows. *Ann. of Math. 102*, 41–54. Problem: 3.9.

[50] Atiyah, M. F. (1969). The signature of fibre-bundles. In D. C. Spencer & S. Iyanaga (Eds.), *Global Analysis. Papers in honor of K. Kodaira*, pp. 73–84. University of Tokyo PressPrinceton Univ. Press. Problem: 2.18.

[51] Atiyah, M. F. (1990). On framings of 3-manifolds. *Topology 29*, 1–8. Problem: 5.20.

[52] Auckly, D. R. (1993). An irreducible homology sphere which is not Dehn surgery on a knot. Preprint. Problem: 3.6.

[53] Auckly, D. R. (1994). Topological methods to compute Chern–Simons invariants. *Math. Proc. Cambridge Philos. Soc. 115*, 229–251. Problem: 3.104.

[54] Auckly, D. R. (1995). Surgery numbers of 3-manifolds: a hyperbolic example. In W. H. Kazez (Ed.), *Geometric Topology, Proc. of the 1993 Georgia International Topology Conference.* International Press. Problem: 3.102.

[55] Axelrod, S. & Singer, I. M. (1992). Chern–Simons perturbation theory. In S. Catto & A. Rocha (Eds.), *International Conference on Differential Geometric Methods in Theoretical Physics*, pp. 3–45. World Scientific. Problems: 3.108, 5.20.

[56] Babenko, I. K. & Bogatyi, S. A. (1989). К отображению сферы в евклидово пространство. *Математисческие Заметки 46*, 3–8. English translation in [57]. Problem: 5.24.

[57] Babenko, I. K. & Bogatyi, S. A. (1990). Mapping a sphere into Euclidean space. *Math. Notes of the Acad. of Sci. USSR 46*, 683–686. Original in Russian in [56].

[58] Baker, M. (1989). Covers of Dehn fillings on once-punctured torus bundles. *Proc. Amer. Math. Soc. 105*, 747–754. Problem: 3.2.

[59] Baker, M. (1990). Covers of Dehn fillings on once-punctured torus bundles II. *Proc. Amer. Math. Soc. 110*, 1099–1108. Problem: 3.2.

[60] Baker, M. (1991). On coverings of figure eight knot surgeries. *Pacific J. Math. 150*, 215–228. Problem: 3.2.

[61] Bar-Natan, D. (1991). *Perturbative aspects of the Chern–Simons topological quantum field theory.* Ph. D. thesis, Princeton Univ. Problem: 5.20.

[62] Bar-Natan, D. (1995a). On the Vassiliev knot invariants. *Topology 34*, 423–472. Problems: 1.89, 1.92, 5.20.

[63] Bar-Natan, D. (1995b). Vassiliev homotopy string link invariants. To appear in *J. Knot Theory Ramifications*. Problem: 1.96.

[64] Barbot, T. (1992). *Géométrie transverse des flots d'Anosov.* Ph. D. thesis, Lyon. Problem: 3.53.

[65] Barbot, T. (1994). Generalizations of the Bonatti-Langevin example of Anosov flow and their classification up to topological equivalence. Preprint, Université de Bourgogne, 51. Problem: 3.53.

[66] Barbot, T. (1995). Caractérisation des flots d'Anosov en dimension 3 par leurs feuilletages faibles. To appear in *Ergod. Th. & Dynam. Sys.*. Problem: 3.53.

[67] Barge, J. & Ghys, E. (1992). Cocycles d'Euler et de Maslov. *Math. Ann. 294*, 235–265. Problem: 2.18.

[68] Barth, W., Peters, C., & de Ven, A. V. (1984). *Compact Complex Surfaces*, Volume 4 of *Ergeb. Math. Grenzgeb. (3)*. Springer-Verlag. Problem: 4.105.

[69] Bass, H. & Morgan, J. W. (1984a). *The Smith Conjecture*, Volume 112 of *Pure and Applied Mathematics.* Academic Press. Problems: 3.38, 3.46.

[70] Bauer, S. (1988). The homotopy type of a 4-manifold with finite fundamental group. In T. tom Dieck (Ed.), *Algebraic Topology and Transformation Groups*, Volume 1361 of *Lecture Notes in Math.*, pp. 1–6. Springer-Verlag. Proc. of Conf. in Gottingen, 1987. Problem: 4.53.

[71] Bauer, S. (1994). Diffeomorphism types of elliptic surfaces with $p_g = 1$. *J. Reine Angew. Math. 451*, 89–148. Problem: 4.11.

[72] Baues, H. J. (1991). *Combinatorial Homotopy and 4-Dimensional Complexes*. W. de Gruyter. Problem: 4.53.

[73] Baumslag, G. (1967). Groups with the same lower central series as a relatively free group I. *Trans. Amer. Math. Soc. 129*, 308–321. Problems: 3.78, 3.79.

[74] Baumslag, G. (1969). Groups with the same lower central series as a relatively free group II. *Trans. Amer. Math. Soc. 142*, 507–538. Problems: 3.78, 3.79.

[75] Baumslag, G., Dyer, E., & Miller, III, C. F. (1983). On the integral homology of finitely presented groups. *Topology 22*, 27–46. Problem: 4.17.

[76] Bennequin, D. (1983). Entrelacements et equations de Pfaff. *Astérisque 107 and 108*, 83–161. Problems: 1.98, 4.142.

[77] Bentzen, S. (1987). Some numerical results on space forms. *Proc. London Math. Soc. 54*, 559–576. Problem: 3.37.

[78] Berge, J. (1991). The knots in $D^2 \times S^1$ with non-trivial Dehn surgeries yielding $D^2 \times S^1$. *Topology Appl. 38*, 1–19. Problem: 1.81.

[79] Berge, J. (1995). Some knots with surgeries yielding lens spaces. Preprint. Problem: 1.78.

[80] Berger, M. (1963). Sur les variétés $\frac{4}{23}$-pincées de dimension 5. *C.R. Acad. Sci. Paris Sér. I Math. 257*, 4122–4125. Problem: 4.67.

[81] Berger, M. (1966). Trois remarques sur les variétés riemanniennes à courbure positive. *C.R. Acad. Sci. Paris Sér. I Math. 263*, A76–A78. Problem: 4.67.

[82] Besse, A. L. (1987). *Einstein Manifolds*, Volume 10 of *Ergeb. Math. Grenzgeb. (3)*. Springer-Verlag. Problem: 4.68.

[83] Bestvina, M. & Handel, M. (1995). Train tracks for surface automorphisms. *Topology 34*, 109–140. Problem: 1.84.

[84] Bestvina, M. & Mess, G. (1991). The boundary of negatively curved groups. *J. Amer. Math. Soc. 4*, 469–481. Problem: 3.7.

[85] Betley, S., Przytycki, J. H., & Zukowski, T. (1986). Hyperbolic structures on Dehn fillings of some punctured torus bundles over S^1. *Kobe J. Math. 3*, 117–147. Problem: 3.60.

[86] Bieri, R. (1976). Normal subgroups in duality groups and in groups of cohomological dimension 2. *J. Pure Appl. Algebra 7*, 35–51. Problem: 5.11.

[87] Bieri, R. (1980). A connection between the integral homology and the centre of a rational linear group. *Math. Zeit. 170*, 263–266. Problem: 5.11.

[88] Bing, R. H. (1964). Inequivalent families of periodic homeomorphisms of E^3. *Ann. of Math. 80*, 78–93. Problem: 3.38.

[89] Bing, R. H. & Martin, J. (1971). Cubes with knotted holes. *Trans. Amer. Math. Soc. 155*, 217–231. Problem: 1.15.

[90] Birman, J. S. (1974). *Braids, Links, and Mapping Class Groups*, Volume 82 of *Ann. of Math. Stud.* Princeton Univ. Press. Problems: 1.84, 4.100.

[91] Birman, J. S. & Craggs, R. (1978). The μ-invariant of 3-manifolds and certain structural properties of the group of the homeomorphisms of a closed, oriented 2-manifold. *Trans. Amer. Math. Soc. 237*, 283–309. Problem: 2.2.

[92] Birman, J. S., Gonzalez-Acuña, E., & Montesinos, J. M. (1976). Heegaard splittings of prime 3-manifolds are not unique. *Michigan Math. J. 23*, 97–103. Problems: 3.25, 3.86.

[93] Birman, J. S. & Hilden, H. (1975). Heegaard splittings of branched coverings of S^3. *Trans. Amer. Math. Soc. 213*, 315–352. Problem: 3.25.

[94] Birman, J. S. & Lin, X.-S. (1993). Knot polynomials and Vassiliev's invariants. *Invent. Math. 111*, 225–270. Problem: 1.89.

[95] Birman, J. S. & Menasco, W. W. (1990). Studying links via closed braids IV, Composite links and split links. *Invent. Math. 102*, 115–139. Problem: 1.84.

[96] Birman, J. S. & Menasco, W. W. (1992). Studying links via closed braids V, The unlink. *Trans. Amer. Math. Soc. 329*, 585–606. Problem: 1.84.

[97] Birman, J. S. & Menasco, W. W. (1993). Studying links via closed braids III, Classifying links which are closed 3-braids. *Pacific J. Math. 161*, 25–113. Problem: 1.84.

[98] Birman, J. S. & Rubinstein, J. H. (1984). One-sided Heegaard splittings and homotopy groups of some 3-manifolds. *Proc. London Math. Soc. 49*, 517–536. Problem: 3.95.

[99] Bižaca, Ž. (1994). A handle decomposition of an exotic $\mathbb{R}^4$. *J. Differential Geom. 39*, 491–508. Problem: 4.43.

[100] Bižaca, Ž. (1995). An explicit family of exotic Casson handles. *Proc. Amer. Math. Soc. 123*, 1297–1302. Problems: 1.39, 4.76.

[101] Bižaca, Ž. & Gompf, R. E. (1995). Elliptic surfaces and some simple exotic $\mathbb{R}^4$'s. To appear in *J. Differential Geom.*. Problem: 4.43.

[102] Björner, A. & Ziegler, G. M. (1992). Combinatorial stratification of complex arrangements. *J. Amer. Math. Soc. 5*, 105–149. Problem: 4.144.

[103] Bleiler, S. A. (1994). Two generator cable knots are tunnel one. *Proc. Amer. Math. Soc. 122*, 1285–1287. Problem: 1.10.

[104] Bleiler, S. A. & Hodgson, C. D. (1995). Spherical space forms and Dehn filling. To appear in *Topology*. Problem: 1.77.

[105] Bleiler, S. A., Hodgson, C. D., & Weeks, J. (1995). In preparation. Problem: 1.81.

[106] Bleiler, S. A. & Jones, A. (1995a). On non simple two generator 3-manifolds. In preparation. Problem: 1.10.

[107] Bleiler, S. A. & Jones, A. (1995b). The free product of groups with amalgamated subgroup malnormal in a single factor. Preprint. Problem: 1.10.

[108] Bleiler, S. A. & Scharlemann, M. G. (1986). Tangles, property P and a problem of J. Martin. *Math. Ann. 273*, 215–225. Problem: 1.18.

[109] Bleiler, S. A. & Scharlemann, M. G. (1988). A projective plane in $\mathbb{R}^4$ with three critical points is standard. Strongly invertible knots have property P. *Topology 27*, 519–540. Problems: 1.2, 4.30.

[110] Boéchat, J. & Haefliger, A. (1970). Plongements différentiables des variétés orientées de dimension 4 dans $\mathbb{R}^7$. In A. Haefliger & R. Narasimhan (Eds.), *Essays on Topology and Related Topics. Memoires dédiés à Georges de Rham.* Springer-Verlag. Problem: 4.19.

[111] Bogle, M. G. V., Hearst, J. E., Jones, V. F. R., & Stoilov, L. (1994). Lissajous knots. *J. Knot Theory Ramifications 3*, 121–140. Problem: 1.62.

[112] Bogomolov, F. A. (1976). Classification of surfaces of class VII_0 with $b_2 = 0$. *Math. USSR-Izv. 10*, 255–269. Problem: 4.102.

[113] Bogomolov, F. A. (1983). Surfaces of class VII_0 and affine geometry. *Math. USSR-Izv. 21*, 31–71. Problem: 4.102.

[114] Boileau, M. (1985). Nœuds rigidement inversables. In R. A. Fenn (Ed.), *Low Dimensional Topology (Sussex, 1982)*, Volume 95 of *London Math. Soc. Lect. Note Ser.*, pp. 1–18. Cambridge Univ. Press. Problem: 1.6.

[115] Boileau, M., Collins, D., & Zieschang, H. (1986). Scindements de Heegaard des petites variétés de Seifert. *C.R. Acad. Sci. Paris Sér. I Math. 305*, 557–560. Problem: 3.86.

[116] Boileau, M., Domergue, & Mathieu, Y. (1995). Chirurgies de Dehn le long de noeuds nul homotopes. Preprint. Problem: 1.80.

[117] Boileau, M. & Flapan, E. L. (1995). On π-hyperbolic knots which are determined by their 2-fold and 4-fold cyclic branched coverings. *Topology Appl. 61*, 229–240. Problem: 1.75.

[118] Boileau, M., Gonzalez-Acuña, E., & Montesinos, J. M. (1987). Surgery on double knots and symmetries. *Math. Ann. 276*, 323–340. Problems: 1.23, 3.24.

[119] Boileau, M. & Otal, J.-P. (1986). Groupe de difféotopies de certaines variétés de Seifert. *C.R. Acad. Sci. Paris Sér. I Math. 303*, 19–22. Problem: 3.95.

[120] Boileau, M. & Otal, J.-P. (1990). Sur les scindements de Heegaard du tore T^3. *J. Differential Geom. 32*, 209–233. Problems: 3.22, 3.85.

[121] Boileau, M. & Otal, J.-P. (1991). Scindements de Heegaard et groupe des homeotopies des petites variétés de Seifert. *Invent. Math. 106*, 85–107. Problems: 3.22, 3.40, 3.85.

[122] Boileau, M. & Otal, J.-P. (1994). Scindements de Heegaard des variétés de Seifert elliptiques et Euclidiennes. Preprint. Problem: 3.85.

[123] Boileau, M., Rost, M., & Zieschang, H. (1988). On Heegaard decompositions on torus knot exteriors and related Seifert fibre spaces. *Math. Ann. 279*, 553–581. Problem: 1.71.

[124] Boileau, M. & Rudolph, L. (1995). Noeuds non concordants à un noeud C-transverse. Preprint. Problem: 1.100.

[125] Boileau, M. & Wang, S. C. (1995). Non-zero degree maps and surface bundles over S^1. Preprint. Problems: 1.80, 3.51.

[126] Boileau, M. & Weber, C. (1983). Le problème de J. W. Milnor sur le nombre gordien des nœuds albébriques. *L'Enseign. Math. 31*, 49–98. Problem: 1.100.

[127] Boileau, M. & Zieschang, H. (1984). Heegaard genus of closed orientable Seifert 3-manifolds. *Invent. Math. 76*, 455–468. Problems: 1.10, 3.15, 3.92.

[128] Bonahon, F. (1983a). Cobordism of automorphisms of surfaces. *Ann. Sci. École Norm. Sup. (4) 16*, 237–270. Problem: 2.3.

[129] Bonahon, F. (1983b). Difféotopies des espaces lenticulaires. *Topology 22*, 305–314. Problems: 3.35, 3.40, 3.49, 3.95.

[130] Bonahon, F. (1986). Bouts des variétés hyperboliques de dimension 3. *Ann. of Math. 124*, 71–158. Problems: 3.8, 3.14, 3.52.

[131] Bonahon, F. & Otal, J.-P. (1983). Scindements de Heegaard des espaces lenticulaires. *Ann. Sci. École Norm. Sup. (4) 16*, 451–466. Problems: 3.22, 3.85.

[132] Bonahon, F. & Siebenmann, L. C. (1987). The characteristic toric splitting of irreducible compact 3-orbifolds. *Math. Ann. 278*, 441–479. Problem: 3.46.

[133] Bonatti, C. (1989). Un point fixe commun pours des difféomorphismes commutants de S^2. *Ann. of Math. 129*, 61–69. Problem: 2.20.

[134] Bonatti, C. & Langevin, R. (1994). Un exemple de flot d'Anosov transitif transverse à un tore et non conjugué à une suspension. *Ergod. Th. & Dynam. Sys. 14*, 633–644. Problem: 3.53.

[135] Borel, A. (1963). Compact Clifford–Klein forms of symmetric spaces. *Topology 2*, 111–122. Problem: 4.39.

[136] Borel, A. (1974). Stable real cohomology of arithmetic groups. *Ann. Sci. École Norm. Sup. (4) 7*, 235–272. Problem: 3.63.

[137] Borel, A. (1981). Commensurability classes and volumes of hyperbolic 3-manifolds. *Ann. Scuola Norm. Sup. Pisa Cl. Sci. (4) 8*, 1–33. Problems: 3.60, 3.62.

[138] Bott, R. & Taubes, C. H. (1994). On the self linking of knots. *J. Math. Phys. 35*, 5247–5287. Problem: 5.20.

[139] Bourguignon, J. P. (1975). Some constructions related to H. Hopf's conjecture on product manifolds. In S.-S. Chern & R. Osserman (Eds.), *Differential Geometry*, Volume 27 of *Proc. Sympos. Pure Math.*, pp. 33–37. American Math. Soc. Proc. of a Conf. at Stanford Univ. Stanford - 1973. Problem: 4.67.

[140] Bourguignon, J. P. (1981a). La conjecture de Hopf sur $S^2 \times S^2$. In L. Bérard-Bergery, M. Berger, & C. Houzel (Eds.), *Géometrie Riemannienne en Dimension 4, Seminaire Arthur Besse.* Problem: 4.67.

[141] Bourguignon, J. P. (1981b). Les variétés de dimension 4 à signature non rulle dont la courbure est harmonique sont d'Einstein. *Invent. Math. 63*, 263–286. Problem: 4.68.

[142] Bourguignon, J. P., Deschamps, A., & Sentenac, P. (1972). Conjecture de H. Hopf sur les produits de variétés. *Ann. Sci. École Norm. Sup. (4) 5*, 277–302. Problem: 4.67.

[143] Bourguignon, J. P., Deschamps, A., & Sentenac, P. (1973). Quelques variations particulières d'un produit de métriques. *Ann. Sci. École Norm. Sup. (4) 6*, 1–16. Problem: 4.67.

[144] Boyer, S. & Zhang, X. (1994a). Exceptional surgery on knots. *Bull. Amer. Math. Soc. 31*, 197–203. Problem: 1.77.

[145] Boyer, S. & Zhang, X. (1994b). The semi-norm and Dehn surgery. Preprint. Problem: 1.77.

[146] Boyer, S. & Zhang, X. (1995). Finite surgery on knots. To appear in *J. Amer. Math. Soc.*. Problem: 1.77.

[147] Brick, S. & Mihalik, M. (1995). The QSF property for groups and spaces. To appear in *Math. Zeit.*. Problem: 3.7.

[148] Brin, M. G. & Thickstun, T. L. (1987). Open, irreducible 3-manifolds which are end 1-movable. *Topology 26*, 211–233. Problem: 3.7.

[149] Brin, M. G. & Thickstun, T. L. (1989). *3-Manifolds which are end 1-movable*, Volume 411 of *Mem. Amer. Math. Soc.* American Math. Soc. Problem: 3.8.

[150] Browder, W. (1961). Torsion in H-spaces. *Ann. of Math. 74*, 24–51. Problem: 5.28.

[151] Browder, W. (1972). Poincaré spaces, their normal fibrations and surgery. *Invent. Math. 17*, 191–202. Problem: 5.29.

[152] Brown, K. S. (1982). *Cohomology of groups*, Volume 87 of *Graduate Texts in Math.* Springer-Verlag. Problem: 5.29.

[153] Brown, M. (1960). A proof of the generalized Schoenflies theorem. *Bull. Amer. Math. Soc. 66*, 74–76. Problem: 4.88.

[154] Brunner, A. M., Mayland, Jr., E. J., & Simon, J. (1982). Knot groups in S^4 with nontrivial homology. *Pacific J. Math. 103*, 315–324. Problem: 4.29.

[155] Bryant, J., Ferry, S. C., Mio, W., & Weinberger, S. (1993). Topology of homology manifolds. *Bull. Amer. Math. Soc. 28*, 324–328. Problem: 4.69.

[156] Bryant, J., Ferry, S. C., Mio, W., & Weinberger, S. (1995). Topology of homology manifolds. To appear in *Ann. of Math.*. Problem: 4.69.

[157] Buchdahl, N., Kwasik, S., & Schultz, R. (1990). One fixed point actions on low-dimensional spheres. *Invent. Math. 102*, 633–662. Problem: 3.46.

[158] Bullock, D. (1994). Skein related links in 3-manifolds. *Topology Appl. 60*, 235–248. Problem: 1.92.

[159] Bullock, D. (1995a). An integral invariant of 3-manifolds derived from the Kauffman bracket. Preprint. Problem: 1.92.

[160] Bullock, D. (1995b). The $(2,\infty)$–skein module of the complement of a $(2,2p+1)$ torus knot. To appear in *J. Knot Theory Ramifications*. Problem: 1.92.

[161] Burde, G. (1971). On branched coverings of S^3. *Canad. J. Math 23*, 84–89. Problem: 1.11.

[162] Burde, G. & Zieschang, H. (1966). Eine Kennzeichnung der Torusknoten. *Math. Ann. 167*, 169–175. Problem: 1.13.

[163] Burde, G. & Zieschang, H. (1985). *Knots*. W. de Gruyter. Problem: 3.15.

[164] Burghelea, D. & Lashof, R. K. (1974). The homotopy type of the space of diffeomorphisms. II. *Trans. Amer. Math. Soc. 196*, 37–50. Problem: 4.126.

[165] Burns, R. G., Karrass, A., & Solitar, D. (1987). A note on groups with separable finitely generated subgroups. *Bull. Austral. Math. Soc. 36*, 153–160. Problem: 3.76.

[166] Buskirk, J. M. V. (1983). A class of negative-amphicheiral knots and their Alexander polynomials. *Rocky Mountain J. Math. 13*, 413–422. Problem: 1.66.

[167] Callahan, P. (1994). Symmetric surgery on asymmetric knots. Preprint. Problem: 1.23.

[168] Camacho, C. & Lins Neto, A. (1973). *Teoria geométrica das folheações*, Volume 9 of *Projeto Euclides*. IMPA. Translation to English in [169].

[169] Camacho, C. & Lins Neto, A. (1985). *Geometric Theory of Foliations*. Birkhäuser. Original in Portuguese in [168]. Problem: 3.53.

[170] Canary, R. D. (1993a). Ends of hyperbolic 3-manifolds. *J. Amer. Math. Soc. 6*, 1–35. Problem: 3.52.

[171] Canary, R. D. (1993b). Geometrically tame hyperbolic 3-manifolds. In R. Greene & S.-T. Yau (Eds.), *Differential Geometry*, Volume 54 part 3 of *Proc. Sympos. Pure Math.*, pp. 99–109. American Math. Soc. Problem: 3.52.

[172] Canary, R. D. (1994). Covering theorems for hyperbolic 3-manifolds. In K. Johannson (Ed.), *Low-Dimensional Topology, Knoxville, 1992*, pp. 21–30. International Press. Problems: 3.14, 3.52.

[173] Canary, R. D., Epstein, D. B. A., & Green, P. (1987). Notes on notes of Thurston. In D. B. A. Epstein (Ed.), *Analytic and Geometric Aspects of Hyperbolic Space*, Volume 111 of *London Math. Soc. Lect. Note Ser.*, pp. 3–92. Cambridge Univ. Press. Proceedings of the two symposia on hyperbolic geometry, Kleinian groups and 3-dimensional topology held at the Univ. of Warwick and the Univ. of Durham, 1984. Problem: 3.14.

[174] Canary, R. D. & Minsky, Y. (1995). On limits of tame hyperbolic 3-manifolds. To appear in *J. Differential Geom.*. Problem: 3.52.

[175] Candel, A. (1993). Uniformization of surface laminations. *Ann. Sci. École Norm. Sup. (4) 26*, 489–516. Problem: 3.56.

[176] Cannon, J. W. (1979). Shrinking cell-like decompositions of manifolds, codimension 3. *Ann. of Math. 110*, 83–112. Problem: 4.4.

[177] Cannon, J. W. & Thurston, W. P. (1995). Group invariant Peano curves. Preprint. Problem: 3.56.

[178] Cappell, S. E. & Shaneson, J. L. (1971). On four dimensional surgery and applications. *Comment. Math. Helv. 46*, 500–528. Problems: 4.11, 4.15.

[179] Cappell, S. E. & Shaneson, J. L. (1974). The codimension two placement problem and homology equivalent manifolds. *Ann. of Math. 99*, 277–348. Problem: 4.6.

[180] Cappell, S. E. & Shaneson, J. L. (1975). Invariants of 3-manifolds. *Bull. Amer. Math. Soc. 81*, 559–562. Problems: 1.25, 1.35.

[181] Cappell, S. E. & Shaneson, J. L. (1976). Some new four-manifolds. *Ann. of Math. 104*, 61–72. Problems: 4.13, 4.45, 4.85.

[182] Cappell, S. E. & Shaneson, J. L. (1980). Link cobordism. *Comment. Math. Helv. 55*, 20–49. Problem: 1.56.

[183] Cappell, S. E. & Shaneson, J. L. (1984). Linking numbers in branched covers. In C. M. Gordon & R. C. Kirby (Eds.), *Four-Manifold Theory (Durham, NH, 1982)*, Volume 35 of *Contemp. Math.*, pp. 165–179. American Math. Soc. Problems: 1.25, 1.26.

[184] Cappell, S. E. & Weinberger, S. (1988). Which H-spaces are manifolds. *Topology 27*, 377–386. Problem: 5.28.

[185] Carter, J. S. (1986). Surgery on Codimension One Immersions in $(n+1)$-space: Removing n-tuple points. *Trans. Amer. Math. Soc. 298*, 83–102. Problem: 4.61.

[186] Casson, A. J. & Gordon, C. M. (1978). On slice knots in dimension three. In R. J. Milgram (Ed.), *Algebraic and Geometric Topology (Stanford, 1976)*, Volume 32, part II of *Proc. Sympos. Pure Math.*, pp. 39–53. American Math. Soc.

[187] Casson, A. J. & Gordon, C. M. (1983). A loop theorem for duality spaces and fibered ribbon knots. *Invent. Math. 74*, 119–137. Problems: 1.43, 4.22.

[188] Casson, A. J. & Gordon, C. M. (1986). Cobordism of classical knots. In L. Guillou & A. Marin (Eds.), *A la Recherche de la Topologie Perdue*, Volume 62 of *Prog. Math.*, pp. 181–200. Birkhäuser. Problems: 4.5, 4.6, 4.33.

[189] Casson, A. J. & Harer, J. L. (1981). Some homology lens spaces which bound rational homology balls. *Pacific J. Math. 96*, 23–26. Problems: 4.2, 4.97.

[190] Casson, A. J. & Jungreis, D. (1994). Convergence groups and Seifert fibered 3-manifolds. *Invent. Math. 118*, 441–456. Problems: 3.5, 3.71.

[191] Cerf, J. (1968). *Sur les difféomorphismes de la sphère de dimension trois*, Volume 53 of *Lecture Notes in Math.* Springer-Verlag. Problem: 3.34.

[192] Charney, R. & Davis, M. W. (1995). The Euler characteristic of a non-positively curved, piecewise Euclidean manifold. To appear in *Pacific J. Math.*. Problem: 4.10.

[193] Cheeger, J. & Ebin, D. G. (1975). *Comparison Theorems in Riemannian Geometry.* North-HollandElsevier Science Pub. Co. Problem: 4.67.

[194] Cheeger, J. & Gromov, M. (1986). L_2-cohomology and group cohomology. *Topology 25*, 189–2216. Problem: 4.17.

[195] Cheeger, J. & Kister, J. M. (1970). Counting topological manifolds. *Topology 9*, 149–151. Problem: 4.87.

[196] Chen, W. (1995). Counterexamples to Knaster's Conjecture. Preprint. Problem: 5.24.

[197] Chen, Z. (1987). On the geography of surfaces: simply connected minimal surfaces with positive index. *Math. Ann. 277*, 141–164. Problems: 4.39, 4.91.

[198] Chern, S.-S. (1955). On curvature and characteristic classes of a riemannian manifold. *Abh. Math. Sem. Univ. Hamburg 20*, 117–126. Problem: 4.10.

[199] Chern, S.-S. (1971). Differential Geometry; its past and its future. In *Actes Congrès Intern. Math.*, Volume 1, pp. 41–53. Gauthier-Villars. Problem: 4.67.

[200] Chillingworth, D. R. J. (1967). Collapsing 3-dimensional convex polyhedra. *Math. Proc. Cambridge Philos. Soc. 63*, 354–357. Problem: 5.5.

[201] Chillingworth, D. R. J. (1972). Winding numbers on surfaces, II. *Math. Ann. 199*, 131–152. Problem: 2.1.

[202] Chinburg, T. C., Friedman, E. C., Jones, K. N., & Reid, A. W. (1995). The smallest volume arithmetic hyperbolic 3-manifold. In preparation. Problem: 3.60.

[203] Chinburg, T. C. & Friedman, R. D. (1986). The smallest arithmetic hyperbolic 3-orbifold. *Invent. Math. 86*, 507–527. Problem: 3.60.

[204] Chou, C. (1980). Elementary amenable groups. *Illinois J. Math. 24*, 396–407. Problem: 5.9.

[205] Clemens, C. H., Kollar, J., & Mori, S. (1988). *Higher Dimensional Complex Geometry*, Volume 166 of *Astérisque*. Soc. Mat. de France. Problem: 4.139.

[206] Clozel, L. (1987). On the cuspidal cohomology of arithmetic subgroups of $SL(2,n)$ and the first Betti number of arithmetic 3-manifolds. *Duke Math. J. 55*, 475–486. Problem: 3.2.

[207] Cochran, T. D. (1983). Ribbon knots in S^4. *J. London Math. Soc. 28*, 563–576. Problem: 1.105.

[208] Cochran, T. D. (1984a). 4-manifolds which imbed in $\mathbb{R}^6$ but not $\mathbb{R}^5$ and Seifert manifolds for fibered knots. *Invent. Math. 77*, 173–184. Problem: 4.63.

[209] Cochran, T. D. (1984b). Embedding 4-manifolds in S^5. *Topology 23*, 257–269. Problem: 4.63.

[210] Cochran, T. D. (1984c). On an invariant of link cobordism in dimension 4. *Topology Appl. 18*, 97–108. Problem: 1.56.

[211] Cochran, T. D. (1984d). Slice links in the 4-sphere. *Trans. Amer. Math. Soc. 285*, 389–401. Problem: 1.56.

[212] Cochran, T. D. (1985). A topological proof of Stalling's theorem on lower central series of groups. *Math. Proc. Cambridge Philos. Soc. 97*, 465–472. Problem: 3.79.

[213] Cochran, T. D. (1987). Link concordance invariants and homotopy theory. *Invent. Math. 90*, 635–645. Problems: 1.96, 3.79.

[214] Cochran, T. D. (1990). *Derivatives of links: Milnor's concordance invariants and Massey's products*, Volume 84, number 427 of *Mem. Amer. Math. Soc.* American Math. Soc. Problem: 1.96.

[215] Cochran, T. D. & Gompf, R. E. (1988). Applications of Donaldson's theorems to classical knot concordance, homology 3-spheres and property *P*. *Topology 27*, 495–512. Problem: 4.74.

[216] Cochran, T. D. & Orr, K. E. (1993). Not all links are concordant to boundary links. *Ann. of Math. 138*, 519–554. Problem: 3.79.

[217] Cochran, T. D. & Orr, K. E. (1995). Stability of the lower central series of compact 3-manifold groups. Preprint. Problem: 3.78.

[218] Cockroft, W. H. (1954). On 2-dimensional aspherical complexes. *Proc. London Math. Soc. 4*, 375–384. Problem: 5.4.

[219] Cohen, D. & Suciu, A. (1995). On Milnor fibrations of arrangements. *J. London Math. Soc. 51*, 105–119. Problem: 4.145.

[220] Cohen, M. M. (1973). *A Course in Simple Homotopy Theory.* Springer-Verlag. Problem: 3.1.

[221] Connell, E. H. (1963). Approximating stable homeomorphisms by piecewise linear ones. *Ann. of Math. 78*, 326–338. Problem: 4.44.

[222] Conner, P. E. & Raymond, F. (1972). Manifolds with few periodic homeomorphisms. In *Proc. 2nd Conf. on Compact Transformation Groups*, Volume 299 of *Lecture Notes in Math.*, pp. 1–74. Springer-Verlag. Problem: 3.40.

[223] Conner, P. E. & Raymond, F. (1977). Deforming homotopy equivalences to homeomorphisms in aspherical manifolds. *Bull. Amer. Math. Soc. 83*, 36–85. Problem: 5.11.

[224] Connor, A. (1969). *Splittable knots.* Ph. D. thesis, Univ. of Georgia. Problem: 1.15.

[225] Conway, J. H. (1969). An enumeration of knots and links. In J. Leech (Ed.), *Computational Problems in Abstract Algebra*, pp. 329–358. Pergamon Press. Problem: 1.59.

[226] Conway, J. H. & Gordon, C. M. (1983). Knots and links in spatial graphs. *J. Graph Theory 7*, 445–453. Problems: 5.8, 5.15.

[227] Cooper, D., Culler, M., Gillett, H., Long, D. D., & Shalen, P. B. (1994). Plane curves associated to character varieties of 3-manifolds. *Invent. Math. 118*, 47–84. Problem: 1.86.

[228] Coxeter, H. S. M. (1957). Factor groups of the braid group. In *Proc. Fourth Canadian Math. Congress, Banff*, pp. 95–122. Problems: 1.59, 1.92.

[229] Culler, M., Gordon, C. M., Luecke, J. E., & Shalen, P. B. (1987). Dehn surgery on knots. *Ann. of Math. 125*, 237–300. Problem: 1.78.

[230] Culler, M. & Shalen, P. B. (1992). Paradoxical decompositions, 2-generator Kleinian groups, and volumes of hyperbolic 3-manifolds. *J. Amer. Math. Soc. 5*, 231–288. Problem: 3.60.

[231] Culler, M. & Vogtmann, K. (1986). Moduli of graphs and automorphisms of free groups. *Invent. Math. 84*, 91–119. Problem: 5.19.

[232] Curtis, C. L., Freedman, M. H., Hsiang, W.-C., & Stong, R. (1996). A decomposition theorem for h-cobordant smooth simply-connected compact 4-manifolds. *Invent. Math. 123*, 343–348. Problem: 4.90.

[233] Daverman, R. J. (1986). *Decompositions of Manifolds*, Volume 124 of *Pure and Applied Mathematics*. Academic Press. Problems: 4.4, 4.73.

[234] Davis, J. (1994). The Borel/Novikov conjectures and stable diffeomorphisms of 4-manifolds. Preprint. Problem: 4.84.

[235] Davis, M. W. (1984). Coxeter groups and aspherical manifolds. In I. Madsen & R. Oliver (Eds.), *Algebraic Topology, Aarhus, 1982*, Volume 1051 of *Lecture Notes in Math.*, pp. 197–221. Springer-Verlag. Problem: 4.10.

[236] Davis, M. W. (1985). A hyperbolic 4-manifold. *Proc. Amer. Math. Soc. 93*, 325–328. Problem: 4.17.

[237] de la Harpe, P. & Valette, A. (1989). La propriété (T) de Kazhdan pour les groupes localement compacts. *Astérisque 175*, 158pp. Problem: 2.11.

[238] de Souza, P. N. R. (1985). Van der Blij type equalites for unimodular quadratic forms. Available at `http://math.berkeley.edu/~desouza/`. Problem: 4.19.

[239] Deligne, P. (1972). Les immeubles des groupes de tresses généralisés. *Invent. Math. 17*, 273–302. Problem: 4.146.

[240] Deligne, P. (1981). Le groupe fondamental du complément d'une courbe plane n'ayant que des points doubles ordinaires est abélian [d'apres W. Fulton]. In *Séminaire Bourbaki, Exposés 543–560, 1979/80*, Volume 842 of *Lecture Notes in Math.*, pp. 1–10. Springer-Verlag. Problem: 4.40.

[241] Delman, C. I. (1995). Essential lamination and surgery on 2-bridge knots. *Topology Appl. 63*, 201–221. Problem: 1.15.

[242] Ding, F. (1995a). Smooth structures on some open 4-manifolds. To appear in *Topology*. Problem: 4.87.

[243] Ding, F. (1995b). Uncountably many smooth structures on some 4-manifolds. Preprint. Problem: 4.87.

[244] do Carmo, M. P. (1979). *Geometria Riemanniana*, Volume 10 of *Projeto Euclides*. IMPA. English translation in [245].

[245] do Carmo, M. P. (1992). *Riemannian Geometry.* Birkhäuser. Translation of [244]. Problem: 4.67.

[246] Doll, H. (1992). A generalized bridge number for links in 3-manifolds. *Math. Ann. 294*, 701–717. Problem: 1.73.

[247] Donaldson, S. K. (1983). An application of gauge theory to 4-dimensional topology. *J. Differential Geom. 18*, 279–315. Problems: 4.1, 4.6, 4.19, 4.37.

[248] Donaldson, S. K. (1986). Connections, cohomology and the intersection forms of 4-manifolds. *J. Differential Geom. 24*, 275–341. Problems: 4.1, 4.26, 4.92, 4.100.

[249] Donaldson, S. K. (1987a). Irrationality and the h-cobordism conjecture. *J. Differential Geom. 26*, 141–168. Problem: 4.45.

[250] Donaldson, S. K. (1987b). The orientation of Yang–Mills moduli spaces and 4-manifold topology. *J. Differential Geom. 26*, 397–428. Problems: 1.46, 4.1, 4.3, 4.19.

[251] Donaldson, S. K. (1990). Polynomial invariants for smooth four-manifolds. *Topology 29*, 257–315. Problems: 4.11, 4.37, 4.91, 4.130.

[252] Donaldson, S. K. & Kronheimer, P. B. (1990). *Geometry of Four-Manifolds.* Oxford Univ. Press. Problem: 4.45.

[253] Donaldson, S. K. & Sullivan, D. P. (1989). Quasiconformal 4-manifolds. *Acta Math. 163*, 181–252. Problem: 4.44.

[254] DuBois, P. & Michel, F. (1994). The integral Seifert form does not determine the topology of plane curve germs. *J. Algebraic Geom. 3*, 1–38. Problem: 3.29.

[255] Dunbar, W. D. (1988). Hierarchies for 3-orbifolds. *Topology Appl. 29*, 267–283. Problem: 3.46.

[256] Dunbar, W. D. & Meyerhoff, G. R. (1994). Volumes of hyperbolic 3-orbifolds. *Indiana Univ. Math. J. 43*, 611–637. Problem: 3.60.

[257] Dunwoody, M. J. (1976). The homotopy type of a two-dimensional complex. *Bull. London Math. Soc. 8*, 282–285. Problem: 5.1.

[258] Dunwoody, M. J. (1985). Accessibility of finitely presented groups. *Invent. Math. 81*, 449–457. Problems: 1.51, 5.10.

[259] Dunwoody, M. J. (1993). An inaccessible group. In G. A. Niblo & M. A. Roller (Eds.), *Geometric Group Theory I (Sussex, 1991)*, Volume 181 of *London Math. Soc. Lect. Note Ser.*, pp. 75–78. Cambridge Univ. Press. Problem: 5.10.

[260] Dunwoody, M. J. (1995). Folding sequences. Preprint, Univ. of Southampton. Problem: 5.10.

[261] Durfee, A. H. (1974). Fibered knots and algebraic singularities. *Topology 13*, 47–59. Problem: 3.29.

[262] Durfee, A. H. (1975). Knot invariants of singularities. In R. Hartshorne (Ed.), *Algebraic Geometry*, Volume 29 of *Proc. Sympos. Pure Math.*, pp. 441–448. American Math. Soc. Proc. of a Conf. in Humboldt State Univ, Arcata - 1974. Problem: 3.27.

[263] Durfee, A. H. (1978). The signature of smoothings of complex surface singularities. *Math. Ann. 232*, 85–98. Problem: 3.31.

[264] Earle, C. & Sipe, P. (1995). Invariant Teichmüller disks and the monodromy of holomorphic families. Preprint. Problem: 2.7.

[265] Eccles, P. J. (1980). Multiple points of codimension one immersions of oriented manifolds. *Math. Proc. Cambridge Philos. Soc. 87*, 213–220. Problem: 4.61.

[266] Eckmann, B. (1993). Manifolds of even dimension with amenable fundamental group. Preprint, ETH Zürich. Problem: 1.54.

[267] Eckmann, B. & Linnell, P. (1983). Poincaré duality groups of dimension 2, II. *Comment. Math. Helv. 58*, 111–114. Problem: 3.77.

[268] Eckmann, B. & Muller, H. (1980). Poincaré duality groups of dimension 2. *Comment. Math. Helv. 55*, 510–520. Problem: 3.77.

[269] Edelman, P. H. & Reiner, V. (1995). Not all free arrangements are $K(\pi, 1)$. *Bull. Amer. Math. Soc. 32*, 61–65. Problem: 4.146.

[270] Edmonds, A. (1988). Involutions on odd four-manifolds. *Topology Appl. 30*, 43–49. Problem: 4.19.

[271] Edmonds, A. & Livingston, C. (1983). Group actions on fibered 3-manifolds. *Comment. Math. Helv. 58*, 529–542. Problem: 1.28.

[272] Edmonds, A. & Livingston, C. (1984). Symmetric representations of knot groups. *Topology Appl. 18*, 281–312. Problem: 1.24.

[273] Edwards, R. D. (1975). The double suspension of a certain homology 3-sphere is S^5. *Notices Amer. Math. Soc. 22*, A334. Abstract 75T–G33. Problem: 4.4.

[274] Edwards, R. D. (1980). The topology of manifolds and cell-like maps. In *Proc. International Congress Mathematicians, Helsinki 1980*, pp. 111–127. Problem: 4.69.

[275] Eilenberg, S. (1934). Sur les transformations périodiques de la surface de sphère. *Fund. Math. 22*, 28–41. Problem: 2.19.

[276] El-Rifai, E. A. & Morton, H. R. (1994). Algorithms for positive braids. *Quart. J. Math. Oxford Ser. (2) 45*, 479–497. Problem: 1.84.

[277] Eliashberg, Y. (1990a). Filling by holomorphic discs and its applications. In S. K. Donaldson & C. B. Thomas (Eds.), *Geometry of Low Dimensional Manifolds, 2 (Durham, 1989)*, Volume 151 of *London Math. Soc. Lect. Note Ser.*, pp. 45–68. Cambridge Univ. Press. Problem: 4.141.

[278] Eliashberg, Y. (1990b). Topological characterization of Stein manifolds of dimension > 2. *Internat. J. Math. 1*, 29–46. Problem: 4.78.

[279] Eliashberg, Y. (1993). Legendrian and transversal knots in tight contact 3-manifolds. In L. R. Goldberg & A. V. Philips (Eds.), *Topological Methods in Modern Mathematics - Proc. of a Symposium in Honor of John Milnor's 60th Birthday*, pp. 171–194. Publish or Perish. Problems: 1.98, 1.99.

[280] Eliashberg, Y. (1995). Topology of 2-knots in $\mathbb{R}^4$ and symplectic geometry. In H. Hofer, C. H. Taubes, A. Weinstein, & E. Zehnder (Eds.), *The Floer Memorial Volume*, Volume 133 of *Progress in Mathematics*, pp. 335–353. Birkhäuser. Problem: 1.100.

[281] Eliashberg, Y. & Gromov, M. (1991). Convex symplectic manifolds. In E. Bedford (Ed.), *Several Complex Variables and Complex Geometry*, Volume 52, Part 2 of *Proc. Sympos. Pure Math.*, pp. 135–162. American Math. Soc. Problem: 4.141.

[282] Eliashberg, Y. & Polterovich, L. (1995). The problem of lagrangian knots. In W. H. Kazez (Ed.), *Geometric Topology, Proc. of the 1993 Georgia International Topology Conference.* International Press. Problem: 4.111.

[283] Epstein, D. B. A. (1961). Projective planes in 3-manifolds. *Proc. London Math. Soc. 11*, 469–484. Problem: 3.5.

[284] Ernst, C. & Sumners, D. W. (1987). The growth of the number of prime knots. *Math. Proc. Cambridge Philos. Soc. 102*, 303–315. Problem: 1.63.

[285] Eudave-Muñoz, M. (1992). Band sums of links which yield composite links. The cabling conjecture for strongly invertible knots. *Trans. Amer. Math. Soc. 330*, 463–501. Problem: 1.79.

[286] Eudave-Muñoz, M. (1994). Non-hyperbolic manifolds obtained by Dehn surgery on hyperbolic knots. Preprint, Instituto de Matemáticas, UNAM. Problem: 1.77.

[287] Evans, B. & Jaco, W. H. (1977). Compactifying covering spaces. Preprint. Problem: 3.8.

[288] Falk, M. (1993). Homotopy types of line arrangements. *Invent. Math. 111*, 139–150. Problem: 4.144.

[289] Falk, M. (1995). $K(\pi, 1)$ arrangements. *Topology 34*, 141–154. Problem: 4.146.

[290] Farrell, F. T. & Hsiang, W.-C. (1981a). On Novikov's conjecture for nonpositively curved manifolds, I. *Ann. of Math. 113*, 199–209. Problem: 4.83.

[291] Farrell, F. T. & Hsiang, W.-C. (1981b). The Whitehead group of poly-(finite or cyclic) groups. *J. London Math. Soc. 24*, 308–324. Problem: 3.32.

[292] Farrell, F. T. & Jones, L. E. (1986). K-theory and dynamics, I. *Ann. of Math. 124*, 531–569. Problem: 3.32.

[293] Farrell, F. T. & Jones, L. E. (1987). Implications of the geometrization conjecture for the algebraic k-theory of 3-manifolds. In C. McCrory & T. Shifrin (Eds.), *Geometry and Topology: Varieties, Manifolds and Knots (Georgia, 1985)*, Volume 105 of *Lect. Notes in Pure and Appl. Math.*, pp. 109–113. Marcel Dekker. Problem: 3.32.

[294] Farrell, F. T. & Jones, L. E. (1989). A topological analogue of Mostow's rigidity theorem. *J. Amer. Math. Soc. 2*, 257–370. Problems: 4.83, 5.29.

[295] Fathi, A. & Herman, M. R. (1977). Existence de difféomorphisms minimax. In *Dynamical Systems I*, Volume 49 of *Astérisque*, pp. 37–59. Soc. Mat. de France. Problem: 3.111.

[296] Fathi, A., Laudenbach, F., & Poénaru, V. (1979). *Travaux de Thurston sur les surfaces*, Volume 66 and 67 of *Astérisque*. Soc. Mat. de France. Problem: 2.7.

[297] Feighn, M. (1989). Actions of finite groups on homotopy 3-spheres. *Trans. Amer. Math. Soc. 284*, 141–51. Problem: 3.46.

[298] Fenley, S. (1992). Asymptotic properties of depth one foliations in hyperbolic 3-manifolds. *J. Differential Geom. 36*, 269–313. Problem: 3.56.

[299] Fenn, R. A. (1973). Some generalizations of the Borsuk–Ulam theorem and applications to realizing homotopy classes by embedded spheres. *Math. Proc. Cambridge Philos. Soc. 74*, 251–256. Problem: 4.27.

[300] Fenn, R. A. & Rourke, C. P. (1979). On Kirby's calculus of links. *Topology 18*, 1–15. Problem: 3.101.

[301] Ferry, S. C. (1977). The homeomorphism group of a compact Hilbert cube manifold is an ANR. *Ann. of Math. 196*, 101–118. Problem: 5.27.

[302] Ferry, S. C. & Weinberger, S. (1991). Curvature, tangentiality and controlled topology. *Invent. Math. 105*, 401–414. Problem: 4.83.

[303] Feustel, C. D. (1976). On the torus theorem and its applications. *Trans. Amer. Math. Soc. 217*, 1–43. Problem: 1.13.

[304] Feustel, C. D. & Whitten, W. (1978). Groups and complements of knots. *Canad. J. Math 30*, 1284–1295. Problem: 1.13.

[305] Finashin, S. M., Kreck, M., & Viro, O. Y. (1987). Exotic knottings of surfaces in the 4-sphere. *Bull. Amer. Math. Soc. 17*, 287–290. Problem: 4.58.

[306] Finashin, S. M., Kreck, M., & Viro, O. Y. (1988). Non-diffeomorphic but homeomorphic knottings in the 4-sphere. In O. Y. Viro (Ed.), *Topology and Geometry: Rohlin Seminar*, Volume 1346 of *Lecture Notes in Math.*, pp. 157–198. Springer-Verlag. Problem: 4.58.

[307] Fintushel, R. & Pao, P. (1977). Identification of certain 4-manifolds with group action. *Proc. Amer. Math. Soc. 67*, 344–350. Problem: 4.15.

[308] Fintushel, R. & Stern, R. J. (1981). An exotic free involution on S^4. *Ann. of Math. 113*, 357–365. Problems: 4.13, 4.45, 4.55, 4.85.

[309] Fintushel, R. & Stern, R. J. (1984). Another construction of an exotic $S^1 \tilde{\times} S^3 \# S^2 \times S^2$. In C. M. Gordon & R. C. Kirby (Eds.), *Four-Manifold Theory (Durham, NH, 1982)*, Volume 35 of *Contemp. Math.*, pp. 269–275. American Math. Soc. Problems: 4.45, 4.85.

[310] Fintushel, R. & Stern, R. J. (1985). Pseudofree orbifolds. *Ann. of Math. 122*, 335–364. Problem: 1.37.

[311] Fintushel, R. & Stern, R. J. (1987). Rational homology cobordisms of spherical space forms. *Topology 26*, 385–393. Problems: 3.20, 4.5.

[312] Fintushel, R. & Stern, R. J. (1990). Instanton homology of Seifert fibred homology three spheres. *J. London Math. Soc. 61*, 109–137. Problems: 1.37, 4.5.

[313] Fintushel, R. & Stern, R. J. (1993). 2-torsion instanton invariants. *J. Amer. Math. Soc. 6*, 299–339. Problem: 4.132.

[314] Fintushel, R. & Stern, R. J. (1994a). Rational blowdowns of smooth 4-manifolds. Preprint. Problems: 4.11, 4.118, 4.128.

[315] Fintushel, R. & Stern, R. J. (1994b). Surgery in cusp neighborhoods and the geography of irreducible 4-manifolds. *Invent. Math. 117*, 455–523. Problem: 4.130.

[316] Fintushel, R. & Stern, R. J. (1995a). Donaldson invariants of 4-manifolds with simple type. To appear in *J. Differential Geom.*. Problems: 4.36, 4.128.

[317] Fintushel, R. & Stern, R. J. (1995b). Immersed spheres in 4-manifolds and the immersed Thom Conjecture. *Turkish J. Math. 19*, 27–40. Problem: 4.36.

[318] Flapan, E. L. (1985). Necessary and sufficient conditions for certain homology 3-spheres to have smooth $\mathbb{Z}/p$-actions. *Pacific J. Math. 117*, 255–265. Problem: 3.24.

[319] Flapan, E. L., Liang, C., & Mislow, K. (1995). Amphicheiral links with odd crossing number. Preprint. Problem: 1.66.

[320] Floyd, E. E. (1955). Real valued mappings of spheres. *Proc. Amer. Math. Soc. 6*, 957–959. Problem: 5.24.

[321] Fox, R. H. (1948). On the imbedding of polyhedra in 3-space. *Ann. of Math. 49*, 462–470. Problem: 3.11.

[322] Fox, R. H. (1958). On knots whose points are fixed under a periodic transformation of the 3-sphere. *Osaka Math. J. 10*, 31–35. Problem: 3.38.

[323] Fox, R. H. (1962a). A quick trip through knot theory. In M. K. Fort, Jr. (Ed.), *Topology of 3-Manifolds and Related Topics (Georgia, 1961)*, pp. 120–167. Prentice-Hall. Problems: 1.20, 1.49.

[324] Fox, R. H. (1962b). Construction of simply connected 3-manifolds. In M. K. Fort, Jr. (Ed.), *Topology of 3-Manifolds and Related Topics (Georgia, 1961)*, pp. 213–216. Prentice-Hall. Problem: 1.11.

[325] Fox, R. H. (1967). Two theorems about periodic transformations of the 3-sphere. *Michigan Math. J. 14*, 331–334. Problem: 3.38.

[326] Francis, G. K. (1970). Extensions to the disk of properly nested plane immersions of the circle. *Michigan Math. J. 17*, 377–383. Problem: 3.19.

[327] Fraser, M. (1994). *Classification of Legendrian knots in tight contact manifolds.* Ph. D. thesis, Stanford Univ. Problem: 1.99.

[328] Freedman, M. H. (1979). A fake $S^3 \times \mathbb{R}$. *Ann. of Math. 110*, 177–201. Problems: 4.8, 4.85.

[329] Freedman, M. H. (1982). The topology of four-dimensional manifolds. *J. Differential Geom. 17*, 357–453. Problems: 1.55, 3.1, 3.20, 4.1, 4.7, 4.8, 4.11, 4.12, 4.16, 4.33, 4.100, 4.105.

[330] Freedman, M. H. (1984). The disk theorem for four dimensional manifolds. In *Proc. International Congress of Mathematicians, Warsaw 1983*, pp. 647–663. Polish Scientific Publishers, Warsaw. Problems: 1.52, 3.78, 4.6, 4.74.

[331] Freedman, M. H. (1986). A geometric reformulation of four dimensional surgery. *Topology Appl. 24*, 133–141. Problem: 3.71.

[332] Freedman, M. H., Hass, J., & Scott, G. P. (1983). Least area incompressible surfaces in 3-manifolds. *Invent. Math. 71*, 609–642. Problem: 3.67.

[333] Freedman, M. H. & He, Z.-X. (1991). Divergence-free fields: energy and asymptotic crossing number. *Ann. of Math. 134*, 189–229. Problem: 1.68.

[334] Freedman, M. H. & Kirby, R. C. (1978). A geometric proof of Rochlin's theorem. In R. J. Milgram (Ed.), *Algebraic and Geometric Topology (Stanford, 1976)*, Volume 32, part II of *Proc. Sympos. Pure Math.*, pp. 85–97. American Math. Soc. Problem: 4.1.

[335] Freedman, M. H. & Lin, X.-S. (1989). On the (A, B)-slice problem. *Topology 28*, 91–110. Problem: 4.46.

[336] Freedman, M. H. & Quinn, F. S. (1990). *Topology of 4-Manifolds.* Princeton Univ. Press. Problems: 1.36, 4.6, 4.7, 4.15, 4.31, 4.46, 4.52, 4.74.

[337] Freedman, M. H. & Taylor, L. R. (1977). Λ-Splitting 4-manifolds. *Topology 16*, 181–184. Problem: 4.116.

[338] Freedman, M. H. & Taylor, L. R. (1986). A universal smoothing of four-space. *J. Differential Geom. 24*, 69–78. Problem: 4.80.

[339] Freedman, M. H. & Teichner, P. (1995). 4-Manifold topology I: Subexponential groups. To appear in *Invent. Math.*. Problems: 4.6, 5.9.

[340] Freedman, M. H. & Teichner, P. (1996). 4-Manifold topology II: Dwyer's filtration and surgery kernels. To appear in *Invent. Math.*. Problem: 5.9.

[341] Freyd, P. J., Yetter, D. N., Hoste, J., Lickorish, W. B. R., Millet, K. C., & Ocneanu, A. (1985). A new polynomial invariant of knots and links. *Bull. Amer. Math. Soc. 12*, 239–246. Problem: 1.90.

[342] Fried, D. (1983). Transitive Anosov flows and pseudo-Anosov maps. *Topology 22*, 299–304. Problems: 3.53, 3.54.

[343] Friedman, J. L. & Witt, D. M. (1986). Homotopy is not isotopy for homeomorphisms of 3-manifolds. *Topology 25*, 35–44. Problem: 3.95.

[344] Friedman, R. D. (1995). Vector bundles and SO(3)-invariants for elliptic surfaces. *J. Amer. Math. Soc. 8*, 29–139. Problem: 4.11.

[345] Friedman, R. D. & Morgan, J. W. (1994). *Smooth Four-manifolds and Complex Surfaces*, Volume 27 of *Ergeb. Math. Grenzgeb. (3)*. Springer-Verlag. Problems: 4.11, 4.125.

[346] Friedman, R. D. & Qin, Z. (1995). On complex surfaces diffeomorphic to rational surfaces. *Invent. Math. 120*, 81–117. Problem: 4.96.

[347] Frohman, C. (1993). Unitary representations of knot groups. *Topology 32*, 121–144. Problem: 3.107.

[348] Frohman, C. & Meeks, III, W. H. (1990). The topology of complete one-ended minimal surfaces and Heegaard surfaces in $\mathbb{R}^3$. *Bull. Amer. Math. Soc. 23*, 417–421. Problem: 3.44.

[349] Frohman, C. & Nicas, A. J. (1994). An intersection homology invariant for knots in a rational homology 3-sphere. *Topology 33*, 123–158. Problem: 3.107.

[350] Fulton, W. (1980). On the fundamental group of the complement of a node curve. *Ann. of Math. 111*, 407–409. Problem: 4.40.

[351] Fulton, W. & MacPherson, R. D. (1994). A compactification of configuration space. *Ann. of Math. 139*, 183–225. Problem: 5.20.

[352] Furuta, M. (1995). Monopole equation and the 11/8-conjecture. Preprint RIMS, Kyoto. Problem: 4.92.

[353] Gabai, D. (1983). The Murasugi sum is a natural geometric operation. In S. J. Lomonaco, Jr. (Ed.), *Low Dimensional Topology, (San Francisco, 1981)*, Volume 20 of *Contemp. Math.*, pp. 131–144. Problem: 1.83.

[354] Gabai, D. (1986). On 3-manifolds finitely covered by surface bundles. In D. B. A. Epstein (Ed.), *Low-dimensional Topology and Kleinian Groups*, Volume 112 of *London Math. Soc. Lect. Note Ser.*, pp. 145–156. Cambridge Univ. Press. Problem: 3.51.

[355] Gabai, D. (1987a). Foliations and the topology of 3-manifolds, III. *J. Differential Geom. 26*, 479–536. Problems: 1.16, 1.19, 1.41, 1.82, 5.7.

[356] Gabai, D. (1987b). Genus is superadditive under band connected sum. *Topology 26*, 209–210. Problem: 1.1.

[357] Gabai, D. (1989). Surgery on knots in solid tori. *Topology 28*, 1–6. Problems: 1.15, 1.81.

[358] Gabai, D. (1992). Convergence groups are fuchsian groups. *Ann. of Math. 136*, 447–510. Problems: 3.5, 3.71.

[359] Gabai, D. (1994a). Homotopy hyperbolic 3-manifolds are virtually hyperbolic. *J. Amer. Math. Soc. 7*, 193–198. Problems: 3.47, 3.95.

[360] Gabai, D. (1994b). On the geometric and topological rigidity of hyperbolic 3-manifolds. *Bull. Amer. Math. Soc. 31*, 228–232. Problems: 3.40, 3.47, 3.49, 3.71, 4.83.

[361] Gabai, D. (1995a). On the topological and geometric rigidity of hyperbolic 3-manifolds. To appear in *J. Amer. Math. Soc.*. Problems: 3.3, 3.47, 3.49, 3.95.

[362] Gabai, D. (1995b). Valentin Poénaru's program for the Poincaré Conjecture. In S.-T. Yau (Ed.), *Geometry, Topology, and Physics for Raoul Bott*, pp. 139–166. International Press. Problem: 4.88.

[363] Gabai, D. & Kazez, W. H. (1993). 3-Manifolds with essential laminations are covered by solid tori. *J. London Math. Soc. 47*, 557–576. Problem: 3.8.

[364] Gabai, D. & Oertel, U. (1989). Essential laminations in 3-manifolds. *Ann. of Math. 130*, 41–73. Problems: 1.15, 3.7, 3.53.

[365] Galewski, D. E. & Stern, R. J. (1980). Classification of simplicial triangulations of topological manifolds. *Ann. of Math. 111*, 1–34. Problem: 4.4.

[366] Garoufalidis, S. (1995). Applications of TQFT in low dimensional topology. To appear in *Topology*. Problem: 3.108.

[367] Garside, F. A. (1969). The braid group and other groups. *Quart. J. Math. Oxford Ser. (2) 20*, 235–254. Problem: 1.84.

[368] Gehring, F. W. & Martin, G. J. (1991). Inequalities for Möbius transformations and discrete groups. *J. Reine Angew. Math. 418*, 31–76. Problem: 3.60.

[369] Gehring, F. W. & Martin, G. J. (1994). On the minimal volume hyperbolic 3-orbifold. *Math. Res. Lett. 1*, 107–114. Problem: 3.60.

[370] Geoghegan, R. (1972). On spaces of homeomorphisms, embeddings and functions-I. *Topology 11*, 159–177. Problem: 5.27.

[371] Geroch, R. (1976). Positive sectional curvatures does not imply positive Gauss-Bonnet integrand. *Proc. Amer. Math. Soc. 54*, 267–270. Problem: 4.10.

[372] Gersten, S. M. & Stallings, J. R. (1991). Casson's idea about 3-manifolds whose universal cover is $\mathbb{R}^3$. *Internat. J. Algebra Comput. 1*, 395–406. Problem: 3.7.

[373] Gerstenhaber, M. & Rothaus, O. S. (1962). The solution of sets of equations in groups. *Proc. Nat. Acad. Sci. U.S.A. 48*, 1531–1533. Problems: 1.1, 4.18, 5.7.

[374] Ghys, E. (1984). Flots d'Anosov sur les 3-variétés fibrées en cercles. *Ergod. Th. & Dynam. Sys. 4*, 67–80. Problem: 3.53.

[375] Gieseking, H. (1912). *Analytische Untersuchungen über topologische Gruppen.* L. Wiegand. Münster's inaugural dissertation. Problem: 3.60.

[376] Giffen, C. H. (1964). *Fibered knots and periodic transformations.* Ph. D. thesis, Princeton Univ. Problem: 3.38.

[377] Giffen, C. H. (1967). Cyclic branched coverings of doubled curves in 3-manifolds. *Illinois J. Math. 11*, 644–646. Problem: 3.38.

[378] Giffen, C. H. (1977). F-isotopy implies I-equivalence. Preprint. Problem: 4.21.

[379] Giffen, C. H. (1979). Link concordance implies link homotopy. *Math. Scand. 45*, 243–254. Problem: 1.44.

[380] Gillman, D. & Rolfsen, D. P. (1983). The Zeeman conjecture for standard spines is equivalent to the Poincaré Conjecture. *Topology 22*, 315–323. Problem: 5.2.

[381] Gilmer, P. M. (1983). Slice knots in S^3. *Quart. J. Math. Oxford Ser. (2) 34*, 305–322. Problem: 1.61.

[382] Gilmer, P. M. (1993). Classical knot and link concordance. *Comment. Math. Helv. 68*, 1–19. Problem: 1.52.

[383] Gilmer, P. M. & Livingston, C. (1983). Embedding 3-manifolds in S^4. *Topology 22*, 241–252. Problem: 3.20.

[384] Gilmer, P. M. & Livingston, C. (1992). Discriminants of Casson–Gordon invariants. *Math. Proc. Cambridge Philos. Soc. 112*, 127–139. Problem: 1.52.

[385] Giroux, E. (1994). Une structure de contact, même tendue, est plus ou moins tordue. *Ann. Sci. École Norm. Sup. (4) 27*, 694–705. Problem: 4.142.

[386] Glimm, J. (1960). Two cartesian products which are euclidean spaces. *Bull. Soc. Math. France 88*, 131–135. Problem: 3.1.

[387] Gluck, H. R. (1962). The embedding of two-spheres in the four-sphere. *Trans. Amer. Math. Soc. 104*, 308–333. Problem: 4.24.

[388] Goldsmith, D. L. (1979). Concordance implies homotopy for classical links in M^3. *Comment. Math. Helv. 54*, 347–355. Problem: 1.44.

[389] Gompf, R. E. (1985). An infinite set of exotic R^4's. *J. Differential Geom. 21*, 283–300. Problem: 4.42.

[390] Gompf, R. E. (1986). An exotic orientable 4-manifold. *Math. Ann. 274*, 177–180. Problem: 4.85.

[391] Gompf, R. E. (1988). On sums of algebraic surfaces. *Invent. Math. 94*, 171–174. Problem: 4.103.

[392] Gompf, R. E. (1991a). Killing the Akbulut-Kirby 4-sphere, with relevance to the Andrews-Curtis and Schoenflies problems. *Topology 30*, 97–115. Problems: 4.13, 5.2.

[393] Gompf, R. E. (1991b). Nuclei of elliptic surfaces. *Topology 30*, 479–511. Problem: 4.98.

[394] Gompf, R. E. (1991c). Sums of elliptic surfaces. *J. Differential Geom. 34*, 93–114. Problem: 4.103.

[395] Gompf, R. E. (1993). An exotic menagerie. *J. Differential Geom. 37*, 199–223. Problems: 1.102, 4.79, 4.80, 4.87.

[396] Gompf, R. E. (1995). A new construction of symplectic manifolds. To appear in *Ann. of Math.*. Problems: 4.64, 4.118.

[397] Gompf, R. E. & Mrowka, T. S. (1993). Irreducible 4-manifolds need not be complex. *Ann. of Math. 138*, 61–111. Problems: 4.50, 4.130.

[398] Gonzalez-Acuña, E. (1970). Dehn's construction on knots. *Bol. Soc. Mat. Mexicana 15*, 58–79. Problem: 1.15.

[399] Goodman, S. (1983). Dehn surgery and Anosov flows. In J. Palis, Jr. (Ed.), *Geometric Dynamics*, Volume 1007 of *Lecture Notes in Math.*, pp. 300–307. Springer-Verlag. Proc. of an Internat. Symp. held at IMPA, Rio de Janeiro, 1981. Problems: 3.53, 3.54.

[400] Goodrick, R. E. (1968). Non-simplicially collapsible triangulations of I^n. *Math. Proc. Cambridge Philos. Soc. 64*, 31–36. Problem: 5.5.

[401] Gordon, C. M. (1977). Uncountably many stably trivial strings in codimension two. *Quart. J. Math. Oxford Ser. (2) 28*, 369–380. Problems: 1.21, 3.38.

[402] Gordon, C. M. (1981a). Homology of groups of surfaces in the 4-sphere. *Math. Proc. Cambridge Philos. Soc. 89*, 113–117. Problem: 4.29.

[403] Gordon, C. M. (1981b). Ribbon concordance of knots in the 3-sphere. *Math. Ann. 257*, 157–170. Problem: 1.100.

[404] Gordon, C. M. & Luecke, J. E. (1987). Only integral Dehn surgeries can yield reducible manifolds. *Math. Proc. Cambridge Philos. Soc. 102*, 97–101. Problems: 1.77, 1.79, 3.102.

[405] Gordon, C. M. & Luecke, J. E. (1989). Knots are determined by their complements. *J. Amer. Math. Soc. 2*, 371–415. Problems: 1.13, 1.15, 1.81, 3.6.

[406] Gordon, C. M. & Luecke, J. E. (1995). Dehn surgery on knots creating essential tori, I & II. Preprint. Problem: 1.77.

[407] Gordon, C. M. & Sumners, D. W. (1975). Knotted ball pairs whose product with an interval is unknotted. *Math. Ann. 217*, 47–52. Problem: 1.3.

[408] Grasse, P. J. M. (1989). Finite presentation of mapping class groups of certain 3-manifolds. *Topology Appl. 32*, 295–305. Problem: 3.49.

[409] Grayson, M. A. (1989). Shortening embedded curves. *Ann. of Math. 129*, 71–112. Problem: 3.68.

[410] Grigorchuk, R. I. (1984). Степени роста конечно порожденных групп и теория инвариантных средних. *Изв. Акад. Наук СССР 48*, 939–985. English translation in [411].

[411] Grigorchuk, R. I. (1985). Degrees of growth of finitely generated groups, and the theory of invariant means. *Math. USSR-Izv. 25*, 259–300. Original in Russian in [410]. Problem: 5.9.

[412] Gromov, M. (1975). Three remarks on geodesic dynamics and fundamental group. Preprint. Problem: 2.6.

[413] Gromov, M. (1982). Volume and bounded cohomology. *Inst. Hautes Études Sci. Publ. Math. 56*, 5–99. Problem: 3.16.

[414] Gromov, M. (1985). Pseudoholomorphic curves in symplectic manifolds. *Invent. Math. 82*, 307–347. Problems: 4.23, 4.141.

[415] Gromov, M. (1987). Hyperbolic groups. In S. M. Gersten (Ed.), *Essays on Group Theory*, Volume 8 of *Math. Sci. Res. Inst. Publ.* Springer-Verlag. Problem: 3.74.

[416] Gromov, M. (1989). Sur le groupe fondamental d'une variété kählerienne. *C.R. Acad. Sci. Paris Sér. I Math. 308*, 67–70. Problem: 4.37.

[417] Gromov, M. (1993). Asymptotic invariants of infinite groups. In G. A. Niblo & M. A. Roller (Eds.), *Geometric Group Theory II (Sussex, 1991)*, Volume 182 of *London Math. Soc. Lect. Note Ser.*, pp. 1–295. Cambridge Univ. Press. Problem: 3.18.

[418] Grossman, E. K. (1974). On the residual finiteness of certain mapping class groups. *J. London Math. Soc. 9*, 160–164. Problems: 2.2, 2.11.

[419] Gudkov, D. A. (1974a). Топология вещественных проективных алгебраических многообразий. *Успехи Мат. Наук 29(4)*, 3–79. English translation in [420]. Problem: 5.6.

[420] Gudkov, D. A. (1974b). The topology of real projective algebraic varieties. *Russian Math. Surveys 29*, 1–79. Russian original in [419].

[421] Guilbault, C. R. (1995). Some compact, contractible manifolds containing disjoint spines. *Topology 34*, 99–108. Problem: 4.73.

[422] Guillou, L. & Marin, A. (1986a). *A la Recherche de la Topologie Perdue*, Volume 62 of *Prog. Math.* Birkhäuser. Problem: 1.39.

[423] Guillou, L. & Marin, A. (1986c). Une extension d'un théorème de Rohlin sur la signature. In L. Guillou & A. Marin (Eds.), *A la Recherche de la Topologie Perdue*, Volume 62 of *Prog. Math.*, pp. 97–118. Birkhäuser. Problem: 4.6.

[424] Gusarov, M. N. (1994). n-equivalent knots and invariants of finite degree. *Adv. Soviet Math. 18*, 173–192. Topology of manifolds and varieties. Problem: 1.92.

[425] Gutierrez, M. A. (1973). Unlinking up to cobordism. *Bull. Amer. Math. Soc. 79*, 1299–1302. Problem: 1.56.

[426] Haefliger, A. & Hirsch, M. (1963). On the existence and classification of differentiable embeddings. *Topology 2*, 129–135. Problem: 4.19.

[427] Hagiwara, Y. (1994). Reidemeister–Singer distance for unknotting tunnels of a knot. *Kobe J. Math. 11*, 89–100. Problem: 3.89.

[428] Haken, W. (1961). Theorie der Normalflächen. *Acta Math. 105*, 245–375. Problems: 1.34, 3.17, 3.82.

[429] Haken, W. (1962). Über das Homöomorphieproblem der 3-Mannigfaltigkeiten I. *Math. Zeit. 80*, 89–120. Problem: 3.82.

[430] Haken, W. (1968). Some results on surfaces in 3-manifolds. In P. J. Hilton (Ed.), *Studies in Modern Topology*, Volume 5 of *MAA Studies in Mathematics*, pp. 39–98. Math. Assoc. of America. Problems: 3.85, 3.91.

[431] Hall, P. & Hartley, B. I. (1966). The stability group of a series of subgroups. *Proc. London Math. Soc. 16*, 1–39. Problem: 3.78.

[432] Hambleton, I. & Kreck, M. (1988). On the classification of topological 4-manifolds with finite fundamental group. *Math. Ann. 280*, 85–104. Problem: 4.53.

[433] Hambleton, I. & Kreck, M. (1990). Smooth structures on algebraic surfaces with finite fundamental group. *Invent. Math. 102*, 109–114. Problem: 4.86.

[434] Hambleton, I., Kreck, M., & Teichner, P. (1994). Non-orientable 4-manifolds with fundamental group of order 2. *Trans. Amer. Math. Soc. 344*, 649–665. Problems: 4.13, 4.53, 4.82.

[435] Hambleton, I. & Madsen, I. (1986). Actions of finite groups on $\mathbb{R}^{n+k}$ with fixed set $\mathbb{R}^k$. *Canad. J. Math 38*, 781–860. Problem: 4.121.

[436] Hamilton, R. (1986). Four-manifolds with positive curvature operators. *J. Differential Geom. 24*, 153–179. Problem: 3.85.

[437] Handel, M. (1992). Commuting homeomorphisms of S^2. *Topology 31*, 293–303. Problem: 2.20.

[438] Hantzch, W. (1938). Einlagerung von Mannigfaltigkeiten in euklidische Räume. *Math. Zeit. 43*, 38–58. Problem: 3.20.

[439] Harer, J. L. (1978). On handlebody structures for hypersurfaces in $\mathbb{C}^3$ and $\mathbb{CP}^3$. *Math. Ann. 238*, 51–58. Problem: 4.18.

[440] Harer, J. L. (1982a). How to construct all fibered knots and links. *Topology 21*, 263–280. Problem: 1.83.

[441] Harer, J. L. (1982b). Representing elements of $\pi_1(M^3)$ by fibered knots. *Math. Proc. Cambridge Philos. Soc. 92*, 133–138. Problem: 3.13.

[442] Harer, J. L. (1986). The virtual cohomological dimension of the mapping class group of an orientable surface. *Invent. Math. 84*, 157–176. Problem: 3.49.

[443] Harer, J. L. (1988). The cohomology of the moduli space of curves. In E. Sernes (Ed.), *Theory of Moduli*, Volume 1337 of *Lecture Notes in Math.*, pp. 138–221. Springer-Verlag. Problem: 3.49.

[444] Harer, J. L., Kas, A., & Kirby, R. C. (1986). *Handlebody decompositions of complex surfaces*, Volume 62, number 350 of *Mem. Amer. Math. Soc.* American Math. Soc. Problems: 4.18, 4.98.

[445] Harikae, T. & Uchida, Y. (1993). Irregular dihedral branched coverings of knots. In M. E. Bozhüyük (Ed.), *Topics in Knot Theory, NATO Adv. Stud. Inst. series C*, Volume 399, pp. 269–276. Kluwer Academic Publisher. Problem: 1.59.

[446] Hartley, R. I. (1977). Counterexample to a conjecture of Murasugi. Preprint. Problem: 1.26.

[447] Hartley, R. I. (1980). Knots and involutions. *Math. Zeit. 171*, 175–185. Problems: 1.5, 1.6.

[448] Hartley, R. I. (1983). Identifying noninvertible knots. *Topology 22*, 137–145. Problem: 1.61.

[449] Hartley, R. I. & Kawauchi, A. (1979). Polynomials of amphicheiral knots. *Math. Ann. 243*, 63–70. Problem: 1.62.

[450] Hartley, R. I. & Murasugi, K. (1978). Homology invariants. *Canad. J. Math 30*, 655–670. Problems: 1.12, 1.26.

[451] Harvey, W. J. (1979). Geometric structure of surface mapping class groups. In C. T. C. Wall (Ed.), *Homological Group Theory*, Volume 36 of *London Math. Soc. Lect. Note Ser.*, pp. 255–269. Cambridge Univ. Press. Problem: 3.49.

[452] Harvey, W. J. (1981). Boundary structure of the modular group. In I. Kra & B. Maskit (Eds.), *Riemann Surfaces and Related Topics (Stony Brook, 1978)*, Volume 97 of *Ann. of Math. Stud.*, pp. 245–251. Problem: 3.49.

[453] Harvey, W. J. (1993). On certain families of compact Riemann surfaces. In C.-F. Bodigheimer & R. M. Hain (Eds.), *Mapping Class Groups and Moduli Spaces of Riemannian Surfaces (Gottingen 1991/Seattle 1991)*, Volume 150 of *Contemp. Math.*, pp. 137–148. Problem: 2.7.

[454] Hass, J. (1983). The geometry of the slice–ribbon problem. *Math. Proc. Cambridge Philos. Soc. 94*, 101–108. Problem: 1.102.

[455] Hass, J. (1987). Minimal surfaces in manifolds with S^1 actions and the simple loop conjecture for Seifert fibered spaces. *Proc. Amer. Math. Soc. 99*, 383–388. Problem: 3.96.

[456] Hass, J. (1992). Genus two Heegaard splittings. *Proc. Amer. Math. Soc. 114*, 565–570. Problem: 3.90.

[457] Hass, J. & Hughes, J. F. (1985). Immersions of surfaces in 3-manifolds. *Topology 24*, 97–112. Problem: 4.61.

[458] Hass, J. & Rubinstein, J. H. (1986). Shortest 1-sided geodesics. *Michigan Math. J. 33*, 155–168. Problem: 3.67.

[459] Hass, J., Rubinstein, J. H., & Scott, G. P. (1989). Compactifying coverings of closed 3-manifolds. *J. Differential Geom. 30*, 817–832. Problem: 3.7.

[460] Hass, J. & Scott, G. P. (1988). The existence of least area surfaces in 3-manifolds. *Trans. Amer. Math. Soc. 310*, 87–114. Problem: 3.66.

[461] Hatcher, A. E. (1976). Homeomorphisms of sufficiently large, P^2 irreducible 3-manifolds. *Topology 15*, 343–347. Problems: 3.47, 3.48.

[462] Hatcher, A. E. (1978). Concordance spaces, higher simple-homotopy theory, and applications. In R. J. Milgram (Ed.), *Algebraic and Geometric Topology (Stanford, 1976)*, Volume 32, part I of *Proc. Sympos. Pure Math.*, pp. 3–22. American Math. Soc. Problem: 4.35.

[463] Hatcher, A. E. (1981). On the diffeomorphism group of $S^1 \times S^2$. *Proc. Amer. Math. Soc. 83*, 427–430. Problem: 3.47.

[464] Hatcher, A. E. (1983). A proof of a Smale conjecture $\mathrm{Diff}(S^3) \simeq O(4)$. *Ann. of Math. 117*, 553–607. Problems: 3.34, 3.47, 4.126.

[465] Hatcher, A. E. & McCullough, D. (1990). Finite presentation of 3-manifold mapping class groups. In R. Piccinini (Ed.), *Groups of Self-Equivalences and Related Topics*, Volume 1425 of *Lecture Notes in Math.*, pp. 102–113. Springer-Verlag. Problem: 3.49.

[466] Hattori, A. (1975). Topology of $\mathbb{C}^n$ minus a finite number of affine hyperplanes in general position. *J. Fac. Sci. Univ. Tokyo Sect. IA Math. 22*, 205–219. Problem: 4.146.

[467] Hausmann, J.-C. (1987). Geometric Hopfian and non-Hopfian situations. In C. McCrory & T. Shifrin (Eds.), *Geometry and Topology: Varieties, Manifolds and Knots (Georgia, 1985)*, Volume 105 of *Lect. Notes in Pure and Appl. Math.*, pp. 157–166. Marcel Dekker. Problem: 5.26.

[468] Hausmann, J.-C. & Weinberger, S. (1985). Charactéristiques d'Euler et groupes fondamentaux des variétés de dimension 4. *Comment. Math. Helv. 60*, 139–144. Problem: 4.59.

[469] Heath, D. J. (1995). Heegaard splittings of the Figure-8 knot complement are standard. Preprint. Problem: 1.71.

[470] Hemion, G. (1979). On the classification of homeomorphisms of 2-manifolds and the classification of 3-manifolds. *Acta Math. 142*, 123–155. Problem: 3.82.

[471] Hempel, J. P. (1964). A simply connected 3-manifold is S^3 if it is the sum of a solid torus and the complement of a torus knot. *Proc. Amer. Math. Soc. 15*, 154–158. Problem: 1.15.

[472] Hempel, J. P. (1976). *3-Manifolds*, Volume 86 of *Ann. of Math. Stud.* Princeton Univ. Press. Problems: 1.85, 3.5, 3.33.

[473] Hempel, J. P. (1986). Coverings of Dehn fillings of surface bundles. *Topology Appl. 24*, 157–70. Problem: 3.2.

[474] Hempel, J. P. (1987a). Coverings of Dehn fillings of surface bundles II. *Topology Appl. 26*, 163–173. Problem: 3.2.

[475] Hempel, J. P. (1987b). Residual finiteness for 3-manifolds. In S. M. Gersten & J. R. Stallings (Eds.), *Combinatorial Group Theory and Topology*, Volume 111 of *Ann. of Math. Stud.*, pp. 379–396. Princeton Univ. Press. Problems: 3.33, 3.46.

[476] Hempel, J. P. (1987c). The finitely generated intersection property for Kleinian groups. In S. M. Gersten & J. R. Stallings (Eds.), *Combinatorial Group Theory and Topology*, Volume 111 of *Ann. of Math. Stud.*, pp. 18–24. Princeton Univ. Press. Problem: 3.52.

[477] Hempel, J. P. (1990). Branched covers over strongly amphicheiral links. *Topology 29*, 247–255. Problem: 3.2.

[478] Henderson, J. P. (1981). Cellularity in polyhedra. *Topology Appl. 12*, 267–282. Problem: 4.70.

[479] Herbert, R. J. (1981). *Multiple points of immersed manifolds*, Volume 34, number 250 of *Mem. Amer. Math. Soc.* American Math. Soc. Problem: 4.20.

[480] Higman, G. (1951). A finitely generated infinite simple group. *J. London Math. Soc. 26*, 61–64. Problem: 3.105.

[481] Hilden, H. M., Lozano, M. T., & Montesinos, J. M. (1983). All three-manifolds are pullbacks of a branched cover S^3 to S^3. *Trans. Amer. Math. Soc. 279*, 729–735. Problem: 3.10.

[482] Hillman, J. A. (1980). Orientability, asphericity and two-knots. *Houston J. Math. 6*, 67–76. Problem: 1.54.

[483] Hillman, J. A. (1981). *Alexander Ideals of Links*, Volume 895 of *Lecture Notes in Math.* Springer-Verlag. Problems: 1.57, 1.96, 3.79.

[484] Hillman, J. A. (1985). Seifert fibre spaces and Poincaré duality groups. *Math. Zeit. 190*, 365–369. Problem: 3.77.

[485] Hillman, J. A. (1989). *2-Knots and their Groups*, Volume 5 of *Australian Math. Soc. Lect. Series.* Cambridge Univ. Press. Problems: 1.48, 1.54.

[486] Hillman, J. A. (1994). *The Algebraic Characterization of Geometric 4-Manifolds*, Volume 198 of *London Math. Soc. Lect. Note Ser.* Cambridge Univ. Press. Problems: 1.48, 1.54, 4.53, 4.57, 4.59.

[487] Hillman, J. A. (1995). Free products and 4-dimensional connected sums. To appear in *Bull. London Math. Soc. 27.* Problems: 1.55, 4.120.

[488] Hilton, P. J. & Roitberg, J. (1969). On principal S^3-bundles over spheres. *Ann. of Math. 90*, 91–107. Problem: 5.28.

[489] Hironaka, E. (1993). *Abelian coverings of the complex projective plane branched along configurations of real lines*, Volume 502 of *Mem. Amer. Math. Soc.* American Math. Soc. Problem: 4.147.

[490] Hirsch, M. (1961). The imbedding of bounding manifolds in euclidean space. *Ann. of Math. 74*, 494–497. Problem: 4.19.

[491] Hirsch, M. W. (1965). On embedding 4-manifolds in $\mathbb{R}^7$. *Math. Proc. Cambridge Philos. Soc. 61*, 657–658. Problem: 4.19.

[492] Hirsch, M. W. & Smale, S. (1959). Involutions of the 3-sphere. *Amer. J. Math. 81*, 893–900. Problem: 3.4.

[493] Hirzebruch, F. (1966). The topology of normal singularities of an algebraic surface [d'apres D. Mumford]. In *Séminaire Bourbaki; 15ème année, Exposés 241–258 1962/1963.* W. A. Benjamin, Inc. Problems: 3.27, 3.28, 4.3.

[494] Hirzebruch, F. (1978). *Topological Methods in Algebraic Geometry*, Volume 131 of *Grundlehren Math. Wiss.* Springer-Verlag. Problem: 4.39.

[495] Hirzebruch, F. (1983). Arrangements of lines and algebraic surfaces. In *Arithmetic and Geometry*, Volume 36, Number 2 of *Prog. Math.*, pp. 113–140. Birkhäuser. Problem: 4.147.

[496] Hirzebruch, F., Neumann, W. D., & Koh, S. S. (1971). *Differentiable manifolds and quadratic forms*, Volume 4 of *Lect. Notes in Pure and Appl. Math.* Marcel Dekker. Problem: 3.20.

[497] Hitchin, N. (1974). Compact 4-dimensional Einstein manifolds. *J. Differential Geom. 9*, 435–441. Problem: 4.68.

[498] Hodgson, C. D. (1986). *Degeneration and regeneration of geometric structures on 3-manifolds.* Ph. D. thesis, Princeton Univ. Problem: 1.77.

[499] Hodgson, C. D. & Rubinstein, J. H. (1985). Involutions and isotopies of lens spaces. In D. P. Rolfsen (Ed.), *Knot Theory and Manifolds, Proc. Vancouver 1983*, Volume 1144 of *Lecture Notes in Math.*, pp. 60–96. Springer-Verlag. Problems: 3.25, 3.26, 3.35, 3.40, 3.46, 3.95.

[500] Hodgson, C. D. & Weeks, J. (1995). A census of closed hyperbolic 3-manifolds. In preparation. Problem: 1.77.

[501] Hog-Angeloni, C., Metzler, W., & Sieradski, A. J. (1993). *Two-dimensional Homotopy and Combinatorial Group Theory*, Volume 197 of *London Math. Soc. Lect. Note Ser.* Cambridge Univ. Press. Problems: 1.103, 5.1, 5.2, 5.4.

[502] Hopf, H. (1944). Eine Verallgemeinerung bekannter Abbildungs- und Überdeckungssätze. *Portugal. Math. 4*, 129–139. Problem: 5.24.

[503] Horikawa, E. (1976a). Algebraic surfaces with small c_1^2, I. *Ann. of Math. 104*, 357–387. Problem: 4.101.

[504] Horikawa, E. (1976b). Algebraic surfaces with small c_1^2, II. *Invent. Math. 37*, 121–155. Problem: 4.101.

[505] Horikawa, E. (1978). Algebraic surfaces with small c_1^2, III. *Invent. Math. 47*, 209–248. Problem: 4.101.

[506] Horikawa, E. (1979). Algebraic surfaces with small c_1^2, IV. *Invent. Math. 50*, 103–128. Problem: 4.101.

[507] Hoste, J. (1995). Infinite framed link diagrams for open 3-manifolds. In preparation. Problem: 3.103.

[508] Hoste, J. & Przytycki, J. H. (1992). A survey of skein modules of 3- manifolds. In *Knots 90*, pp. 363–379. W. de Gruyter. Problem: 1.92.

[509] Hoste, J. & Przytycki, J. H. (1993). The $(2,\infty)$-skein module of lens spaces; a generalization of the Jones polynomial. *J. Knot Theory Ramifications 2*, 321–333. Problem: 1.92.

[510] Hoste, J. & Przytycki, J. H. (1995). The skein module of genus 1 Whitehead type manifolds. To appear in *Math. Zeit.*. Problem: 1.92.

[511] Hsiang, W.-C. (1984). Geometric applications of algebraic K-theory. In *Proc. International Congress of Mathematicians, Warsaw 1983*, pp. 99–119. Polish Scientific Publishers, Warsaw. Problem: 5.29.

[512] Hsiang, W.-C. & Jahren, B. (1983). A remark on the isotopy classes of diffeomorphisms of lens spaces. *Pacific J. Math. 109*, 411–423. Problem: 3.36.

[513] Hsiang, W.-C. & Szczarba, R. H. (1971). On embedding surfaces in four-manifolds. In A. Liulevicius (Ed.), *Algebraic Topology (Wisconsin, 1970)*, Volume 22 of *Proc. Sympos. Pure Math.*, pp. 97–103. American Math. Soc. Problems: 4.26, 4.36.

[514] Hsiang, W.-Y. & Kleiner, B. (1989). On the topology of positively curved 4-manifolds with symmetry. *J. Differential Geom. 29*, 615–621. Problem: 4.67.

[515] Hu, B. Z. (1993). Whitehead groups of polyhedra with nonpositive curvature. *J. Differential Geom. 38*, 501–517. Problem: 5.29.

[516] Hughes, J. F. (1982). *Invariants of regular homotopy and bordism of low dimensional immersions.* Ph. D. thesis, UC Berkeley. Problem: 4.61.

[517] Hughes, J. F. (1983). Triple points of immersed $2n$-manifolds in $3n$-space. *Quart. J. Math. Oxford Ser. (2) 34*, 427–431. Problem: 4.20.

[518] Hughes, J. F. & Melvin, P. M. (1985). The Smale invariant of a knot. *Comment. Math. Helv. 60*, 615–627. Problem: 4.61.

[519] Inoue, M. (1974). On surfaces of class VII_0. *Invent. Math. 24*, 269–310. Problem: 4.102.

[520] Itenberg, I. & Kharlamov, V. (1993). Towards the maximal number of components of a non-singular surface of degree 5 in $\mathbb{RP}^3$. Preprint LMENS - 93 - 9. École Normale Supérieur - Paris (to appear in *Adv. Soviet Math.*). Problem: 3.69.

[521] Ivanov, N. V. (1976). Группы диффеоморфизмов многообразий Вальдхаузена. *Исследования по Топологии 2, Записки Научных Семинаров Ленинградского Отделения Математический ин-та. им. Стеклова 66*, 172–176. Translation to English in [525]. Problem: 3.47.

[522] Ivanov, N. V. (1979a). Homotopies of automorphism spaces of some three dimensional manifolds. *Докл. Акад. Наук СССР 244*, 274–277. See also the correction in [523]. Translation to English incorporating the correction in [524]. Problems: 3.36, 3.47.

[523] Ivanov, N. V. (1979b). Correction to: Homotopies of automorphism spaces of some three dimensional manifolds. *Докл. Акад. Наук СССР 249*, 1288. Correction to [522]. Problems: 3.36, 3.47.

[524] Ivanov, N. V. (1979c). Homotopy of spaces of automorphisms of some three-dimensional manifolds. *Soviet Math. Dokl. 20*, 47–50. Original in Russian in [522] and correction in [523].

[525] Ivanov, N. V. (1979d). Diffeomorphism groups of Waldhausen manifolds. *J. Soviet Math. 12*, 115–118. Original in Russian in [521].

[526] Ivanov, N. V. (1980). Spaces of surfaces in Waldhausen manifolds. LOMI Preprint P-5-80, 31pp.

[527] Ivanov, N. V. (1982). Гомотопии пространств диффеоморфизмов некоторых трехмерных многообразий. *Исследования по Топологии 4, Записки Научных Семинаров Ленинградского Отделения Математический ин-та. им. Стеклова 122*, 72–103. Translation to English in [528]. Problems: 3.36, 3.47.

[528] Ivanov, N. V. (1984). Homotopy of spaces of diffeomorphisms of some three-dimensional manifolds. *J. Soviet Math. 26*, 1646–1664. Translation of [527].

[529] Ivanov, N. V. (1987a). Complexes of curves and the Teichmüller modular groups. *Russian Math. Surveys 42(3)*, 55–107. Original in Russian in [530].

[530] Ivanov, N. V. (1987b). Комплексы кривых и модулярные группы Тайхмюллера. *Успехи Мат. Наук 42(3)*, 49–91. Translation to English in [529]. Problem: 2.10.

[531] Ivanov, N. V. (1989). Automorphisms of complexes of curves and of Teichmüller spaces. Preprint IHES/M/89/60.

[532] Ivanov, N. V. (1992). *Subgroups of the Teichmüller modular groups*, Volume 115 of *Transl. Math. Monographs.* American Math. Soc. Problem: 2.11.

[533] Jackson, B. (1981). End invariants of group extensions. *Topology 21*, 71–81. Problem: 3.7.

[534] Jaco, W. H. (1970). Surfaces embedded in $M^2 \times S^1$. *Canad. J. Math 22*, 553–568. Problems: 3.17, 3.18.

[535] Jaco, W. H. (1980). *Lectures on Three-manifold Topology*, Volume 43 of *CBMS Regional Conf. Ser. in Math.* American Math. Soc. Problem: 3.52.

[536] Jaco, W. H. & Oertel, U. (1984). An algorithm to decide if a 3-manifold is a Haken manifold. *Topology 23*, 195–209. Problems: 3.17, 3.82.

[537] Jaco, W. H. & Shalen, P. B. (1976). Peripheral structure of 3-manifolds. *Invent. Math. 38*, 55–88. Problem: 3.3.

[538] Jaco, W. H. & Shalen, P. B. (1979). *Seifert fibered spaces in 3-manifolds*, Volume 220 of *Mem. Amer. Math. Soc.* American Math. Soc. Problem: 3.45.

[539] Jaeger, F. (1988). On Tutte polynomials and link polynomials. *Proc. Amer. Math. Soc. 103*, 647–654. Problem: 1.60.

[540] James, I. M. (1976). *The topology of Stiefel manifolds*, Volume 24 of *London Math. Soc. Lect. Note Ser.* Cambridge Univ. Press. Problem: 4.61.

[541] Jiang, T. & Yau, S. S.-T. (1993). Topological invariance of intersection lattices of arrangements in $\mathbb{CP}^2$. *Bull. Amer. Math. Soc. 29*, 88–93. Problem: 4.144.

[542] Johannson, K. (1979). *Homotopy equivalences of 3-manifolds with boundary*, Volume 761 of *Lecture Notes in Math.* Springer-Verlag. Problems: 3.3, 3.39, 3.45, 3.49.

[543] Johannson, K. (1990). Heegaard surfaces in Haken 3-manifolds. *Bull. Amer. Math. Soc. 23*, 91–98. Problem: 3.90.

[544] Johannson, K. (1995). *Topology and Combinatorics of 3-Manifolds*, Volume 1599 of *Lecture Notes in Math.* Springer-Verlag. Problems: 3.89, 3.90.

[545] Johnson, D. (1980). Quadratic forms and the Birman-Craggs homomorphisms. *Trans. Amer. Math. Soc. 261*, 235–254. Problem: 2.2.

[546] Johnson, D. (1983). The structure of the Torelli group I: a finite set of generators for $\mathcal{J}$. *Ann. of Math. 118*, 423–442. Problems: 2.1, 2.9.

[547] Johnson, D. (1985a). The structure of the Torelli group II: a characterization of the group generated by twists on bounding curves. *Topology 24*, 113–126. Problem: 2.1.

[548] Johnson, D. (1985b). The structure of the Torelli group III: the abelianization of $\mathcal{J}$. *Topology 24*, 127–144. Problem: 2.1.

[549] Johnson, F. E. A. & Kotschick, D. (1993). On the signature and Euler characteristic of certain four-manifolds. *Math. Proc. Cambridge Philos. Soc. 114*, 431–437. Problem: 4.59.

[550] Jones, J. D. S. & Westbury, B. W. (1995). Algebraic K-theory, homology spheres, and the η-invariant. To appear in *Topology*. Problem: 3.63.

[551] Jones, V. F. R. (1987). Hecke algebra representations of braid groups and link polynomials. *Ann. of Math. 126*, 335–388. Problem: 1.88.

[552] Jones, V. F. R. (1989). On a certain value of the Kauffman polynomial. *Comm. Math. Phys. 125*, 459–467. Problem: 1.91.

[553] Jost, J. & Yau, S.-T. (1993). Harmonic mappings and algebraic varieties over function fields. *Amer. J. Math. 115*, 1197–1227. Problem: 4.99.

[554] Joyce, D. D. (1995). Compact Riemannian 7-manifolds with holonomy G_2, I & II. Preprint. Problem: 4.68.

[555] Kaiser, U. (1985). Regulaere Homotopien von Immersionen von Sphaeren im Euklidischen Raum. Diplomarbeit, Siegen. Problem: 4.61.

[556] Kaiser, U. (1988). Immersions in codimension 1 up to regular homotopy. *Archiv. Math. (Basel) 51*, 371–377. Problem: 4.61.

[557] Kaiser, U. & Li, B. H. (1991). On the reflection of immersions in Euclidean spaces. *Acta Math. Sci. (English Ed.) 11*, 301–307. Problem: 4.61.

[558] Kan, D. M. & Thurston, W. P. (1976). Every connected space has the homology of a $K(\pi, 1)$. *Topology 15*, 253–258. Problem: 4.17.

[559] Kanenobu, T. (1986). Infinitely many knots with the same polynomial invariant. *Proc. Amer. Math. Soc. 97*, 158–162. Problem: 1.91.

[560] Kanenobu, T. & Murakami, H. (1986). Two-bridge knots with unknotting number one. *Proc. of the AMS 98*, 499–502. Problem: 1.69.

[561] Karoubi, M. (1987). *Homologie cyclique et K-théorie*, Volume 149 of *Astérisque.* Soc. Mat. de France. Problem: 3.63.

[562] Kato, M. (1978). Compact complex manifolds containing global spherical shells, I. In *Proc. of the International Symposium on Algebraic Geometry, Kyoto 1977*, Tokyo, pp. 45–84. Kinokuniya. Problem: 4.37.

[563] Kauffman, L. H. (1987). State models and the Jones polynomial. *Topology 26*, 395–407. Problems: 1.65, 1.66.

[564] Kauffman, L. H. (1988). New invariants in the theory of knots. *Amer. Math. Monthly 95*, 195–242. Problems: 1.65, 1.66.

[565] Kawauchi, A. (1976). On quadratic forms of 3-manifolds, $\tilde{H}$-cobordisms. *Osaka J. Math. 13*, 567–590. Problem: 3.20.

[566] Kawauchi, A. (1980). On a 4-manifold homology equivalent to a bouquet of surfaces. *Trans. Amer. Math. Soc. 262*, 95–112. Problems: 4.25, 4.73.

[567] Kawauchi, A. (1986). Three dualities on the integral homology of infinite cyclic coverings of manifolds. *Osaka J. Math. 23*, 633–651. Problem: 1.51.

[568] Kearton, C. (1989). Mutation of knots. *Proc. Amer. Math. Soc. 105*, 206–208. Problem: 1.53.

[569] Keating, M. E. (1973). On the K-theory of the quaternion group. *Mathematika 20*, 59–62. Problem: 3.37.

[570] Kerckhoff, S. P. (1983). The Nielsen realization problem. *Ann. of Math. 117*, 235–265. Problem: 2.6.

[571] Kervaire, M. A. (1965). Les nœuds de dimensions supérieur. *Bull. Soc. Math. France 93*, 193–201. Problems: 1.48, 1.56, 1.105.

[572] Kervaire, M. A. & Milnor, J. W. (1961). On 2-spheres in 4-manifolds. *Proc. Nat. Acad. Sci. U.S.A. 47*, 1651–1657. Problem: 4.26.

[573] Kim, M. H., Kojima, S., & Raymond, F. (1992). Homotopy invariants of nonorientable 4-manifolds. *Trans. Amer. Math. Soc. 333*, 71–81. Problem: 4.53.

[574] Kim, P. K. & Tollefson, J. L. (1980). Splitting the PL involutions of nonprime 3-manifolds. *Michigan Math. J. 27*, 259–274. Problem: 3.25.

[575] Kinoshita, S. (1957). On Wendt's theorem of knots, I. *Osaka Math. J. 9*, 61–66. Problem: 1.59.

[576] Kinoshita, S. (1958). On knots and periodic transformations. *Osaka Math. J. 10*, 43–52. Problem: 3.38.

[577] Kirby, R. C. (1970). Codimension-two locally flat embeddings have normal bundles. In J. C. Cantrell & C. H. Edwards, Jr. (Eds.), *Topology of Manifolds, (Georgia, 1969)*, pp. 416–423. Markham Pub. Co. Problem: 4.31.

[578] Kirby, R. C. (1989). *Topology of 4-Manifolds*, Volume 1374 of *Lecture Notes in Math.* Springer-Verlag. Problems: 4.43, 4.54.

[579] Kirby, R. C. (1996). Akbulut's corks and h-cobordisms of smooth, simply connected 4-manifolds. In S. Akbulut, T. Onder, & R. J. Stern (Eds.), *Proc. of Gökova Topology–Geometry Conference, 1995*, pp. 85–94. Scientific and Technical Research Council of Turkey.

[580] Kirby, R. C. & Melvin, P. M. (1978). Slice knots and Property R. *Invent. Math. 45*, 57–59. Problems: 1.16, 1.19.

[581] Kirby, R. C. & Melvin, P. M. (1991a). Quantum invariants of lens spaces and a Dehn surgery formula. *Abstracts Amer. Math. Soc. 12*, 435. Problem: 3.108.

[582] Kirby, R. C. & Melvin, P. M. (1991b). The 3-manifold invariants of Witten and Reshetikhin–Turaev for $sl(2,\mathbb{C})$. *Invent. Math. 105*, 473–545. Problem: 3.108.

[583] Kirby, R. C. & Melvin, P. M. (1995). Framings of 3-manifolds. In preparation. Problem: 5.20.

[584] Kirby, R. C. & Siebenmann, L. C. (1974). Normal bundles for codimension 2 locally flat imbeddings. In L. C. Glaser & T. B. Rushing (Eds.), *Geometric Topology - Park City 1974 Conference*, Volume 438 of *Lecture Notes in Math.*, pp. 310–324. Springer-Verlag. Problem: 4.31.

[585] Kirby, R. C. & Siebenmann, L. C. (1977). *Foundational Essays on Topological Manifolds, Smoothings, and Triangulations*, Volume 88 of *Ann. of Math. Stud.* Princeton Univ. Press. Problem: 5.29.

[586] Kirby, R. C. & Taylor, L. R. (1990). *Pin* structures on low dimensional manifolds. In S. K. Donaldson & C. B. Thomas (Eds.), *Geometry of Low Dimensional Manifolds, 2 (Durham, 1989)*, Volume 151 of *London Math. Soc. Lect. Note Ser.*, pp. 177–242. Cambridge Univ. Press. Problem: 4.112.

[587] Klingenberg, W. (1982). *Riemannian Geometry.* W. de Gruyter. Problem: 4.67.

[588] Klyachko, A. A. (1993). A funny property of spheres and equations over groups. *Comm. Algebra. 21*, 2555–2575. Problem: 5.7.

[589] Ko, K. H. & Smolinsky, L. (1992). The framed braid group and 3-manifolds. *Proc. Amer. Math. Soc. 115*, 541–551. Problem: 3.101.

[590] Kobayashi, T. (1984). Structures of the Haken manifolds with Heegaard splittings of genus 2. *Osaka J. Math. 21*, 437–455. Problem: 3.15.

[591] Kobayashi, T. (1988). Casson–Gordon's rectangle condition of Heegaard diagrams and incompressible tori in 3-manifolds. *Osaka J. Math. 25*, 553–573. Problem: 3.89.

[592] Kobayashi, T. (1989). Minimal genus Seifert surface for unknotting number 1 knots. *Kobe J. Math. 6*, 53–62. Problem: 1.69.

[593] Kobayashi, T. (1992). A construction of 3-manifolds whose homeomorphism classes of Heegaard splittings have polynomial growth. *Osaka J. Math. 29*, 653–674. Problem: 3.91.

[594] Kobayashi, T. (1994). A construction of arbitrarily high degeneration of tunnel numbers of knots under connected sum. *J. Knot Theory Ramifications 3*, 179–186. Problem: 1.70.

[595] Kodaira, K. (1963). On compact analytic surfaces III. *Ann. of Math. 78*, 563–626. Problem: 4.11.

[596] Kodaira, K. (1964). On the structure of compact complex analytic surfaces, I. *Amer. J. Math. 86*, 751–798. Problem: 4.37.

[597] Kodaira, K. (1967). A certain type of irregular algebraic surface. *J. Analyse Math. 19*, 207–215. Problem: 4.39.

[598] Kohara, T. & Suzuki, S. (1992). Some remarks on knots and links in spatial graphs. In *Knots 90*, pp. 435–445. W. de Gruyter. Problem: 5.15.

[599] Kohno, T. (1992). Topological invariants for 3-manifolds using representations of mapping class groups I. *Topology 31*, 303–230. Problem: 2.2.

[600] Kojima, S. (1986). Determining knots by branched covers. In D. B. A. Epstein (Ed.), *Low-dimensional Topology and Kleinian Groups*, Volume 112 of *London Math. Soc. Lect. Note Ser.*, pp. 193–207. Cambridge Univ. Press. Problems: 1.27, 1.74.

[601] Kojima, S. & Long, D. D. (1988). Virtual betti numbers of some hyperbolic 3-manifolds. In Y. Matsumoto, T. Mizutani, & S. Morita (Eds.), *A Fete of Topology*, pp. 417–437. Academic Press. Problem: 3.2.

[602] Kojima, S. & Miyamoto, I. (1991). The smallest hyperbolic 3-manifolds with totally geodesic boundary. *J. Differential Geom. 34*, 175–192. Problem: 3.60.

[603] Kontsevich, M. (1994). Feynman diagrams and low-dimensional topology. In A. Joseph, F. Mingnot, F. Murat, B. Prum, & R. Rentschler (Eds.), *First European Congress of Mathematics - 1992 Paris, vol 2*, Volume 120 of *Prog. Math.*, pp. 98–121. Birkhäuser. Problems: 3.108, 5.19, 5.20.

[604] Koschorke, U. (1985). Higher order homotopy invariants for higher dimensional link maps. In A. Dold & B. Eckmann (Eds.), *Algebraic Topology*, Volume 1172 of *Lecture Notes in Math.*, pp. 116–129. Springer-Verlag. Göttingen, 1984. Problem: 1.97.

[605] Kotschick, D. (1989). On manifolds homeomorphic to $\mathbb{CP}^2\#8\overline{\mathbb{CP}}^2$. *Invent. Math. 95*, 591–600. Problems: 4.45, 4.96.

[606] Kotschick, D. (1992a). All fundamental groups are almost complex. *Bull. London Math. Soc. 24*, 377–378. Problem: 5.21.

[607] Kotschick, D. (1992b). Orientation-reversing homeomorphisms in surface geography. *Math. Ann. 292*, 375–381. Problems: 4.39, 4.133.

[608] Kotschick, D. (1992c). Remarks on geometric structures on compact complex surfaces. *Topology 31*, 317–321. Problem: 4.39.

[609] Kotschick, D. (1993). On connected sum decompositions of algebraic surfaces and their fundamental groups. *Internat. Math. Res. Notices,* bound within *Duke Math. J. 6*, 179–182. Problem: 4.37.

[610] Kotschick, D. (1994). Four-manifold invariants of finitely presentable groups. In K. Fukaya, M. Furuta, T. Kohno, & D. Kotschick (Eds.), *Topology, Geometry and Field Theory*, pp. 89–99. World Scientific. Proc. of the 31st Taniguchi Symposium, 1993 - Kobe. Problems: 4.59, 4.120.

[611] Kotschick, D. (1995a). On irreducible four-manifolds. Preprint. Problems: 4.37, 4.138.

[612] Kotschick, D. (1995b). On the blowups of numerical Godeaux surfaces. *C.R. Acad. Sci. Paris Sér. I Math. 320*, 577–580. Problem: 4.96.

[613] Kotschick, D. & Lisca, P. (1995). Instanton invariants of $\mathbb{CP}^2$ via topology. To appear in *Math. Ann.*. Problem: 4.129.

[614] Kotschick, D. & Matic, G. (1995). Embedded surfaces in four-manifolds, branched covers and **so**(3)-invariants. *Math. Proc. Cambridge Philos. Soc. 117*, 275–286. Problems: 4.26, 4.36.

[615] Kowng, H.-Z. (1994). *Straightening tori in Heegaard splittings*. Ph. D. thesis, UC Santa Barbara. Problem: 1.70.

[616] Kramer, R. (1983). Dehn twists and handlebodies of genus 2. In S. J. Lomonaco, Jr. (Ed.), *Low Dimensional Topology, (San Francisco, 1981)*, Volume 20 of *Contemp. Math.*, pp. 201–211. Problem: 2.4.

[617] Kranjc, M. (1991). Embedding a 2-complex K in $\mathbb{R}^4$ when $H^2(K)$ is a cyclic group. *Pacific J. Math. 150*, 329–339. Problem: 5.3.

[618] Kreck, M. (1984). Some closed 4-manifolds with exotic differentiable structure. In I. Madsen & R. Oliver (Eds.), *Algebraic Topology, Aarhus, 1982*, Volume 1051 of *Lecture Notes in Math.*, pp. 246–262. Springer-Verlag. Problem: 4.45.

[619] Kreck, M. (1995). Surgery and duality. To be published by Vieweg–Verlag. Problem: 4.84.

[620] Kreck, M., Lück, W., & Teichner, P. (1995). Counterexamples to the Kneser Conjecture in dimension four. To appear in *Comment. Math. Helv.*. Problem: 4.120.

[621] Kreck, M. & Teichner, P. (1995). 4-manifolds with infinite cyclic fundamental group. In preparation. Problem: 4.30.

[622] Kreher, R. & Metzler, W. (1983). Simpliziale transformationen von Polyedern und die Zeeman-Vermutung. *Topology 22*, 19–26. Problem: 5.2.

[623] Kronheimer, P. B. & Mrowka, T. S. (1993). Gauge theory for embedded surfaces, I. *Topology 32*, 773–826. Problems: 1.4, 1.40, 1.98, 1.100.

[624] Kronheimer, P. B. & Mrowka, T. S. (1994a). Recurrence relations and asymptotics for four-manifold invariants. *Bull. Amer. Math. Soc. 30*, 215–221. Problems: 4.36, 4.128.

[625] Kronheimer, P. B. & Mrowka, T. S. (1994b). The genus of embedded surfaces in the projective plane. *Math. Res. Lett. 1*, 797–808. Problem: 1.100.

[626] Kronheimer, P. B. & Mrowka, T. S. (1995a). Embedded surfaces and the structure of Donaldson's polynomial invariants. *J. Differential Geom. 41*, 573–734. Problem: 4.128.

[627] Kronheimer, P. B. & Mrowka, T. S. (1995b). Gauge theory for embedded surfaces, II. *Topology 34*, 37–97. Problems: 1.98, 1.100.

[628] Kuga, K. (1984). Representing homology classes of $S^2 \times S^2$. *Topology 23*, 133–137. Problem: 1.42.

[629] Kuiper, N. H. (1974). The quotient space of $\mathbb{CP}^2$ by complex conjugation is the 4-sphere. *Math. Ann. 208*, 175–177. Problem: 4.113.

[630] Kuperberg, G. (1995). A volume-preserving counterexample to the Seifert conjecture. To appear in *Comment. Math. Helv.*, AMS Preprint Server #199407-58-002. Problem: 3.113.

[631] Kuperberg, G. & Kuperberg, K. (1995). Generalized counterexamples to the Seifert conjecture. To appear in *Ann. of Math.*, AMS Preprint Server # 199407-58-001. Problem: 3.113.

[632] Kuperberg, K. (1994). A smooth counterexample to the Seifert conjecture. *Ann. of Math. 140*, 723–732. Problem: 3.113.

[633] Kwasik, S. (1987). Low dimensional concordances, Whitney towers and isotopies. *Math. Proc. Cambridge Philos. Soc. 102*, 103–119. Problem: 4.35.

[634] Kwasik, S. & Schultz, R. (1992a). Icosahedral group actions on $\mathbb{R}^3$. *Invent. Math. 108*, 385–402. Problem: 3.46.

[635] Kwasik, S. & Schultz, R. (1992b). Vanishing of Whitehead torsion in dimension four. *Topology 31*, 735–756. Problem: 4.9.

[636] Kwasik, S. & Schultz, R. (1995). Pseudo-isotopies of 3-manifolds. To appear in *Topology*. Problem: 4.35.

[637] Lambert, H. W. (1977). Longitude surgery on genus 1 knots. *Proc. Amer. Math. Soc. 63*, 359–362. Problem: 1.16.

[638] Lambropoulou, S. S. F. & Rourke, C. P. (1996). On Markov's theorem in 3-manifolds. To appear in *Topology Appl.*. Problem: 3.101.

[639] Lashof, R. K. & Taylor, L. R. (1984). Smoothing theory and Freedman's work on four manifolds. In I. Madsen & R. Oliver (Eds.), *Algebraic Topology, Aarhus, 1982*, Volume 1051 of *Lecture Notes in Math.*, pp. 271–292. Springer-Verlag. Problem: 4.85.

[640] Lawson, T. (1984). Detecting the standard embedding of $\mathbb{RP}^2$ in S^4. *Math. Ann. 267*, 439–448. Problem: 4.58.

[641] Lê Dũng Tráng (1973). Un critère d'équisingularité. In *Singularités à Cargèse*, Volume 7 and 8 of *Astérisque*, pp. 183–192. Soc. Mat. de France. Problem: 3.30.

[642] Lê Dũng Tráng & Ramanujam, C. P. (1976). The invariance of Milnor's number implies the invariance of the topological type. *Amer. J. Math. 98*, 67–78. Problem: 3.30.

[643] le Dimet, J.-Y. (1988). Cobordisme d'enlacement de disques. *Mém. Soc. Math. de France (N.S.) 32*, 1–92. Problem: 1.56.

[644] Lee, R. (1973). Semicharacteristic classes. *Topology 12*, 183–199. Problem: 3.37.

[645] Lee, R. & Thomas, C. B. (1973). Free finite group actions on S^3. *Bull. Amer. Math. Soc. 79*, 211–215. Problem: 3.37.

[646] Lee, R. & Wilczynski, D. M. (1995). Representing homology classes by locally flat surfaces of minimum genus. Preprint. Problem: 4.36.

[647] Lehmer, D. H. (1933). Factorizations of certain cyclotomic functions. *Ann. of Math. 34*, 461–479. Problem: 5.12.

[648] Levine, J. (1995). Finitely presented groups with long lower central series. Preprint. Problem: 3.78.

[649] Li, B. H. & Peterson, F. P. (1985). On the reflection of n-manifolds in $(n+1)$-manifolds. *Kexue Tongbao (Chinese) 20*, 166–169. Problem: 4.61.

[650] Li, J., Yau, S.-T., & Zheng, F. (1990). A simple proof of Bogomolov's theorem on class VII_0 surfaces with $b_2 = 0$. *Illinois J. Math. 34*, 217–220. Problem: 4.102.

[651] Li, J., Yau, S.-T., & Zheng, F. (1994). On projectively flat hermitian manifolds. *Comm. Anal. Geom. 2*, 103–110. Problem: 4.102.

[652] Lickorish, W. B. R. (1977). Surgery on knots. *Proc. Amer. Math. Soc. 60*, 296–298. Problem: 3.6.

[653] Lickorish, W. B. R. (1985). The unknotting number of a classical knot. *Contemp. Math. 44*, 117–121. Problem: 1.69.

[654] Lickorish, W. B. R. (1988). The panorama of polynomials for knots, links and skeins. In J. S. Birman & A. L. Libgober (Eds.), *Braids (Santa Cruz, 1986)*, Volume 78 of *Contemp. Math.*, pp. 399–414. American Math. Soc. Problem: 1.91.

[655] Lickorish, W. B. R. & Millett, K. C. (1986). Some evaluations of link polynomials. *Comment. Math. Helv. 61*, 349–359. Problem: 1.91.

[656] Lickorish, W. B. R. & Thistlethwaite, M. B. (1988). Some links with nontrivial polynomials and their crossing-numbers. *Comment. Math. Helv. 63*, 527–539. Problem: 1.68.

[657] Lima, E. L. (1963). Commuting vector fields on 2-manifolds. *Bull. Amer. Math. Soc. 69*, 366–368. Problem: 2.20.

[658] Lima, E. L. (1964a). Common singularities of commuting vector fields on 2-manifolds. *Comment. Math. Helv. 39*, 97–110. Problem: 2.20.

[659] Lima, E. L. (1964b). Commuting vector fields in S^2. *Proc. Amer. Math. Soc. 15*, 138–141. Problem: 2.20.

[660] Lin, X.-S. (1992). Milnor's link invariants are all of finite type. Preprint. Problem: 1.96.

[661] Lin, X.-S. (1995). Power series expansions and invariants of links. In W. H. Kazez (Ed.), *Geometric Topology, Proc. of the 1993 Georgia International Topology Conference.* International Press. Problem: 1.96.

[662] Linnell, P. (1983). On accessibility of groups. *J. Pure Appl. Algebra 30*, 39–46. Problem: 5.10.

[663] Lisca, P. (1994). On the Donaldson polynomials of elliptic surfaces. *Math. Ann. 299*, 624–639. Problem: 4.11.

[664] Lisca, P. (1995). Computations of instanton invariants using Donaldson–Floer Theory. *Invent. Math. 119*, 347–359. Problem: 4.11.

[665] Litherland, R. A. (1979). Knots in solid tori. *Proc. London Math. Soc. 39*, 311–331. Problem: 1.15.

[666] Litherland, R. A. (1981). The second homology of the group of a knotted surface. *Quart. J. Math. Oxford Ser. (2) 32*, 425–434. Problem: 4.29.

[667] Livesay, G. R. (1954). On a theorem of F. J. Dyson. *Ann. of Math. 59*, 227–229. Problem: 5.24.

[668] Livesay, G. R. (1960). Fixed point free involutions on the 3-sphere. *Ann. of Math. 72*, 603–611. Problems: 3.37, 3.99.

[669] Livesay, G. R. (1963). Involutions with two fixed points on the three-sphere. *Ann. of Math. 78*, 582–593. Problems: 3.4, 3.99.

[670] Livingston, C. (1983). Knots which are not concordant to their reverses. *Quart. J. Math. Oxford Ser. (2) 34*, 323–328. Problem: 1.61.

[671] Livingston, C. (1987). Knots with finite weight commutator subgroups. *Proc. Amer. Math. Soc. 101*, 195–198. Problem: 1.14.

[672] Livingston, C. (1988). The free genus of doubled knots. *Proc. Amer. Math. Soc. 104*, 329–333. Problem: 1.20.

[673] Long, D. D. (1983). *Cobordism of knots and surface automorphisms.* Ph. D. thesis, Cambridge Univ. Problem: 1.62.

[674] Long, D. D. (1987). Immersions and embeddings of totally geodesic surfaces. *Bull. London Math. Soc. 19*, 481–484. Problem: 3.2.

[675] Long, D. D. & Niblo, G. A. (1991). Subgroup separability and 3-manifold groups. *Math. Zeit. 207*, 209–215. Problem: 3.76.

[676] Long, D. D. & Reid, A. W. (1995). Simple quotients of hyperbolic 3-manifold groups. Preprint. Problem: 3.58.

[677] Los, J. E. (1993). Pseudo-Anosov maps and invariant train tracks in the disc: a finite algorithm. *Proc. London Math. Soc. 66*, 400–430. Problem: 1.84.

[678] Los, J. E. (1994). Knots, braid index and dynamical type. *Topology 33*, 257–270. Problem: 1.84.

[679] Lubotzky, A. & Magid, A. R. (1985). *Varieties of representation of finitely generated groups*, Volume 58, number 336 of *Mem. Amer. Math. Soc.* American Math. Soc. Problem: 3.105.

[680] Lubotzky, A., Moses, S., & Raghunathan, M. S. (1993). Cyclic subgroups of exponential growth and metrics on discrete groups. *C.R. Acad. Sci. Paris Sér. I Math. 317*, 735–740. Problem: 2.16.

[681] Lück, W. (1994). L^2-Betti numbers of mapping tori and groups. *Topology 33*, 203–214. Problem: 4.59.

[682] Luecke, J. E. & Wu, Y.-Q. (1995). Relative Euler number and finite covers of graph manifolds. In W. H. Kazez (Ed.), *Geometric Topology, Proc. of the 1993 Georgia International Topology Conference*. International Press. Problem: 3.51.

[683] Luft, E. & Zhang, X. (1994). Symmetric knots and the cabling conjecture. *Math. Ann. 298*, 489–496. Problem: 1.79.

[684] Luke, R. & Mason, W. K. (1972). The space of homeomorphisms on a compact two-manifold is an absolute neighborhood retract. *Trans. Amer. Math. Soc. 164*, 275–285. Problem: 5.27.

[685] Luo, F. (1988). The existence of $K(\pi_1, 1)$ 4-manifolds which are rational homology 4-spheres. *Proc. Amer. Math. Soc. 104*, 1315–1321. Problem: 4.17.

[686] Lustig, M. & Moriah, Y. (1991). Nielsen equivalence in Fuchsian groups and Seifert fibered spaces. *Topology 30*, 191–204. Problems: 3.86, 3.88, 5.1.

[687] Lyon, H. C. (1972). Knots without unknotted incompressible spanning surfaces. *Proc. Amer. Math. Soc. 35*, 617–620. Problem: 1.20.

[688] Maeda, T. (1977). On a composition of knot groups. ii. algebraic bridge index. *Math. Sem. Notes Kobe Univ. 5*, 457–464. Problem: 4.29.

[689] Maeda, T. & Yajima, T. (1976). On a composition of knot groups. *Math. Sem. Notes, Kwansei Gakuin Univ. 25*, 105–109. Problem: 4.29.

[690] Mandelbaum, R. (1979). Decomposing analytic surfaces. In J. C. Cantrell (Ed.), *Geometric Topology (Georgia, 1977)*, pp. 147–218. Academic Press. Problem: 4.103.

[691] Mandelbaum, R. & Moishezon, B. (1976). On the topological structure of non-singular algebraic surfaces in $\mathbb{CP}^3$. *Topology 15*, 23–40. Problems: 4.12, 4.103.

[692] Mandelbaum, R. & Moishezon, B. (1983). Numerical invariants of links in 3-manifolds. In S. J. Lomonaco, Jr. (Ed.), *Low Dimensional Topology, (San Francisco, 1981)*, Volume 20 of *Contemp. Math.*, pp. 285–304. Problem: 4.18.

[693] Marden, A. (1974). The geometry of finitely generated Kleinian groups. *Ann. of Math. 99*, 383–462. Problem: 3.52.

[694] Margulis, G. A. (1967). Appendix to D. Anosov & Ya. Sinai, Some smooth ergodic systems. *Успехи Мат. Наук 22*, 107–172. Problem: 3.53.

[695] Margulis, G. A. (1991). *Discrete Subgroups of Semisimple Lie Groups*, Volume 17 of *Ergeb. Math. Grenzgeb. (3)*. Springer-Verlag. Problem: 2.15.

[696] Marin, A. (1977). La transversalité topologique. *Ann. of Math. 106*, 269–294. Problem: 4.52.

[697] Martin, G. J. (1988). Infinite group actions on spheres. *Rev. Mat. Iberoamericana. 4*, 407–451. Problem: 3.71.

[698] Martin, G. J. (1995). Triangle subgroups of Kleinian groups. To appear in *Comment. Math. Helv.*. Problem: 3.70.

[699] Martin, G. J. & Skora, R. (1989). Group actions on the two sphere. *Amer. J. Math. 111*, 387–402. Problems: 3.71, 5.10.

[700] Martin, G. J. & Tukia, P. (1988). Convergence and Möbius groups. In D. Drasin (Ed.), *Holomorphic Functions and Moduli (MSRI, 1986)*, Volume 10 of *Math. Sci. Res. Inst. Publ.*, pp. 113–140. Springer-Verlag. Problem: 3.71.

[701] Maruyama, N. (1984). Knot surgery descriptions of some closed orientable 3-manifolds. *J. Tsuda College 16*, 1–13. Problem: 4.28.

[702] Massey, W. S. (1960). On the Stiefel-Whitney class of a manifold. *Amer. J. Math. 82*, 92–102. Problem: 4.19.

[703] Massey, W. S. (1962). On the Stiefel-Whitney class of a manifold II. *Proc. Amer. Math. Soc. 13*, 938–942. Problem: 4.19.

[704] Massey, W. S. & Peterson, F. P. (1963). On the dual Stiefel-Whitney classes of a manifold. *Bol. Soc. Mat. Mexicana 8*, 1–13. Problem: 4.19.

[705] Masur, H. (1986). Closed trajectories for quadratic differentials with an application to billiards. *Duke Math. J. 53*, 307–314. Problem: 2.7.

[706] Mathieu, Y. (1990). *Sur des nœds qui ne sont pas determinés par leur complement et problems de chirurgie dans les variétés de dimension 3.* Ph. D. thesis, Marseille. Problem: 1.81.

[707] Matic, G. (1988). $SO(3)$-connections and rational homology cobordisms. *J. Differential Geom. 28*, 277–307. Problem: 4.5.

[708] Matsumoto, Y. (1975). A 4-manifold which admits no spine. *Bull. Amer. Math. Soc. 81*, 467–470. Problems: 4.25, 4.73.

[709] Matsumoto, Y. (1982a). On 4-manifolds fibered by tori. *Proc. Japan Acad. Ser. A Math. Sci. 58*, 298–301. Problem: 4.115.

[710] Matsumoto, Y. (1982b). On the bounding genus of homology 3-spheres. *J. Fac. Sci. Univ. Tokyo Sect. IA Math. 29*, 287–318. Problems: 4.2, 4.28.

[711] Matsumoto, Y. (1984). Good torus fibrations. In C. M. Gordon & R. C. Kirby (Eds.), *Four-Manifold Theory (Durham, NH, 1982)*, Volume 35 of *Contemp. Math.*, pp. 375–397. American Math. Soc. Problem: 4.115.

[712] Matsumoto, Y. (1986). An elementary proof of Rohlin's signature theorem and its extension by Guillou and Marin. In L. Guillou & A. Marin (Eds.), *A la Recherche de la Topologie Perdue*, Volume 62 of *Prog. Math.*, pp. 119–139. Birkhäuser. Problem: 4.112.

[713] Matsumoto, Y. (1989). Topology of torus fibrations. *Sugaku Expositions 2*, 55–73. Problem: 4.115.

[714] Matsumoto, Y. & Montesinos-Amilibia, J. M. (1991). A proof of Thurston's uniformization theorem of geometric orbifolds. *Tokyo J. Math. 14*, 181–196. Problem: 3.46.

[715] Matsumoto, Y. & Montesinos-Amilibia, J. M. (1994). Pseudo-periodic homeomorphisms and degeneration of Riemann surfaces. *Bull. Amer. Math. Soc. 30*, 70–75. Problems: 2.7, 4.99.

[716] Matsumoto, Y. & Montesinos-Amilibia, J. M. (1995). Pseudo-periodic maps and degeneration of Riemann surfaces, I and II. Preprint. Problem: 4.99.

[717] Matsumoto, Y. & Venema, G. A. (1979). Failure of the Dehn lemma on contractible 4-manifolds. *Invent. Math. 51*, 205–218. Problem: 4.25.

[718] Matumoto, T. (1976). *Variétés simpliciales d'homologie et variétés topologiques métrisables.* Ph. D. thesis, Orsay. Problem: 4.4.

[719] Matumoto, T. (1992). Homologically trivial smooth involutions on K3-surfaces. In Y. Matsumoto & S. Morita (Eds.), *Aspects of Low-Dimensional Manifolds*, Volume 20 of *Adv. Studies in Pure Math.*, pp. 365–374. Kinokuniya. Problem: 4.124.

[720] Matveev, S. V. (1987). Zeeman conjecture for unthickenable special polyhedra is equivalent to the Andrews–Curtis conjecture. *Sibirsk. Mat. Zh. 28*, 66–80. Problem: 5.2.

[721] Matveyev, R. (1995). A decomposition of smooth simply-connected h-cobordant 4-manifolds. Preprint. Problem: 4.90.

[722] Mayland, Jr., E. J. (1977). A class of 2-bridge knots with Property P. Preprint. Problem: 1.15.

[723] Mazur, B. (1959). On embeddings of spheres. *Bull. Amer. Math. Soc. 65*, 59–65. Problem: 4.88.

[724] Mazur, B. (1961). A note on some contractible 4-manifolds. *Ann. of Math. 73*, 221–228. Problem: 4.73.

[725] McCullough, D. (1985). Twist groups of compact 3-manifolds. *Topology 24*, 461–474. Problems: 2.4, 3.49.

[726] McCullough, D. (1986). Mappings of reducible 3-manifolds. In H. Toruńczyk (Ed.), *Proc. of the Semester in Geometric and Algebraic Topology of the Banach International Mathematical Center*, Warsaw, pp. 61–76. Banach Center Publ. Problem: 3.49.

[727] McCullough, D. (1990). Topological and algebraic automorphisms of 3-manifolds. In R. Piccinini (Ed.), *Groups of Self-Equivalences and Related Topics*, Volume 1425 of *Lecture Notes in Math.*, pp. 102–113. Springer-Verlag. Problem: 3.49.

[728] McCullough, D. (1991). Virtually geometrically finite mapping class groups of 3-manifolds. *J. Differential Geom. 33*, 1–65. Problem: 3.49.

[729] McCullough, D. & Miller, A. (1986a). *Homeomorphisms of 3-manifolds with compressible boundary*, Volume 60, number 344 of *Mem. Amer. Math. Soc.* American Math. Soc. Problems: 3.46, 3.49.

[730] McCullough, D. & Miller, A. (1986b). The genus 2 Torelli group is not finitely generated. *Topology Appl. 22*, 43–49. Problem: 2.9.

[731] McCullough, D. & Miller, A. (1989). Manifold covers of 3-orbifolds with geometric pieces. *Topology Appl. 31*, 169–185. Problem: 3.46.

[732] McCullough, D. & Rubinstein, J. H. (1989). The generalized Smale Conjecture for 3-manifolds with genus 2 one-sided Heegaard splittings. Preprint. Problem: 3.37.

[733] McDuff, D. (1987). Examples of symplectic structures. *Invent. Math. 89*, 13–36. Problem: 4.135.

[734] McDuff, D. (1990). The structure of rational and ruled symplectic 4-manifolds. *J. Amer. Math. Soc. 3*, 679–712. erratum *J. Amer. Math. Soc.*, 5 (1992), 987–988. Problem: 4.139.

[735] McDuff, D. (1991). Immersed spheres in symplectic 4-manifolds. *Ann. Inst. Fourier (Grenoble) 42*, 651–671. Problem: 4.139.

[736] McMillan, Jr., D. R. (1961). Cartesian products of contractible open manifolds. *Bull. Amer. Math. Soc. 67*, 510–514. Problems: 3.1, 3.7.

[737] McMillan, Jr., D. R. (1964). A criterion for cellularity in a manifold. *Ann. of Math. 79*, 327–337. Problem: 1.39.

[738] McMillan, Jr., D. R. (1970). Acyclicity in 3-manifolds. *Bull. Amer. Math. Soc. 76*, 942–964. Problem: 3.1.

[739] McMullen, C. T. (1989). Iteration on Teichmüller space. *Invent. Math. 97*, 95–127. Problem: 3.14.

[740] Meeks, III, W. H. (1979). Circles invariant under diffeomorphisms of finite order. *J. Differential Geom. 14*, 377–383. Problem: 2.8.

[741] Meeks, III, W. H. & Scott, G. P. (1986). Finite group actions on 3-manifolds. *Invent. Math. 86*, 287–346. Problems: 3.46, 3.47.

[742] Meeks, III, W. H., Simon, L., & Yau, S.-T. (1982). Embedded minimal surfaces, exotic spheres, and manifolds with positive Ricci curvature. *Ann. of Math. 116*, 621–659. Problems: 3.7, 3.67.

[743] Meeks, III, W. H. & Yau, S.-T. (1980). Topology of three dimensional manifolds and the embedding problems in minimal surface theory. *Ann. of Math. 112*, 441–484. Problems: 1.21, 3.12, 3.67.

[744] Meeks, III, W. H. & Yau, S.-T. (1982). The classical Plateau problem and the topology of three-dimensional manifolds. The embedding of the solution given by Douglas–Morrey and an analytic proof of Dehn's lemma. *Topology 21*, 409–442. Problems: 3.2, 3.39, 3.67.

[745] Melvin, P. M. (1977). *Blowing up and down in 4-manifolds.* Ph. D. thesis, UC Berkeley. Problems: 1.105, 4.24.

[746] Melvin, P. M. (1981). On 4-manifolds with singular torus actions. *Math. Ann. 256*, 255–276. Problem: 4.56.

[747] Melvin, P. M. (1984). 4-Dimensional oriented bordism. In C. M. Gordon & R. C. Kirby (Eds.), *Four-Manifold Theory (Durham, NH, 1982)*, Volume 35 of *Contemp. Math.*, pp. 399–405. American Math. Soc. Problem: 4.54.

[748] Menasco, W. W. & Reid, A. W. (1992). Totally geodesic surfaces in hyperbolic link complements. In B. N. Apanasov, W. D. Neumann, A. W. Reid, & L. C. Siebenmann (Eds.), *Topology 90, (Ohio St. U., 1990)*, pp. 215–226. W. de Gruyter. Problem: 1.76.

[749] Menasco, W. W. & Thistlethwaite, M. B. (1992). Surfaces with boundary in alternating knot exteriors. *J. Reine Angew. Math. 426*, 47–65. Problem: 1.79.

[750] Menasco, W. W. & Thistlethwaite, M. B. (1993). The classification of alternating links. *Ann. of Math. 138*, 113–171. Problems: 1.66, 1.64.

[751] Menasco, W. W. & Thompson, A. A. (1989). Compressing handlebodies with holes. *Topology 28*, 485–494. Problem: 3.11.

[752] Mess, G. (1992). The Torelli groups for genus 2 and 3 surfaces. *Topology 31*, 775–790. Problems: 2.1, 2.9.

[753] Mess, G. (1995a). The Seifert fibered space conjecture and groups which are coarse quasi-isometric to planes. Preprint. Problem: 3.5.

[754] Mess, G. (1995b). Unit tangent bundle subgroups of the mapping class group. Preprint. Problems: 2.14, 3.7.

[755] Metzler, W. (1976). Über den Homotopietyp zweidimensionaler CW-Komplexe und Elementartransformationen bei Darstellungen von Gruppen durch Erzeugende und definierende Relationen. *J. Reine Angew. Math. 285*, 7–23. Problem: 5.1.

[756] Metzler, W. (1990). Die Unterscheidung von Homotopietyp und einfachen Homotopietyp bei zweidimensionalen Komplexen. *J. Reine Angew. Math. 403*, 201–219. Problem: 5.1.

[757] Meyerhoff, G. R. (1985). The cusped hyperbolic 3-orbifold of minimum volume. *Bull. Amer. Math. Soc. 13*, 154–156. Problem: 3.60.

[758] Mihalik, M. (1995a). Artin and Coxeter groups are semistable at infinity. Preprint. Problem: 3.7.

[759] Mihalik, M. (1995b). Compactifying coverings of 3-manifolds. Preprint. Problem: 3.8.

[760] Mihalik, M. & Tschantz, S. (1995). Tame combings and the quasi-simply-filtered condition for groups. Preprint. Problem: 3.7.

[761] Mikasa, J. (1993). *On projections of spatial theta-curves.* Ph. D. thesis, Kwansei Gakuin Univ. In Japanese. Problem: 5.16.

[762] Milgram, R. J. (1985). Evaluating the Swan finiteness obstruction for periodic groups. In *Algebraic and Geometric Topology*, Volume 1126 of *Lecture Notes in Math.*, pp. 127–158. Springer-Verlag. Problem: 3.37.

[763] Milnor, J. W. (1954). Link groups. *Ann. of Math. 59*, 177–195. Problem: 1.97.

[764] Milnor, J. W. (1957a). Groups which act on S^n without fixed points. *Amer. J. Math. 79*, 623–630. Problems: 3.37, 4.121.

[765] Milnor, J. W. (1957b). Isotopy of links. In R. H. Fox, D. C. Spencer, & A. W. Tucker (Eds.), *Algebraic Geometry and Topology - A Symposium in Honor of Solomon Lefschetz*, Volume 12, pp. 280–306. Princeton Univ. Press. Problem: 1.96.

[766] Milnor, J. W. (1968a). Infinite cyclic coverings. In J. G. Hocking (Ed.), *Conference on the Topology of Manifolds (Michigan St. U., 1967)*, pp. 115–133. Prindle, Weber and Schmidt. Problem: 3.20.

[767] Milnor, J. W. (1968b). *Singular Points of Complex Hypersurfaces*, Volume 61 of *Ann. of Math. Stud.* Princeton Univ. Press. Problem: 1.4.

[768] Milnor, J. W. (1983). On polylogarithms, Hurwitz zeta functions, and their Kubert identities. *L'Enseign. Math. 29*, 281–322. Problem: 3.62.

[769] Milnor, J. W. & Husemoller, D. H. (1973). *Symmetric Bilinear Forms*, Volume 73 of *Ergeb. Math. Grenzgeb. (3)*. Springer-Verlag. Problem: 4.1.

[770] Milnor, J. W. & Thurston, W. P. (1977). Characteristic numbers of 3-manifolds. *L'Enseign. Math. 23*, 249–254. Problem: 3.16.

[771] Minsky, Y. (1994). On rigidity, limit sets, and end invariants of hyperbolic 3-manifolds. *J. Amer. Math. Soc. 7*, 539–588. Problems: 3.52, 3.56.

[772] Miyamoto, Y. (1994). Volumes of hyperbolic manifolds with geodesic boundary. *Topology 33*, 613–630. Problem: 3.60.

[773] Miyaoka, Y. (1974). Kähler metrics on elliptic surfaces. *Proc. Japan Acad. Ser. A Math. Sci. 50*, 533–536. Problem: 4.37.

[774] Miyaoka, Y. (1977). On the Chern numbers of surfaces of general type. *Invent. Math. 42*, 225–237. Problems: 4.39, 5.21.

[775] Moise, E. E. (1977). *Geometric Topology in Dimensions 2 and 3*, Volume 47 of *Graduate Texts in Math.* Springer-Verlag. Problem: 3.97.

[776] Moise, E. E. (1979). Statically tame periodic homeomorphisms of compact connected 3-manifolds, I. Homeomorphisms conjugate to rotations of the 3-sphere. *Trans. Amer. Math. Soc. 252*, 1–47. Problem: 3.42.

[777] Moise, E. E. (1980). Statically tame periodic homeomorphisms of compact connected 3-manifolds, II. Statically tame implies tame. *Trans. Amer. Math. Soc. 259*, 255–280. Problem: 3.42.

[778] Moishezon, B. G. & Teicher, M. (1987). Simply-connected algebraic surfaces of positive index. *Invent. Math. 89*, 601–643. Problems: 4.39, 4.91.

[779] Montesinos, J. M. (1978). 4-manifolds, 3-fold covering spaces and ribbons. *Trans. Amer. Math. Soc. 245*, 453–467. Problem: 4.113.

[780] Montesinos, J. M. (1985). A note on moves and on irregular covers of S^4. In J. R. Harper & R. Mandelbaum (Eds.), *Combinatorial Methods in Topology and Algebraic Geometry (Rochester, 1982)*, Volume 44 of *Contemp. Math.*, pp. 345–349. Problem: 4.113.

[781] Montesinos, J. M. & Whitten, W. (1986). Constructions of two-fold branched covering spaces. *Pacific J. Math. 125*, 415–446. Problems: 3.25, 3.86.

[782] Montgomery, D. C. & Yang, C. T. (1972). Differentiable pseudo-free circle actions on homotopy seven spheres. In *Conference on Compact Transformation Groups*, Lecture Notes in Math., pp. 41–101. Springer-Verlag. Proc. of a Conference in Univ. of Massachusetts, Amherst. Problem: 4.123.

[783] Montgomery, D. C. & Zippin, L. (1954). Examples of transformation groups. *Proc. Amer. Math. Soc. 5*, 460–465. Problem: 3.38.

[784] Moore, G. W. & Seiberg, N. (1989). Classical and quantum conformal field theory. *Comm. Math. Phys. 123*, 177–254. Problem: 2.2.

[785] Morgan, J. W. (1979). Non-singular Morse–Smale flows on 3-dimensional manifolds. *Topology 18*, 41–54. Problem: 3.9.

[786] Morgan, J. W. (1984). On Thurston's uniformization theorem for three-dimensional manifolds. In H. Bass & J. W. Morgan (Eds.), *The Smith Conjecture*, pp. 37–125. Academic Press. Problem: 3.14.

[787] Morgan, J. W. & Mrowka, T. S. (1992). A note on Donaldson's polynomial invariants. *Internat. Math. Res. Notices,* bound within *Duke Math. J. 68*, 223–230. Problem: 4.130.

[788] Morgan, J. W. & Mrowka, T. S. (1993). On the diffeomorphism classification of regular elliptic surfaces. *Internat. Math. Res. Notices,* bound within *Duke Math. J. 70*, 183–184. Problem: 4.11.

[789] Morgan, J. W. & O'Grady, K. G. (1994). *Differential Topology of Complex Surfaces*, Volume 1545 of *Lecture Notes in Math.* Springer-Verlag. Problem: 4.11.

[790] Morgan, J. W. & Shalen, P. B. (1984). Valuations, trees, and degenerations of hyperbolic structures, I. *Ann. of Math. 120*, 401–476. Problem: 3.14.

[791] Morgan, J. W. & Shalen, P. B. (1988a). Degenerations of hyperbolic structures, II: Measured laminations, trees, and 3-manifolds. *Ann. of Math. 127*, 403–456. Problem: 3.14.

[792] Morgan, J. W. & Shalen, P. B. (1988b). Degenerations of hyperbolic structures, III: Actions of 3-manifold groups on trees and Thurston's compactness theorem. *Ann. of Math. 127*, 457–520. Problem: 3.14.

[793] Morgan, J. W., Szabó, Z., & Taubes, C. H. (1995). The generalized Thom Conjecture. Preprint. Problem: 4.36.

[794] Moriah, Y. (1987). On the free genus of knots. *Proc. Amer. Math. Soc. 99*, 373–380. Problem: 1.20.

[795] Moriah, Y. (1988). Heegaard splittings of Seifert fibered spaces. *Invent. Math. 91*, 465–481. Problem: 3.86.

[796] Moriah, Y. & Rubinstein, J. H. (1993). Heegaard structures of negatively curved 3-manifolds. Preprint. Problem: 1.70.

[797] Moriah, Y. & Schultens, J. (1995). Irreducible Heegaard splittings of Seifert fibered spaces are either vertical or horizontal. Preprint. Problem: 3.86.

[798] Morimoto, K. (1995). There are knots whose tunnel numbers go down under connected sum. To appear in *Proc. Amer. Math. Soc.*. Problem: 1.70.

[799] Morimoto, K. & Sakuma, M. (1991). On unknotting tunnels for knots. *Math. Ann. 289*, 143–167. Problem: 1.71.

[800] Morimoto, K., Sakuma, M., & Yokota, Y. (1994). Concrete examples of tunnel number one knots which have the property $1+1=3$. Preprint. Problems: 1.70, 1.73.

[801] Morita, S. (1986). Finite coverings of punctured torus bundles and the first Betti number. *Sci. Papers College Arts Sci. Univ. Tokyo 35*, 109–121. Problem: 3.2.

[802] Morita, S. (1987). Characteristic classes of surface bundles. *Invent. Math. 90*, 551–577. Problem: 2.6.

[803] Morrey Jr., C. B. (1966). *Multiple Integrals in the Calculus of Variations.* Springer-Verlag. Problem: 3.66.

[804] Morton, H. R. (1977). A criterion for a surface in $\mathbb{R}^3$ to be unknotted. *Notices Amer. Math. Soc. 24*, A303. Abstract #77T–G56. Problem: 3.11.

[805] Morton, H. R. (1978). Infinitely many fibered knots having the same Alexander polynomial. *Topology 17*, 101–104. Problems: 1.7, 4.61.

[806] Morton, H. R. (1979). A criterion for an embedded surface in $\mathbb{R}^3$ to be unknotted. In R. A. Fenn (Ed.), *Topology of Low-Dimensional Manifolds (Sussex, 1977)*, Volume 722 of *Lecture Notes in Math.*, pp. 93–98. Springer-Verlag. Problem: 3.11.

[807] Morton, H. R. (1985). Exchangeable braids. In R. A. Fenn (Ed.), *Low Dimensional Topology (Sussex, 1982)*, Volume 95 of *London Math. Soc. Lect. Note Ser.*, pp. 86–105. Cambridge Univ. Press. Problem: 1.7.

[808] Morton, H. R. (1986). Threading knot diagrams. *Math. Proc. Cambridge Philos. Soc. 99*, 247–260. Problem: 1.84.

[809] Morton, H. R. (1988). Problems. In J. S. Birman & A. L. Libgober (Eds.), *Braids (Santa Cruz, 1986)*, Volume 78 of *Contemp. Math.*, pp. 557–574. American Math. Soc. Problems: 1.59, 1.91.

[810] Morton, H. R. & Traczyk, P. (1988). The Jones polynomial of satellite links around mutants. In J. S. Birman & A. L. Libgober (Eds.), *Braids (Santa Cruz, 1986)*, Volume 78 of *Contemp. Math.*, pp. 587–592. American Math. Soc. Problem: 1.91.

[811] Moser, L. E. (1974). On the impossibility of obtaining $S^2 \times S^1$ by elementary surgery along a knot. *Pacific J. Math. 53*, 519–524. Problem: 1.16.

[812] Mosher, L. E. & Oertel, U. (1995). Spaces which are not negatively curved. Preprint. Problem: 3.74.

[813] Mostow, G. D. (1967). Quasi-conformal mappings in n-space and the rigidity of hyperbolic space forms. *Inst. Hautes Études Sci. Publ. Math. 34*, 53–104. Problem: 3.95.

[814] Mostow, G. D. (1973). *Strong Rigidity of Locally Symmetric Spaces*, Volume 78 of *Ann. of Math. Stud.* Princeton Univ. Press. Problems: 3.39, 3.47.

[815] Motto, M. (1992). Maximal triads and prime decompositions of surfaces embedded in 3-manifolds. *Trans. Amer. Math. Soc. 331*, 851–867. Problem: 3.11.

[816] Mumford, D. (1961). The topology of normal singularities of an algebraic surface and a criterion for simplicity. *Inst. Hautes Études Sci. Publ. Math. 9*, 5–22. Problem: 3.28.

[817] Murakami, H. (1994). Quantum SU(2)-invariants dominate Casson's SU(2)-invariant. *Math. Proc. Cambridge Philos. Soc. 115*, 253–281. Problem: 3.108.

[818] Murasugi, K. (1963). On a certain subgroup of the group of an alternating link. *Amer. J. Math. 85*, 544–550. Problem: 1.83.

[819] Murasugi, K. (1965). On a certain numerical invariant of link types. *Trans. Amer. Math. Soc. 117*, 387–342. Problem: 1.26.

[820] Murasugi, K. (1987). Jones polynomials and classical conjectures in knot theory. *Topology 26*, 187–194. Problem: 1.66.

[821] Myers, R. (1975). Open book decompositions of 3-manifolds. *Notices Amer. Math. Soc. 22*, A651. Abstract 75T–G94. Problem: 3.13.

[822] Myers, R. (1981a). Free involutions on lens spaces. *Topology 20*, 313–318. Problem: 3.37.

[823] Myers, R. (1981b). Homology 3-spheres which admit no PL involutions. *Pacific J. Math. 94*, 379–384. Problem: 3.24.

[824] Myers, R. (1988). Contractible open 3-manifolds which are not covering spaces. *Topology 27*, 27–35. Problem: 3.7.

[825] Naimi, R. (1994). Foliations transverse to fibers of Seifert manifolds. *Comment. Math. Helv. 69*, 155–162. Problem: 2.19.

[826] Naimi, R. (1995). Constructing essential laminations in 2-bridge knot surgered 3-manifolds. Preprint. Problem: 1.15.

[827] Nakagawa, Y. (1975). A new class of knots with Property P. *Publ. Res. Inst. Math. Sci. 10*, 445–455. Problem: 1.15.

[828] Nakamura, I. (1984). On surfaces of class VII_0 with curves. *Invent. Math. 78*, 393–443. Problem: 4.102.

[829] Nakanishi, Y. (1981). A note on unknotting number. *Math. Sem. Notes Kobe Univ. 9*, 99–108. Problem: 1.69.

[830] Nakanishi, Y. (1984). Fox's congruence modulo (2,1). *Sūrikaisekikenkyūsho Kōkyūroku 518*, 96–101. Problem: 1.59.

[831] Nakanishi, Y. (1992). Unknotting operations. *Sūrikaisekikenkyūsho Kōkyūroku 813*, 102–110. Problem: 1.59.

[832] Nakanishi, Y. (1994). On generalized unknotting operations. *J. Knot Theory Ramifications 3*, 197–209. Problem: 1.59.

[833] Nakanishi, Y. & Suzuki, S. (1987). On Fox's congruence classes of knots. *Osaka J. Math. 24*, 217–225. Problem: 1.59.

[834] Neumann, W. D. (1981). A calculus for plumbing applied to the topology of complex algebraic surface singularities and degenerating complex curves. *Trans. Amer. Math. Soc. 268*, 299–343. Problem: 3.28.

[835] Neumann, W. D. (1989). Complex algebraic plane curves via their links at infinity. *Invent. Math. 198*, 445–489. Problem: 1.100.

[836] Neumann, W. D. (1995). Commensurability and virtual fibration for graph manifolds. Preprint. Problem: 3.51.

[837] Neumann, W. D. & Raymond, F. (1978). Seifert manifolds, plumbing, μ-invariants and orientation reversing maps. In K. C. Millett (Ed.), *Algebraic and Geometric Topology (Santa Barbara, 1977)*, Volume 664 of *Lecture Notes in Math.*, pp. 162–194. Springer-Verlag. Problem: 3.51.

[838] Neumann, W. D. & Reid, A. W. (1992a). Arithmetic of hyperbolic manifolds. In B. N. Apanasov, W. D. Neumann, A. W. Reid, & L. C. Siebenmann (Eds.), *Topology 90, (Ohio St. U., 1990)*, pp. 273–310. W. de Gruyter. Problems: 3.61, 3.64, 5.12.

[839] Neumann, W. D. & Reid, A. W. (1992b). Notes on Adams' small volume orbifolds. In B. N. Apanasov, W. D. Neumann, A. W. Reid, & L. C. Siebenmann (Eds.), *Topology 90, (Ohio St. U., 1990)*, pp. 311–314. W. de Gruyter. Problem: 3.60.

[840] Neumann, W. D. & Rudolph, L. (1987). Unfoldings in knot theory. *Math. Ann. 278*, 409–439. Problems: 1.83, 1.102.

[841] Neumann, W. D. & Rudolph, L. (1988). The enhanced Milnor number in higher dimensions. In U. Koschorke (Ed.), *Topology Symposium, Siegen 1987*, Volume 1350 of *Lecture Notes in Math.*, pp. 109–121. Springer-Verlag. Problem: 1.83.

[842] Neumann, W. D. & Rudolph, L. (1990). Difference index of vectorfields and the enhanced Milnor number. *Topology 29*, 83–100. Problem: 1.83.

[843] Neumann, W. D. & Yang (1995). Conjectures in K-theory and rationality of Chern–Simons invariants of hyperbolic 3-manifolds. To appear in *L'Enseign. Math.*. Problem: 3.62.

[844] Neuwirth, L. P. (1965). *Knot Groups*, Volume 56 of *Ann. of Math. Stud.* Princeton Univ. Press. Problem: 1.85.

[845] Neuzil, J. P. (1975). Surgery on a curve in a solid torus. *Trans. Amer. Math. Soc. 204*, 385–406. Problem: 1.15.

[846] Newman, M. H. A. (1931). A theorem on periodic transformations of spaces. *Quart. J. Math. Oxford Ser. (2) 2*, 1–8. Problem: 3.95.

[847] Nicas, A. J. (1986). An infinite family of non-Haken hyperbolic 3-manifolds with vanishing Whitehead groups. *Math. Proc. Cambridge Philos. Soc. 99*, 239–246. Problem: 3.2.

[848] Nicas, A. J. & Stark, C. W. (1984). Whitehead groups of certain hyperbolic manifolds. *Math. Proc. Cambridge Philos. Soc. 95*, 299–308. Problem: 3.2.

[849] Nicas, A. J. & Stark, C. W. (1987). Whitehead groups of certain hyperbolic manifolds. In S. M. Gersten & J. R. Stallings (Eds.), *Combinatorial Group Theory and Topology*, Volume 111 of *Ann. of Math. Stud.*, pp. 415–432. Princeton Univ. Press. Problem: 3.2.

[850] Noga, D. (1967). Über den Außenraum von Produktknoten und die Bedeutung der Fixgruppen. *Math. Zeit. 101*, 131–141. Problem: 1.15.

[851] Novikov, S. P. (1963). Topology of foliations. *Trans. Moscow Math. Soc. 14*, 268–305. Problem: 3.53.

[852] Odlyzko, A. M. & Sloane, N. J. A. (1981). On exceptions of integral quadratic forms. *J. Reine Angew. Math. 321*, 212–216. Problem: 4.19.

[853] Ohshika, K. (1995a). Kleinian groups which are limits of geometrically finite groups. Preprint. Problem: 3.52.

[854] Ohshika, K. (1995b). Rigidity and topological conjugates of topologically tame Kleinian groups. To appear in *Proc. Amer. Math. Soc.*. Problem: 3.52.

[855] Oliver, R. (1988). *Whitehead Groups of Finite Groups*, Volume 132 of *London Math. Soc. Lect. Note Ser.* Cambridge Univ. Press. Problem: 3.32.

[856] Ol'shanskii, A. Y. (1991). Периодические фактор-группы гиперболических групп. *Матхематический Сборник 182*, 543–567. English translation in [857].

[857] Ol'shanskii, A. Y. (1992). Periodic factor groups of hyperbolic groups. *Math. USSR.-Sb 72*, 519–541. Russian original in [856]. Problem: 3.73.

[858] Orlik, P. & Raymond, F. (1974). Actions of the torus on 4-manifolds, II. *Topology 13*, 89–112. Problem: 4.57.

[859] Orlik, P. & Solomon, L. (1980). Combinatorics and topology of complements of hyperplanes. *Invent. Math. 56*, 167–189. Problem: 4.145.

[860] Orlik, P. & Terao, H. (1992). *Arrangements of Hyperplanes*, Volume 300 of *Grundlehren Math. Wiss.* Springer-Verlag. Problem: 4.144.

[861] Orr, K. E. (1987). New link invariants and applications. *Comment. Math. Helv. 62*, 542–560. Problem: 1.56.

[862] Otal, J.-P. (1995). Le theoréme d'hyperbolisation pour les variétés fibrée de dimension 3. To appear in *Astérisque*. Problem: 3.14.

[863] Otal, J.-P. & Paulin, F. (1995). Géométrie hyperbolique et groupes kleiniens. In preparation. Problem: 3.14.

[864] Pachner, U. (1991). Homeomorphic manifolds are equivalent by elementary shellings. *European. J. Combin. 12*, 129–145. Problem: 5.13.

[865] Page, D. (1979). A compact rotating gravitational instanton. *Phys. Lett. B 79*, 235–238. Problem: 4.68.

[866] Palis, Jr., J. & de Melo, W. (1982a). *Geometric Theory of Dynamical systems, An Introduction.* Springer-Verlag. Translation of [867]. Problem: 3.109.

[867] Palis, Jr., J. & de Melo, W. (1982b). *Introdução aos Sistemas Dinâmicos*, Volume 6 of *Projeto Euclides.* Springer-Verlag. Translation to English in [866].

[868] Palmeira, C. F. B. (1978). Open manifolds foliated by planes. *Ann. of Math. 107*, 109–131. Problem: 3.7.

[869] Paterson, A. L. T. (1988). *Amenability*, Volume 29 of *Math. Surveys and Monographs.* American Math. Soc. Problem: 5.9.

[870] Peixoto, M. M. (1962). Structural stability on two-dimensional manifolds. *Topology 1*, 101–120. Problem: 3.109.

[871] Perko, K. A. (1975). Octahedral knot covers. In L. P. Neuwirth (Ed.), *Knots, Groups, and 3-Manifolds*, Volume 84 of *Ann. of Math. Stud.*, pp. 47–50. Princeton Univ. Press. Papers Dedicated to the Memory of R. H. Fox. Problem: 1.24.

[872] Perko, K. A. (1976). On dihedral covering spaces of knots. *Invent. Math. 34*, 77–82. Problems: 1.25, 1.26.

[873] Perron, B. (1986). Pseudo-isotopies. *Topology 25*, 381–397. Problem: 4.35.

[874] Petrie, T. (1975). Equivariant quasi-equivalence, transversality and normal cobordism. In R. D. James (Ed.), *Proc. International Congress of Mathematicians, Vancouver 1974*, pp. 537–541. Problem: 4.123.

[875] Piergallini, R. (1995). Four-manifolds as 4-fold branched covers of S^4. *Topology 34*, 497–508. Problem: 4.113.

[876] Plante, J. (1981). Anosov flows, transversally affine foliations, and a conjecture of Verjovsky. *J. London Math. Soc. 2*, 359–362. Problem: 3.53.

[877] Plotnick, S. P. (1980). Vanishing of Whitehead groups for Seifert manifolds with infinite fundamental group. *Comment. Math. Helv. 55*, 654–667. Problem: 3.32.

[878] Plotnick, S. P. (1983). Homotopy type of 4-dimensional knot complements. *Math. Zeit. 183*, 447–471. Problem: 1.55.

[879] Plotnick, S. P. (1986). Equivariant intersection forms, knots in S^4, and rotation in 2-spheres. *Trans. Amer. Math. Soc. 296*, 543–575. Problem: 4.84.

[880] Poénaru, V. (1969). Extension des immersions en codimension 1 (d'après Samuel Blank). In *Séminaire Bourbaki, Exposés 341-346 1967/1968*. W. A. Benjamin, Inc. Problem: 3.19.

[881] Poénaru, V. (1974). A remark on simply-connected 3-manifolds. *Bull. Amer. Math. Soc. 80*, 1203–1204. Problem: 3.1.

[882] Poénaru, V. (1991). Killing handles of index one stably and π_1^∞. *Duke Math. J. 63*, 431–447. Problem: 3.7.

[883] Poénaru, V. (1992). Almost convex groups, Lipschitz combing, and π_1^∞ for universal covering spaces of closed 3-manifolds. *J. Differential Geom. 35*, 103–130. Problem: 3.7.

[884] Poénaru, V. (1994). Geometry "a la Gromov" for the fundamental group of a closed 3-manifold M^3 and the simple connectivity at infinity of $\tilde{M}^3$. *Topology 33*, 181–196. Problem: 3.7.

[885] Poénaru, V. & Tanasi, C. (1993). k-Weakly almost convex groups and $\pi_1^\infty(\tilde{M}^3)$. *Geom. Dedicata 48*, 57–81. Problem: 3.7.

[886] Poincaré, H. (1892). *Les Methodes Nouvelles de la Mecanique Céleste*, Volume 1. Gauthier-Villars. Problem: 3.109.

[887] Polterovich, L. (1991). The surgery of Lagrange manifolds. *Geom. Funct. Anal. 1*, 198–210. Problem: 4.111.

[888] Powell, J. (1978). Two theorems on the mapping class group of a surface. *Proc. Amer. Math. Soc. 68*, 347–350. Problems: 2.1, 2.13.

[889] Prasad, G. (1973). Strong rigidity of $\mathbb{Q}$-rank 1 lattices. *Invent. Math. 21*, 255–286. Problems: 3.39, 3.47.

[890] Przytycka, T. M. & Przytycki, J. H. (1987). Signed dichromatic graphs of oriented link diagrams and matched diagrams. Preprint, Univ. of British Columbia. Problem: 1.60.

[891] Przytycka, T. M. & Przytycki, J. H. (1992). Subexponentially computable truncations of Jones-type polynomials, with appendix on Vertigan's algorithm. Preprint, Odense Univ. Problem: 1.90.

[892] Przytycka, T. M. & Przytycki, J. H. (1993). Subexponentially computable truncations of Jones-type polynomials. In *Graph Structure Theory*, Volume 147 of *Contemp. Math.*, pp. 63–108. American Math. Soc. Proc. AMS-IMS-SIAM Joint Summer Res. Conf. on Graph Minors, 1991. Problems: 1.60, 1.90.

[893] Przytycki, J. H. (1986). Survey on recent invariants in classical knot theory. Uniwersytet Warszawski, Preprints 6,8,9. Problem: 1.91.

[894] Przytycki, J. H. (1988). t_k-moves on links. In J. S. Birman & A. L. Libgober (Eds.), *Braids (Santa Cruz, 1986)*, Volume 78 of *Contemp. Math.*, pp. 615–656. American Math. Soc. Problem: 1.59.

[895] Przytycki, J. H. (1989). Equivalence of cables of mutants of knots. *Canad. J. Math 26*, 250–478. Problem: 1.91.

[896] Przytycki, J. H. (1990). $t_3, \bar{t}_4$ moves conjecture for oriented links with matched diagrams. *Math. Proc. Cambridge Philos. Soc. 108*, 55–61. Problem: 1.59.

[897] Przytycki, J. H. (1991). Skein modules of 3-manifolds. *Bull. Polish Acad. Sci. Math. 39*, 91–100. Problem: 1.92.

[898] Przytycki, J. H. (1992a). Quantum group of links in a handlebody. In M. Gerstenhaber & J. D. Stasheff (Eds.), *Deformation Theory and Quantum Groups with Applications to Mathematical Physics*, Volume 134 of *Contemp. Math.*, pp. 235–245. Problem: 1.92.

[899] Przytycki, J. H. (1992b). Skein module of links in a handlebody. In B. N. Apanasov, W. D. Neumann, A. W. Reid, & L. C. Siebenmann (Eds.), *Topology 90, (Ohio St. U., 1990)*, pp. 315–342. W. de Gruyter. Problem: 1.92.

[900] Przytycki, J. H. (1993). Elementary conjectures in classical knot theory. In L. H. Kauffman & R. A. Baadhio (Eds.), *Quantum Topology*, Volume 3 of *Ser. Knots and Everything*, pp. 292–320. World Scientific. Problems: 1.59, 1.60.

[901] Przytycki, J. H. (1994a). 3-coloring and other elementary invariants of knots. Odense preprint # 26. Problem: 1.59.

[902] Przytycki, J. H. (1994b). $H_1(M^3)$ and its skein module quantizations. *Abstracts Amer. Math. Soc. 15*, 322. Abstract 891–57–180. Problem: 1.92.

[903] Przytycki, J. H. (1994c). Vassiliev-Gusarov skein modules of 3-manifolds and criteria for periodicity of knots. In K. Johannson (Ed.), *Low-Dimensional Topology, Knoxville, 1992*, pp. 157–176. International Press. Problem: 1.91.

[904] Przytycki, J. H. & Traczyk, P. (1987). Invariants of links of Conway type. *Kobe J. Math. 4*, 115–139. Problem: 1.90.

[905] Pugh, C. C. (1967). The Closing Lemma. *Amer. J. Math. 89*, 956–1009. Problem: 3.109.

[906] Pugh, C. C. & Robinson, C. (1983). The C^1 Closing Lemma, including Hamiltonians. *Ergod. Th. & Dynam. Sys. 3*, 261–313. Problem: 3.109.

[907] Quillen, D. (1970). Cohomology of groups. In *Actes du Congrès International de Mathématiciens - Nice*, pp. 47–51. Gauthier-Villars. Problem: 3.63.

[908] Quillen, D. (1975). Higher algebraic k-theory. In R. D. James (Ed.), *Proc. International Congress of Mathematicians, Vancouver 1974*, pp. 171–176. Problem: 3.63.

[909] Quinn, F. S. (1982). Ends of maps, III: dimensions 4 and 5. *J. Differential Geom. 17*, 503–521. Problem: 4.52.

[910] Quinn, F. S. (1983). Resolutions of homology manifolds and the topological characterization of manifolds. *Invent. Math. 72*, 267–284. Problem: 4.69.

[911] Quinn, F. S. (1986). Isotopy of 4-manifolds. *J. Differential Geom. 24*, 343–372. Problem: 4.35.

[912] Quinn, F. S. (1987). An obstruction to the resolution of homology manifolds. *Michigan Math. J. 34*, 285–291. Problem: 4.69.

[913] Quinn, F. S. (1988). Topological transversality holds in all dimensions. *Bull. Amer. Math. Soc. 18*, 145–148. Problems: 4.7, 4.52.

[914] Ramakrishnan, D. (1989). Regulators, algebraic cycles, and values of L-functions. In M. R. Stein & R. K. Dennis (Eds.), *Algebraic K-Ttheory and Algebraic Number Theory*, Volume 83 of *Contemp. Math.*, pp. 183–310. Problem: 3.62.

[915] Ramanujam, C. P. (1971). A topological characterization of the affine plane as an algebraic variety. *Ann. of Math. 94*, 69–88. Problem: 4.78.

[916] Randall, A. D. & Schweitzer, P. A. (1994). On foliations, concordance spaces, and the Smale conjectures. In *Differential Topology, Foliations, and Group Actions*, Contemp. Math., pp. 235–258. Proc. of Conf. in Rio de Janeiro, 1992. Problem: 4.126.

[917] Randell, R. C. (1982). The fundamental group of the complement of a union of complex hyperplanes. *Invent. Math. 69*, 103–108. See also the correction in [918]. Problem: 4.145.

[918] Randell, R. C. (1985). The fundamental group of the complement of a union of complex hyperplanes: correction. *Invent. Math. 80*, 467–468. A correction to [917]. Problem: 4.145.

[919] Raymond, F. & Scott, L. (1977). Failure of Nielsen's Theorem in higher dimensions. *Archiv. Math. (Basel) 29*, 643–654. Problem: 3.40.

[920] Reid, A. W. (1990). A note on trace fields of Kleinian groups. *Bull. London Math. Soc. 22*, 349–352. Problem: 3.61.

[921] Reid, A. W. (1991). Arithmeticity of knot complements. *J. London Math. Soc. 43*, 171–184. Problem: 3.64.

[922] Reid, A. W. (1995). A non-Haken hyperbolic 3-manifold covered by a surface bundle. *Pacific J. Math. 167*, 163–182. Problem: 3.51.

[923] Reshetikhin, N. Y. & Turaev, V. G. (1991). Invariants of 3-manifolds via link polynomials and quantum groups. *Invent. Math. 103*, 547–597. Problem: 3.108.

[924] Reznikov, A. (1995a). All regulators of flat bundles are torsion. *Ann. of Math. 141*, 373–386. Problem: 3.63.

[925] Reznikov, A. (1995b). Rationality of secondary classes. To appear in *J. Differential Geom.*. Problem: 3.63.

[926] Rice, P. M. (1968). Free actions of $\mathbb{Z}_4$ on S^3. *Duke Math. J. 36*, 749–751. Problem: 3.37.

[927] Riley, R. (1974). Knots with the parabolic property P. *Quart. J. Math. Oxford Ser. (2) 25*, 273–283. Problem: 1.15.

[928] Ritter, G. X. (1975). Free Z_8 actions on S^3. *Trans. Amer. Math. Soc. 181*, 195–212. Problem: 3.37.

[929] Roberts, R. (1995). Constructing taut foliations. To appear in *Comment. Math. Helv.*. Problem: 1.15.

[930] Robertson, N., Seymour, P. D., & Thomas, R. (1993). Linkless embeddings of graphs in 3-space. *Bull. Amer. Math. Soc. 28*, 84–89. Problem: 5.15.

[931] Rochlin, V. A. (1971). Two-dimensional submanifolds of 4-dimensional manifolds. *Functional Anal. Appl. 5*, 39–48. Problems: 4.18, 4.36.

[932] Rolfsen, D. P. (1976). *Knots and Links.* Publish or Perish. Problem: 4.6.

[933] Rong, Y. W. (1991). The Kauffman polynomial and the two-fold cover of a link. *Indiana Univ. Math. J. 40*, 321–331. Problem: 1.91.

[934] Rong, Y. W. (1992). Degree one maps between geometric 3-manifolds. *Trans. Amer. Math. Soc. 332*, 411–436. Problem: 3.100.

[935] Rong, Y. W. (1995a). Link polynomials of higher order. Preprint. Problem: 1.92.

[936] Rong, Y. W. (1995b). Some knots not determined by their complements. To appear in *Topology Appl.*. Problem: 1.81.

[937] Rong, Y. W. & Wang, S. C. (1992). The pre-images of submanifolds. *Math. Proc. Cambridge Philos. Soc. 112*, 271–279. Problem: 3.100.

[938] Rosenberg, H. (1974). Un contre-exemple à la conjecture de Seifert. In *Séminaire Bourbaki, Exposés 418–435, 1972/73*, Volume 383 of *Lecture Notes in Math.*, pp. 294–305. Springer-Verlag. Problem: 2.20.

[939] Ruan, Y. (1992). A universal constraint for Donaldson invariants. *J. Differential Geom. 35*, 711–722. Problem: 4.132.

[940] Ruan, Y. (1993). Symplectic topology and extremal rays. *Geom. Funct. Anal. 3*, 395–439. Problem: 4.135.

[941] Ruan, Y. (1994). Symplectic topology and algebraic 3-folds. *J. Differential Geom. 39*, 215–227. Problem: 4.136.

[942] Ruan, Y. & Tian, G. (1995). Higher genus symplectic invariants and σ-model coupled with gravity. Preprint. Problem: 4.136.

[943] Ruberman, D. (1982). Imbedding four-manifolds and slicing links. *Math. Proc. Cambridge Philos. Soc. 91*, 107–110. Problem: 4.19.

[944] Ruberman, D. (1988). Rational homology cobordisms of rational space forms. *Topology 27*, 401–414. Problem: 4.5.

[945] Ruberman, D. (1990). Seifert surfaces of knots in S^4. *Pacific J. Math. 145*, 97–116. Problems: 4.60, 4.116.

[946] Ruberman, D. (1995). Involutions on spin 4-manifolds. *Proc. Amer. Math. Soc. 123*, 593–596. Problem: 4.124.

[947] Rubinstein, J. H. (1976). Heegaard splittings and a theorem of Livesay. *Proc. Amer. Math. Soc. 60*, 317–320. Problems: 3.4, 3.99.

[948] Rubinstein, J. H. (1979a). Free actions of some finite groups on S^3, I. *Math. Ann. 240*, 165–175. Problem: 3.37.

[949] Rubinstein, J. H. (1979b). On 3-manifolds that have finite fundamental group and contain Klein bottles. *Trans. Amer. Math. Soc. 251*, 129–137. Problems: 3.37, 3.95.

[950] Rubinstein, J. H. (1994). An algorithm to recognize S^3. In *Proc. International Congress Mathematicians, Zurich 1994*. Problem: 3.37.

[951] Rubinstein, J. H. (1995). Polyhedral minimal surfaces, Heegaard splittings and decision problems for 3-dimensional manifolds. In W. H. Kazez (Ed.), *Geometric Topology, Proc. of the 1993 Georgia International Topology Conference*. International Press. Problem: 3.82.

[952] Rubinstein, J. H. & Scharlemann, M. G. (1995). Comparing Heegaard splittings of non-Haken 3-manifolds. Preprint. Problem: 3.89.

[953] Rubinstein, J. H. & Wang, S. C. (1995). π_1-injective surfaces in graph manifolds. Preprint. Problem: 3.2.

[954] Rudolph, L. (1976). A Morse function for a surface in $\mathbb{CP}^3$. *Topology 14*, 301–303. Problem: 4.18.

[955] Rudolph, L. (1982). Embeddings of the line in the plane. *J. Reine Angew. Math. 337*, 113–118. Problem: 1.102.

[956] Rudolph, L. (1983). Algebraic functions and closed braids. *Topology 22*, 191–202. Problem: 1.40.

[957] Rudolph, L. (1984). Some topologically locally-flat surfaces in the complex projective plane. *Comment. Math. Helv. 59*, 592–599. Problem: 4.36.

[958] Rudolph, L. (1987). Isolated critical points of mapping from $\mathbb{R}^4$ to $\mathbb{R}^2$ and a natural splitting of the Milnor number of a classical fibered link I, Basic theory, Examples. *Comment. Math. Helv. 62*, 630–645. Problem: 1.83.

[959] Rudolph, L. (1990). A congruence between link polynomials. *Math. Proc. Cambridge Philos. Soc. 107*, 319–327. Problem: 1.40.

[960] Rudolph, L. (1993). Quasipositivity as an obstruction to sliceness. *Bull. Amer. Math. Soc. 29*, 51–59. Problems: 1.4, 1.38, 1.100, 4.74.

[961] Rudolph, L. (1995). An obstruction to sliceness via "classical" gauge theory and contact geometry. *Invent. Math. 119*, 155–163. Problem: 1.38.

[962] Sacks, J. & Uhlenbeck, K. K. (1981). The existence of minimal immersions of 2-spheres. *Ann. of Math. 113*, 1–24. Problem: 3.12.

[963] Sato, N. (1984). Cobordism of semi-boundary links. *Topology Appl. 18*, 225–234. Problem: 1.56.

[964] Sato, Y. (1994). Scharlemann's 4-manifolds and smooth 2-knots in $S^2 \times S^2$. *Proc. Amer. Math. Soc. 121*, 1289–1294. Problem: 4.15.

[965] Scharlemann, M. G. (1976a). Constructing strange manifolds with the dodecahedral space. *Duke Math. J. 43*, 33–40. Problems: 4.15, 4.85.

[966] Scharlemann, M. G. (1976b). Transversality theories at dimension four. *Invent. Math. 33*, 1–14. Problem: 4.7.

[967] Scharlemann, M. G. (1984). The four-dimensional Schoenflies conjecture is true for genus two imbeddings. *Topology 23*, 211–217. Problem: 4.32.

[968] Scharlemann, M. G. (1985a). Smooth spheres in $\mathbb{R}^4$ with four critical points are standard. *Invent. Math. 79*, 125–141. Problem: 1.2.

[969] Scharlemann, M. G. (1985b). Unknotting number one knots are prime. *Invent. Math. 82*, 37–55. Problem: 1.69.

[970] Scharlemann, M. G. (1989). Sutured manifolds and generalized Thurston norms. *J. Differential Geom. 29*, 557–614. Problem: 1.1.

[971] Scharlemann, M. G. (1990). Producing reducible 3-manifolds by surgery on a knot. *Topology 29*, 481–500. Problem: 1.79.

[972] Scharlemann, M. G. & Thompson, A. A. (1989). Link genus and the Conway moves. *Comment. Math. Helv. 64*, 527–535. Problem: 1.58.

[973] Schoen, R. & Yau, S.-T. (1979). Existence of incompressible minimal surfaces and the topology of three-dimensional manifolds with non-negative scalar curvature. *Ann. of Math. 110*, 127–142. Problem: 3.67.

[974] Schultens, J. (1993). The classification of Heegaard splittings for (compact orientable surface) $\times S^1$. *Proc. London Math. Soc. 67*, 425–448. Problems: 3.22, 3.86, 3.88.

[975] Schultens, J. (1995). Weakly reducible Heegaard splittings of Seifert fibered spaces. Preprint. Problems: 3.86, 3.88, 3.89.

[976] Schweitzer, P. A. (1974). Counterexamples to the Seifert conjecture and opening closed leaves of foliations. *Ann. of Math. 100*, 386–400. Problem: 3.113.

[977] Scott, G. P. (1978). Subgroups of surface groups are almost geometric. *J. London Math. Soc. 17*, 555–565. See correction in [982]. Problem: 3.76.

[978] Scott, G. P. (1980). A new proof of the annulus and torus theorems. *Amer. J. Math. 102*, 241–277. Problem: 3.45.

[979] Scott, G. P. (1983a). The geometries of 3-manifolds. *Bull. London Math. Soc. 15*, 401–487. Problem: 3.45.

[980] Scott, G. P. (1983b). There are no fake Seifert fibre spaces with infinite π_1. *Ann. of Math. 117*, 35–70. Problems: 3.3, 3.5.

[981] Scott, G. P. (1985a). Homotopy implies isotopy for some Seifert fibre spaces. *Topology 24*, 341–352. Problems: 3.76, 3.95.

[982] Scott, G. P. (1985b). Subgroups of surface groups are almost geometric. *J. London Math. Soc. 32*, 217–220. Correction to [977]. Problem: 3.76.

[983] Seaman, W. (1987). On four manifolds which are positively pinched. *Ann. Global Anal. Geom. 5*, 193–198. Problem: 4.67.

[984] Seaman, W. (1988). Some remarks on positively curved 4-manifolds. *Michigan Math. J. 35*, 179–183. Problem: 4.67.

[985] Seaman, W. (1989). A pinching theorem for four-manifolds. *Geom. Dedicata 31*, 37–40. Problem: 4.67.

[986] Seifert, H. (1933). Topologie dreidimensionaler gefaserte Raüme. *Acta Math. 60*, 147–238. English translation in [988]. Problems: 1.77, 4.123.

[987] Seifert, H. (1936). Algebraische approximation von Mannigfaltigkeiten. *Math. Zeit. 41*, 1–17. Problem: 3.11.

[988] Seifert, H. (1980). Topology of 3-dimensional fibered spaces. In *A Textbook of Topology*, Volume 89 of *Pure and Applied Mathematics*. Academic Press. Transl. of [986], published as an Appendix [991, pages 359-422].

[989] Seifert, H. & Threlfall, W. (1934). *Lehrbuch der Topologie*. Leipzig: B. G. Teubner. English and Spanish transl. in [991], [990], also available in Russian and Chinese.

[990] Seifert, H. & Threlfall, W. (1951). *Lecciones de Topologia*, Volume 1 of *Coleccion de textos de matemática moderna*. Instituto Jorge Juan de Matemáticas. Transl. of [989].

[991] Seifert, H. & Threlfall, W. (1980). *A Textbook of Topology*, Volume 89 of *Pure and Applied Mathematics*. Academic Press. Transl. of [989].

[992] Shiga, H. & Tanigawa, H. (1989). On the Maskit coordinates of Teichmüller space and modular transformations. *Kodai Math. J. 12*, 437–43. Problem: 2.7.

[993] Shimabara, M. (1988). Knots in certain spatial graphs. *Tokyo J. Math. 11*, 405–413. Problem: 5.15.

[994] Shubert, H. (1953). Knoten und Vollringe. *Acta Math. 90*, 131–286. Problem: 1.6.

[995] Siebenmann, L. C. (1970). Are nontriangulable manifolds triangulable? In J. C. Cantrell & C. H. Edwards, Jr. (Eds.), *Topology of Manifolds, (Georgia, 1969)*, pp. 77–84. Markham Pub. Co. Problem: 3.1.

[996] Siebenmann, L. C. (1980). On vanishing of the Rohlin invariant and nonfinitely amphicheiral homology. In U. Koschorke & W. D. Neumann (Eds.), *Topology Symposium, Siegen 1979*, Volume 788 of *Lecture Notes in Math.*, pp. 172–222. Problem: 3.24.

[997] Simon, J. (1970). Some classes of knots with Property P. In J. C. Cantrell & C. H. Edwards, Jr. (Eds.), *Topology of Manifolds, (Georgia, 1969)*, pp. 195–199. Markham Pub. Co. Problem: 1.15.

[998] Simon, J. (1971). On knots with nontrivial interpolating manifolds. *Trans. Amer. Math. Soc. 160*, 467–473. Problem: 1.15.

[999] Simon, J. (1973). An algebraic classification of knots in S^3. *Ann. of Math. 97*, 1–13. Problem: 1.13.

[1000] Simon, J. (1975). On the problems of determining knots by their complements and knot complements by their groups. *Proc. Amer. Math. Soc. 57*, 140–142. Problem: 1.13.

[1001] Simon, J. (1976). Compactification of covering spaces of compact 3-manifolds. *Michigan Math. J. 23*, 245–256. Problem: 3.8.

[1002] Siu, Y.-T. (1983). Every $K3$ surface is Kähler. *Invent. Math. 73*, 139–150. Problem: 4.37.

[1003] Skinner, A. (1994). The word problem in a class of non-Haken 3-manifolds. *Topology 33*, 215–239. Problem: 3.81.

[1004] Smale, S. (1959). Diffeomorphisms of the 2-sphere. *Proc. Amer. Math. Soc. 10*, 621–626. Problem: 4.126.

[1005] Smale, S. (1967). Differential dynamical systems. *Bull. Amer. Math. Soc. 73*, 747–819. Problem: 3.109.

[1006] Smillie, J. (1995). Teichmüller disks in moduli space. Preprint. Problem: 2.7.

[1007] Smith, P. A. (1939). Transformations of finite period, II. *Ann. of Math. 40*, 690–711. Problem: 3.38.

[1008] Smythe, N. (1966). Boundary links. In R. H. Bing & R. J. Bean (Eds.), *Topology Seminar, Wisconsin 1965*, Volume 60 of *Ann. of Math. Stud.*, pp. 69–72. Princeton Univ. Press. Problem: 1.36.

[1009] Stallings, J. R. (1965). Homology and central series of groups. *J. Algebra 2*, 170–181. Problem: 3.79.

[1010] Stallings, J. R. (1978). Construction of fibered knots and links. In R. J. Milgram (Ed.), *Algebraic and Geometric Topology (Stanford, 1976)*, Volume 32, part II of *Proc. Sympos. Pure Math.*, pp. 55–60. American Math. Soc. Problem: 1.83.

[1011] Stern, R. J. (1978). Some more Brieskorn spheres which bound contractible manifolds. *Notices Amer. Math. Soc. 25*, A448. Abstract 78T–G75. Problem: 4.2.

[1012] Stipsicz, A. & Szabó, Z. (1994a). Classification of complex surfaces. *Turkish J. Math. 18*, 42–52. Problem: 4.128.

[1013] Stipsicz, A. & Szabó, Z. (1994b). The smooth classification of elliptic surfaces with $b^+ > 1$. *Duke Math. J. 75*, 1–50. Problem: 4.11.

[1014] Stolz, S. (1988). Exotic structures on 4-manifolds detected by spectral invariants. *Invent. Math. 94*, 149–162. Problem: 4.13.

[1015] Stong, R. (1968). *Notes on Cobordism Theory*, Volume 7 of *Math. Notes*. Princeton Univ. Press. Problem: 4.54.

[1016] Sullivan, D. P. (1975). On the intersection ring of compact three manifolds. *Topology 14*, 275–278. Problem: 3.27.

[1017] Sullivan, D. P. (1979). Hyperbolic geometry and homeomorphisms. In J. C. Cantrell (Ed.), *Geometric Topology*, pp. 543–555. Academic Press. Proc. of the 1977 Georgia Topology Conf. Problems: 4.44, 5.25.

[1018] Sullivan, D. P. (1981). Travaux de Thurston sur les groupes quasi-fuchsien et les variétés hyperboliques de dimension 3 fibrées sur S^1. In *Séminaire Bourbaki, Exposés 543–560, 1979/80*, Volume 842 of *Lecture Notes in Math.*, pp. 196–214. Springer-Verlag. Problem: 3.14.

[1019] Suslin, A. A. (1984). On the k-theory of local fields. *J. Pure Appl. Algebra 34*, 301–318. Problem: 3.63.

[1020] Suslin, A. A. (1987). Algebraic K-theory of fields. In A. M. Gleason (Ed.), *Proc. International Congress Mathematicians, Berkeley 1986*, pp. 222–244. American Math. Soc. Problem: 3.63.

[1021] Suzuki, S. (1975a). On surfaces in 3-sphere: prime decompositions. *Hokkaido Math. J. 4*, 179–195. Problem: 3.11.

[1022] Suzuki, S. (1975b). On surfaces in 3-sphere: stable congruence. *Math. Japon. 20*, 65–83. Problem: 3.11.

[1023] Suzuki, S. (1977). On homeomorphisms of a 3-dimensional handlebody. *Canad. J. Math 29*, 111–124. Problem: 2.4.

[1024] Swan, R. G. (1960). Periodic resolutions for finite groups. *Ann. of Math. 72*, 267–291. Problem: 3.37.

[1025] Swarup, G. A. (1975). An unknotting criterion. *J. Pure Appl. Algebra 6*, 291–296. Problem: 1.51.

[1026] Swarup, G. A. (1977). Relative version of a theorem of Stallings. *J. Pure Appl. Algebra 11*, 75–82. Problem: 1.55.

[1027] Swarup, G. A. (1980). Two finiteness properties in 3-manifolds. *Bull. London Math. Soc. 12*, 296–302. Problem: 3.3.

[1028] Swarup, G. A. (1986). A remark on cable knots. *Bull. Amer. Math. Soc. 18*, 401–402. Problem: 1.13.

[1029] Swarup, G. A. (1993). A note on accessibility. In G. A. Niblo & M. A. Roller (Eds.), *Geometric Group Theory I (Sussex, 1991)*, Volume 181 of *London Math. Soc. Lect. Note Ser.*, pp. 204–207. Cambridge Univ. Press. Problem: 1.51.

[1030] Tait, P. G. (1876-77). On knots. *Trans. Royal Soc. Edinburgh 28*, 145–190. Republished in [1031]. Problem: 1.66.

[1031] Tait, P. G. (1898). *Scientific papers*, Volume 1. Cambridge Univ. Press. Republished in 1971.

[1032] Takahashi, M. (1981). *Two-bridge knots have Property P*, Volume 239 of *Mem. Amer. Math. Soc.* American Math. Soc. Problem: 1.15.

[1033] Takeuchi (1991). Waldhausen's classification theorem for finitely uniformizable 3-orbifolds. *Trans. Amer. Math. Soc. 328*, 151–200. Problem: 3.46.

[1034] Takeuchi, Y. (1988). A classification of a class of 3-branchfolds. *Trans. Amer. Math. Soc. 307*, 481–502. Problem: 3.46.

[1035] Taniyama, K. (1989). A partial order of knots. *Tokyo J. Math. 12*, 205–229. Problem: 5.16.

[1036] Taubes, C. H. (1987). Gauge theory on asymptotically periodic 4-manifolds. *J. Differential Geom. 25*, 363–430. Problems: 4.41, 4.42, 4.49, 4.72, 4.95.

[1037] Taubes, C. H. (1994). The Seiberg–Witten invariants and symplectic forms. *Math. Res. Lett. 1*, 809–822. Problems: 4.130, 4.138, 4.140.

[1038] Taubes, C. H. (1995a). Homology cobordism and the simplest perturbative Chern-Simons 3-manifold invariant. In S.-T. Yau (Ed.), *Geometry, Topology, and Physics for Raoul Bott*, pp. 429–538. International Press. Problems: 5.19, 5.20.

[1039] Taubes, C. H. (1995b). Metabolic cobordisms and the simplest perturbative Chern-Simons invariant. Preprint. Problem: 5.20.

[1040] Taubes, C. H. (1995c). The Seiberg–Witten and the Gromov invariants. *Math. Res. Lett. 2*, 221–238. Problems: 4.128, 4.138.

[1041] Teichner, P. (1992). *Topological Four-Manifolds with Finite Fundamental Group*. Verlag Shaker. Problem: 4.84.

[1042] Teichner, P. (1995). On the star-construction for topological 4-manifolds. In W. H. Kazez (Ed.), *Geometric Topology, Proc. of the 1993 Georgia International Topology Conference.* International Press. Problem: 4.82.

[1043] Thistlethwaite, M. B. (1987). A spanning tree expansion of the Jones polynomial. *Topology 26*, 297–309. Problems: 1.65, 1.66.

[1044] Thistlethwaite, M. B. (1988). On the Kauffman polynomial of an adequate link. *Invent. Math. 93*, 285–296. Problem: 1.68.

[1045] Thom, R. (1960). Les intégrales premières d'un système différentiel sur une variété compacte. Preprint. Problem: 3.109.

[1046] Thomas, C. B. (1977). On Poincaré 3-complexes with binary polyhedral fundamental group. *Math. Ann. 226*, 207–221. Problem: 3.37.

[1047] Thomas, C. B. (1984). Splitting theorems for certain PD^3-groups. *Math. Zeit. 186*, 201–209. Problem: 3.77.

[1048] Thomas, C. B. (1995). 3-Manifolds and $PD(3)$-groups. In S. C. Ferry, A. A. Ranicki, & J. M. Rosenberg (Eds.), *Novikov Conjectures, Index Theorems and Rigidity, Volume 1 (Oberwolfach, 1993)*, Volume 226 of *London Math. Soc. Lect. Note Ser.*, pp. 301–308. Cambridge Univ. Press. Problem: 3.77.

[1049] Thompson, A. A. (1994). Thin position and the recognition problem for S^3. *Math. Res. Lett. 1*, 613–630. Problem: 3.80.

[1050] Thurston, W. P. (1977). Geometry and topology of 3-manifolds. Lectures Notes from Princeton Univ. 1977, available from MSRI - for information order e-mail `shaw@msri.org`, or write to Nancy Shaw - MSRI, 1000 Centennial Drive, Berkeley CA 94720 - USA. Problems: 1.77, 1.86, 3.14, 3.15.

[1051] Thurston, W. P. (1982). Three dimensional manifolds, Kleinian groups and hyperbolic geometry. *Bull. Amer. Math. Soc. 6*, 357–382. Problems: 3.33, 3.46.

[1052] Thurston, W. P. (1986a). A norm for the homology of 3-manifolds. *Mem. Amer. Math. Soc. 59, number 339*, 99–130. Problem: 3.16.

[1053] Thurston, W. P. (1986b). Hyperbolic structures on 3-manifolds I: Deformation of acylindrical 3-manifolds. *Ann. of Math. 124*, 203–246. Problem: 3.14.

[1054] Thurston, W. P. (1986c). Hyperbolic structures on 3-manifolds, II: Surface groups and manifolds which fiber over S^1. Preprint. Problem: 3.14.

[1055] Thurston, W. P. (1986d). Hyperbolic structures on 3-manifolds, III: Deformation of 3-manifolds with incompressible boundary. Preprint. Problem: 3.14.

[1056] Thurston, W. P. (1988). On the geometry and dynamics of diffeomorphisms of surfaces. *Bull. Amer. Math. Soc. 19*, 417–431. Problem: 1.84.

[1057] Todorov, A. N. (1980). Applications of the Kähler-Einstein-Calabi-Yau metric to moduli of $K3$ surfaces. *Invent. Math. 61*, 251–265. Problem: 4.37.

[1058] Tollefson, J. (1976). Periodic homeomorphisms of 3-manifolds fibered over S^1. *Trans. Amer. Math. Soc. 223*, 223–234. See erratum in Trans. Amer. Math. Soc. 243 (1978) 309–310. Problem: 3.40.

[1059] Toruńczyk, H. (1973). Compact absolute retracts as factors of the Hilbert space. *Fund. Math. 83*, 75–84. Problem: 5.27.

[1060] Trace, B. (1980). A class of 4-manifolds which have 2-spines. *Proc. Amer. Math. Soc. 79*, 155–156. Problem: 4.18.

[1061] Trace, B. (1982). On attaching 3-handles to a 1-connected 4-manifold. *Pacific J. Math. 99*, 175–181. Problem: 4.18.

[1062] Tristram, A. G. (1969). Some cobordism invariants for links. *Math. Proc. Cambridge Philos. Soc. 66*, 251–264. Problems: 4.18, 4.26.

[1063] Trotter, H. F. (1964). Non-invertible knots exist. *Topology 2*, 275–280. Problem: 1.61.

[1064] Trotter, H. F. (1975). Some knots spanned by more than one unknotted surface of minimal genus. In L. P. Neuwirth (Ed.), *Knots, Groups, and 3-Manifolds*, Volume 84 of *Ann. of Math. Stud.*, pp. 51–62. Princeton Univ. Press. Papers Dedicated to the Memory of R. H. Fox. Problem: 1.20.

[1065] Tsukui, Y. (1970). On surfaces in 3-space. *Yokohama Math. J. 18*, 93–104. Problem: 3.11.

[1066] Tsukui, Y. (1975). On a prime surface of genus 2 and homeomorphic splitting of 3-sphere. *Yokohama Math. J. 23*, 63–75. Problem: 3.11.

[1067] Tukia, P. (1988). Homeomorphic conjugates of Fuchsian groups. *J. Reine Angew. Math. 391*, 1–54. Problem: 3.5.

[1068] Turaev, V. G. (1988). Модули Конвея и Кауффмана полнотория. *Исследования по Топологии 6, Записки Научных Семинаров Ленинградского Отделения Математический ин-та. им. Стеклова 167*, 79–89. English translation in [1069]. Problem: 1.92.

[1069] Turaev, V. G. (1990). The Conway and Kauffman modules of a solid torus. *J. Soviet Math. 52*, 2799–2805. Russian original in [1068].

[1070] Turaev, V. G. (1991). Skein quantization of Poisson algebras of loops on surfaces. *Ann. Sci. École Norm. Sup. (4) 24*, 635–704. Problem: 1.92.

[1071] Turaev, V. G. & Wenzl, H. (1993). Quantum invariants of 3-manifolds associated with classical simple Lie algebras. *Internat. J. Math. 4*, 323–358. Problem: 3.108.

[1072] Vassiliev, V. A. (1990). Cohomology of knot spaces. In V. I. Arnold (Ed.), *Theory of Singularities and its Applications*, Volume 1 of *Adv. Soviet Math.* American Math. Soc. Problem: 1.89.

[1073] Veech, W. A. (1989). Teichmüller curves in moduli space, Eisenstein series and an application to triangular billiards. *Invent. Math. 97*, 553–583. Problem: 2.7.

[1074] Veech, W. A. (1992). The billiard in a regular polygon. *Geom. Funct. Anal. 2*, 341–379. Problem: 2.7.

[1075] Vertigan, D. L. (1995). On the computational complexity of Tutte, Homfly and Kauffman invariants. To appear. Problem: 1.90.

[1076] Ville, M. (1989). Les variétés riemanniennes de dimension 4 4/19-pincées. *Ann. Inst. Fourier (Grenoble) 39*, 149–154. Problem: 4.67.

[1077] Viro, O. J. (1979a). Построение многокомпонентных вещественных алгебраических поверхностей . *Докл. Акад. Наук СССР 248*, 279–282. English translation in [1078].

[1078] Viro, O. J. (1979b). Construction of multicomponent real algebraic surfaces. *Soviet Math. Dokl. 20*, 991–995. Russian original in [1077]. Problem: 3.69.

[1079] Vogel, P. (1995). Algebraic structures on modules of diagrams. Preprint, Université Paris VII. Problem: 1.61.

[1080] von Kerékjártó, B. (1934a). Bemerkung über reguläre Abbildungen von Flächen. *Acta Sci. Math. (Szeged) 7*, 206. Problem: 2.19.

[1081] von Kerékjártó, B. (1934b). Über reguläre Abbildungen von Flächen auf sich. *Acta Sci. Math. (Szeged) 7*, 65–75. Problem: 2.19.

[1082] Wagreich, P. (1972). Singularities of complex surfaces with solvable local fundamental group. *Topology 11*, 51–72. Problem: 3.28.

[1083] Wajnryb, B. (1983). A simple presentation for the mapping class group of an orientable surface. *Israel J. Math. 45*, 157–174. Problem: 2.18.

[1084] Waldhausen, F. (1967a). Eine Klasse von 3-dimensionalen Mannigfaltigkeiten, I. *Invent. Math. 3*, 308–333. Problems: 3.27, 3.28.

[1085] Waldhausen, F. (1967b). Eine Klasse von 3-dimensionalen Mannigfaltigkeiten, II. *Invent. Math. 4*, 87–117. Problems: 3.27, 3.28.

[1086] Waldhausen, F. (1967c). Gruppen mit Zentrum und 3-Dimensionale Mannigfaltigkeiten. *Topology 6*, 505–517. Problem: 3.5.

[1087] Waldhausen, F. (1968a). Heegaard-Zerlegungen der 3-Sphäre. *Topology 7*, 195–203. Problem: 3.85.

[1088] Waldhausen, F. (1968b). On irreducible 3-manifolds which are sufficiently large. *Ann. of Math. 87*, 56–88. Problems: 3.1, 3.7, 3.95.

[1089] Waldhausen, F. (1968c). The word problem in fundamental groups of sufficiently large irreducible 3-manifolds. *Ann. of Math. 88*, 272–280. Problems: 3.47, 3.81.

[1090] Waldhausen, F. (1969). Über Involutionen der 3-Sphäre. *Topology 8*, 81–91. Problems: 1.6, 3.38.

[1091] Waldhausen, F. (1978a). Algebraic K-theory of generalized free products. *Ann. of Math. 108*, 135–256. Problem: 3.32.

[1092] Waldhausen, F. (1978b). Recent results on sufficiently large 3-manifolds. In R. J. Milgram (Ed.), *Algebraic and Geometric Topology (Stanford, 1976)*, Volume 32, part II of *Proc. Sympos. Pure Math.*, pp. 21–38. American Math. Soc. Problem: 3.49.

[1093] Wall, C. T. C. (1964a). Diffeomorphisms of 4-manifolds. *J. London Math. Soc. 39*, 131–140. Problem: 4.36.

[1094] Wall, C. T. C. (1964b). On simply-connected 4-manifolds. *J. London Math. Soc. 39*, 141–149. Problem: 4.11.

[1095] Wall, C. T. C. (1966). Classification problems in differential topology V: On certain 6-manifolds. *Invent. Math. 1*, 355–374. Problem: 5.21.

[1096] Wall, C. T. C. (1971). Pairs of relative cohomological dimension one. *J. Pure Appl. Algebra 1*, 141–154. Problem: 5.10.

[1097] Wang, S. (1995). A vanishing theorem for Seiberg–Witten invariants. *Math. Res. Lett. 2*, 305–310. Problem: 4.104.

[1098] Wang, S. C. (1990). The virtual $\mathbb{Z}$-representability of 3-manifolds which admit orientation reversing involutions. *Proc. Amer. Math. Soc. 110*, 499–503. Problem: 3.2.

[1099] Wedel, C. (1994). Neuere Resultate zur Zeeman–Vermutung. Diplomarbeit, Frankfurt/Main. Problem: 5.14.

[1100] Weeks, J. (1985). *Hyperbolic structures on 3-manifolds.* Ph. D. thesis, Princeton Univ. Problem: 3.60.

[1101] Weeks, J. (1995). SnapPea. The program is available by **anonymous ftp**, from `geom.umn.edu`. Problem: 1.77.

[1102] Weinberger, S. (1987). On fibering four- and five-manifolds. *Israel J. Math. 59*, 1–7. Problem: 4.121.

[1103] Weinberger, S. (1994). *The Topological Classification of Stratified Spaces.* Chicago Lectures in Math. Univ. of Chicago Press. Problem: 5.29.

[1104] Weinstein, A. (1970). Positively curved n-manifolds in $\mathbb{R}^{n+2}$. *J. Differential Geom. 4*, 1–4. Problem: 4.67.

[1105] Wells, R. (1966). Cobordism groups of immersions. *Topology 5*, 281–294. Problem: 4.61.

[1106] Welsh, D. J. A. (1992). On the number of knots and links. In *Sets, graphs and numbers (Budapest, 1991)*, Volume 60 of *Colloq. Math. Soc. János Bolyai*, pp. 713–718. North-Holland. Problem: 1.63.

[1107] Welsh, D. J. A. (1993). *Complexity: Knots, Colouring and Counting*, Volume 186 of *London Math. Soc. Lect. Note Ser.* Cambridge Univ. Press. Problem: 1.90.

[1108] Wenzl, H. (1993). Braids and invariants of 3-manifolds. *Invent. Math. 114*, 235–275. Problem: 3.108.

[1109] White, J. H. (1975). Twist invariants and the Pontryagin numbers of immersed manifolds. In S.-S. Chern & R. Osserman (Eds.), *Differential Geometry*, Volume 27 of *Proc. Sympos. Pure Math.*, pp. 429–437. American Math. Soc. Proc. of a Conf. at Stanford Univ. Stanford - 1973. Problem: 4.20.

[1110] Whitehead, J. H. C. (1962). *Mathematical Works.* Pergamon Press and Macmillan. Ed. I. M. James. Problem: 1.39.

[1111] Whitney, H. (1941). On the topology of differentiable manifolds. In R. L. Wilder & W. L. Ayres (Eds.), *Lectures in Topology (Michigan Conf., 1940)*, pp. 101–141. The Univ. of Michigan. Press. Problem: 4.111.

[1112] Whitney, H. (1944). The self-intersections of a smooth n-manifold in 2n-space. *Ann. of Math. 45*, 220–246. Problem: 4.19.

[1113] Whitney, H. (1957). *Geometric Integration Theory*, Volume 21 of *Princeton Math. Seri.* Princeton Univ. Press. Problem: 3.110.

[1114] Whitten, W. (1981). Inverting double knots. *Pacific J. Math. 97*, 209–216. Problem: 1.6.

[1115] Whitten, W. (1986). Rigidity among prime-knot complements. *Bull. Amer. Math. Soc. 14*, 299–300. Problem: 1.13.

[1116] Witten, E. (1989). Quantum Field Theory and the Jones polynomial. *Comm. Math. Phys. 121*, 351–399. Problems: 3.108, 5.20.

[1117] Witten, E. (1994). Monopoles and four-manifolds. *Math. Res. Lett. 1*, 769–796. Problem: 4.143.

[1118] Wolf, J. A. (1967). *Spaces of Constant Curvature.* McGraw Hill. Republished by Publish or Perish in 1974, 1977 and 1984. Problem: 3.37.

[1119] Wright, D. G. (1992). Contractible open manifolds which are not covering spaces. *Topology 31*, 281–291. Problems: 3.7, 4.73.

[1120] Wright, G. (1994). The Reshetikhin–Turaev representation of the mapping class group. *J. Knot Theory Ramifications 3*, 547–574. Problem: 2.2.

[1121] Wright, P. (1975). Group presentations and formal deformations. *Trans. Amer. Math. Soc. 208*, 161–169. Problem: 5.1.

[1122] Wu, W.-T. (1952). Sur les classes caractéristiques des structures fibrées sphériques. *Publ. Inst. Math. Univ. Strasbourg 11*, 5–89, 155–156. Problem: 5.21.

[1123] Wu, W.-T. (1963). On the imbedding of orientable manifolds in a euclidean space. *Sci. Sinica 12*, 25–33. Problem: 4.19.

[1124] Xu, Y.-J. & Yau, S. S.-T. (1993). Durfee conjecture and coordinate free characterization of homogeneous singularities. *J. Differential Geom. 37*, 375–396. Problem: 3.31.

[1125] Yajima, T. (1964). On simply-knotted spheres in $\mathbb{R}^4$. *Osaka J. Math. 1*, 133–152. Problem: 1.105.

[1126] Yajima, T. (1969). On a characterization of knot groups of some spheres in $\mathbb{R}^4$. *Osaka J. Math. 6*, 435–446. Problem: 4.29.

[1127] Yajima, T. (1970). Wirtinger presentation of knot groups. *Proc. Japan Acad. Ser. A Math. Sci. 46*, 997–1000. Problem: 4.29.

[1128] Yamabe, H. & Yujobo, Z. (1950). On the continuous functions defined on a sphere. *Osaka Math. J. 2*, 19–22. Problem: 5.24.

[1129] Yamada, S. (1995). The absolute value of the Chern–Simons–Witten invariants of lens spaces. *J. Knot Theory Ramifications 4*, 319–327. Problem: 3.108.

[1130] Yang, C. T. (1957). On maps from spheres to Euclidean spaces. *Amer. J. Math. 79*, 725–732. Problem: 5.24.

[1131] Yau, S.-T. (1976). Parallelizable manifolds without complex structure. *Topology 15*, 51–53. Problem: 5.21.

[1132] Yau, S.-T. (1977). Calabi's conjecture and some new results in algebraic geometry. *Proc. Nat. Acad. Sci. U.S.A. 74*, 1798–1799. Problems: 4.68, 5.21.

[1133] Yau, S.-T. (1978). On the Ricci curvature of a compact Kähler manifold and the complex Monge-Ampère equation I. *Comm. Pure Appl. Math. 31*, 339–411. Problems: 4.39, 4.106.

[1134] Yoshikawa, K. (1982). On a 2-knot group with trivial center. *Bull. Austral. Math. Soc. 25*, 321–326. Problem: 1.57.

[1135] Zabrodsky, A. (1970). Homotopy associativity and finite CW-complexes. *Topology 9*, 121–216. Problem: 5.28.

[1136] Zariski, O. (1929). On the problem of existence of algebraic functions of two variables processing a given branch curve. *Amer. J. Math. 51*, 305–328. Problem: 4.40.

[1137] Zariski, O. (1935). *Algebraic Surfaces*, Volume 3 of *Ergeb. Math. Grenzgeb. (3)*. Springer-Verlag. Problems: 4.38, 4.40.

[1138] Zeeman, E. C. (1964). On the dunce hat. *Topology 2*, 341–358. Problems: 5.2, 5.14.

[1139] Zhou, Q. (1990). *Three-dimensional geometric cone structures.* Ph. D. thesis, UCLA. Problem: 3.46.

[1140] Zieschang, H. (1981). *Finite Groups of Mapping Classes of Surfaces*, Volume 875 of *Lecture Notes in Math.* Springer-Verlag. Problem: 3.40.

[1141] Zimmermann, B. (1982). Das Nielsensche Realisieungsproblem für hinreichend große 3-Mannigfaltigkeiten. *Math. Zeit. 180*, 349–359. Problems: 3.38, 3.40, 3.42.